BASIC COMMUNICATION SKILLS
FOR TECHNOLOGY
2nd Edition

Andrea J. Rutherfoord, Ph.D.
formerly of DeVry Institute of Technology
Atlanta, Georgia

Prentice
Hall

Upper Saddle River, New Jersey
Columbus, Ohio

Library of Congress Cataloging-in-Publication Data

Rutherfoord, Andrea J.
    Basic communication skills for technology / Andrea J. Rutherfoord.
       p. cm.
    Includes bibliographical references and index.
    ISBN 0-13-087822-7
      1. Technical writing. I. Title.
  T11 .T83 2001
  808.0666—dc21   00-027426

Vice President and Publisher: Dave Garza
Editor in Chief: Stephen Helba
Executive Editor: Debbie Yarnell
Production Editor: Tricia Huhn
Photo Coordinator: Ryan Lamb
Design Coordinator: Robin G. Chukes
Text Designer: Carlisle Communications, Ltd.
Cover art: Elizabeth Shippell
Cover Designer: Thomas Borah
Production Manager: Brian Fox
Marketing Manager: Chris Bracken

This book was set in Century Schoolbook by Carlisle Communications, Ltd. and was printed and bound by R .R. Donnelley & Sons Company. The cover was printed by R.R. Donnelley & Sons Company.

10 9 8 7 6 5 4 3 2 1
ISBN 0-13-087822-7

Contents

Preface to the Student

Your technical skills will be an important factor in your career. However, if the technical skills of two candidates are equal, the decision for hiring (or promoting) is usually based on the ability to communicate. For some positions, communications skills are so vital that poor writers or speakers are passed over, no matter what level of expertise they have in technical skills. To confirm the importance of communication skills in technology, read "Messages from Industry". Leaders of major corporations state their beliefs in the value of good communication.

This book presents the types of writing skills you need to have for a career in technology. Much of the information is based on my career as an educator and professional technical writer. During the course of my career, particularly in the business world, I have witnessed the growing demands and increased expectations for written and spoken communication that employees at every level face on a daily basis.

A few years ago, employees could rely on a department secretary or administrative assistant for help with writing, editing, and distribution. Today, all employees, except those in the highest levels of upper management, are personally responsible for their communication tasks. And the results depend as much on their ability to master a computer and software as on their ability to write effectively.

Companies still hire or contract professional technical writers to produce critical documents, such as proposals, user's guides, and online help. But the majority of employees must write their own e-mail, letters, reports, and presentation aids, sometimes without time for internal review. This can be an intimidating challenge to someone who is unfamiliar with basic writing skills. For those who take the time to practice and experiment, it can be an opportunity to shine.

The tools for communication continue to become more powerful and sophisticated. Most businesses expect workers to operate desktop computers, laptops, personal digital assistants (PDAs), and other equipment for word processing and messaging, as well as peripheral devices such as high-quality printers, scanners, and digital cameras to enhance the quality and detail of documents and graphics.

Writing software has become more powerful, too, ranging from easy-to-use word processors, database programs, and graphic tools, to ever-more-powerful desktop-publishing programs and presentation systems, capable of producing manuals, documents, online help, and presentations in any language and any format, and available on all types of computer platforms. Many companies have adopted strategies such as telecommuting, teleconferencing, and even telework as alternatives to expansion, making it even more important for employees to maintain and manage their own equipment and software. This was driven home to me recently when a service technician came to my home to provide an estimate for a repair. During the process of testing my air conditioning ducts, he also set up his laptop computer, external zip drive, and portable printer. His entire office was now operational on my kitchen table, and he was able to complete the diagnostic pricing and contract, authorize it, and print it in his portable office. This scenario is sometimes called telework. The technician doesn't have an office at his employer's site; instead, he brings his office with him to each customer site.

And of course, no business can afford to overlook the value of e-business, e-commerce, and e-recruiting: conducting all types of transactions using the Internet. Progressive companies know that technology provides the most effective way to communicate with customers, employees and potential employees, business partners, and customers. It also provides immediate access to information, products, and services. Employees often use the Internet as their primary source of company information, research and development, training, and communication.

All of these trends should impress upon you the value of learning more about information technology and how to use the computer as your basic communication tool. Your entire work environment might one day be contained within the casing of your computer and a few peripheral devices.

Many exercises in this text require the use of a word processor. Some require the use of the Internet to research specific topics.

Before you get started, I want to tell you how to use this book. I wrote it using a systems approach to communication. A system is an arrangement of related, individual elements that, together, form a unity. Language has several individual elements, as demonstrated in each chapter of this book. Each chapter has five sections. The first section is an article to READ, followed by WRITING, SPELLING, VOCABULARY, and WORD WATCH sections.

The READING articles for each chapter either discuss or demonstrate the topic of the chapter and consist of facts and ideas presented in an interesting way. If you encounter new words, underline the words and keep reading. Sometimes you will figure out new words by their context—the words around them. Look up the words that you have not figured out after you have finished reading, then reread the sentence. Finally, answer the comprehension questions at the end of the article. They will help you interpret, organize, and respond to what you have read.

The WRITING sections present the skills most useful to technical people and deal with one primary writing skill at a time. Part 1 presents the foundations of technical writing: audience, language and style, and organization. Part 2 presents chapters on specific writing tasks, such as writing a technical definition, technical description, summary, graphics, comparison and contrast, and instructions. Part 3 presents chapters on longer or more complicated tasks, including reports, presentations, and searching for a job.

The SPELLING sections review some spelling patterns that are reliable. They will also give you a helpful aid to remember correct spellings of tricky words. Poor spelling is unprofessional. Spelling checks find and correct some errors, but not all. Since writing takes time, thought, and effort, it seems senseless to degrade our own work with misspellings—it's almost like wearing an expensive shirt inside-out.

The VOCABULARY sections attempt to bring some order to the haphazard collection of foreign roots, prefixes, and suffixes that form technical words. Latin and Greek, particularly, form the roots for most of the difficult words that we encounter in technology. These sections will help you analyze words to determine their meanings.

Finally, the WORD WATCH sections review groups of easily confused and misused words. Sometimes the placement of one letter completely changes the meaning of a word, as in *tough* and *though*. Other times, two related words, such as *affect* and *effect,* have completely different meanings and uses.

Following the chapters, two sections review the fundamental rules of composing clear and correct sentences: The Grammar and Mechanics Units. The Grammar Units deal with the components of language: individual words, groups of words, and sentence structure and their functions in communication. Just as science is guided by a limited set of theories and principles, English is guided by a limited set of rules. *Limited* does not mean a small number, just a learnable number. The Grammar Units provide a review of these rules.

The Mechanics Units deal with the tools of our language: symbols, abbreviations, numbers, and punctuation. The longest unit reviews the comma, often the most troublesome punctuation mark for writers. The Mechanics Units provide a review of the rules surrounding punctuation, numbers, and symbols.

At the very end of the book, six appendices include information that supplements other chapters. Look through them to see what is offered. Learning the techniques of effective writing requires thought and practice. Whatever effort you put in, however, will pay off—in this course and in your career. Consider this book as a ladder to career advancement. Good luck!

Andrea J. Rutherfoord

ACKNOWLEDGMENTS

The second edition is the result of many years of learning from instructors, colleagues, students, and more recently, from my workteams, co-workers, managers, and business associates. I would like to thank them all collectively for their individual contributions, large and small.

I would also like to acknowledge the people who shared their professional experiences with business communication. They each supplied examples and suggestions for applying general writing practices to specific careers. They include Tony Cook and Wayne Wofford, Gwinnett Area Institute of Technology; Sonny Cox, Sonny's Service Center; Earl Friedell and Shirley McCree, DeKalb Technical Institute; David Homback, Southern College of Technology; David Hurst, State Farm Corporation; Jean Burns and Shelley Fischer-Wylie, human resources; Jim Brown and Patrick Williams, Colder Products Company; and Steve Hookstra, paramedic. My family also provided many experiences from their careers—formally and informally. Thank you to Dick Ottum, Duane Ottum, Lara Nichols, Chris Sorlie, and Vermont Rutherfoord.

I also want to thank the faculty and staff at DeVry Institute of Technology (Atlanta) for their insights into electronics and computer science. For this edition, I particularly want to thank Kyle Jones and Jack Griffin.

Despite my many resources, however, any technical inaccuracies are solely my own.

My Prentice Hall editor, Steve Helba, and his assistant, Nancy Kesterson, provided many of the tools and the encouragement I needed to get this into production.

Many thanks to my reviewers for helping to improve the text: Nancy M. Staub, Lamar University; Deborah L. Lamm, Lenior Community College; Kent Harrelson, Dalton State College; Robin Newcomer, Olympic College; and F. C. Campbell, Ph.D., Devry–Chicago.

And of course, I want to express appreciation to my family for letting me spend my weekends at the library or computer, miss a few holidays, continue to type while we talked on the phone, and put off many events (except becoming a grandmother) for a few months. And Tom, my husband, learned how to fend for himself, again, as I disappeared into the office for days on end. Without their patience and understanding, I could not have continued. I dedicate this book to my family: Tom, Chris, Vermont, Lara and Joel, and baby Sintra Autumn.

Messages from Industry

Tomorrow's problem will not be the communication of data but more importantly, 'information.' Data abounds. Useful information is still unfortunately sparse.

Matthew A. Kenny, President, Racal-Milgo, Inc., Sunrise, Florida

The best ideas in the world are useless unless they can be communicated to others. By the word 'communication' I mean that there's not only reception, but understanding of the information conveyed. Miscommunication of information not only destroys many great ideas, but causes untold waste in daily business activities.

Richard W. Oliver, Assistant Vice-President, Northern Telecom Limited, Nashville, Tennessee

The ability to speak and write clearly is not only important to the communication of technical concepts, it is an essential part of the innovation process itself. Translating an idea into the written word is one of the better ways of validating the soundness of one's thinking.

Ian M. Ross, President, AT&T Bell Laboratories, Holmdel, New Jersey

Your knowledge is only as valuable as your ability to communicate it to someone else.

Gerald E. Schultz, President, Bell & Howell Company, Skokie, Illinois

... Communications skills are the second most important [skill] that any technical person can learn, and the first is learning to learn. Without communication skills, however, doing the most important is much more difficult than it needs to be. Someone who has already attained communication skills is ready to move ahead more quickly, for he must learn them to move ahead at all.

Court Skinner, Manager, Advanced Technology, National Semiconductor, Santa Clara, California

One is not evaluated on technical skills alone, but also on the image one presents while communicating.

Keith R. Welker, Personnel, Administrator, Hughes Aircraft Company, El Segundo, California

Foundations

CHAPTER 1

Technical Communication

- Identify the factors to consider in technical communication.
- Analyze the "audience."
- Rewrite passages for a different audience.
- Spell plurals correctly.
- Use *pseudo* and *quasi* correctly.
- Use *its* and *it's* correctly.

READING: Technical Communication

The way we communicate reveals many things about us: our emotions, our view of the world, our interests, and our experience with the different methods of communication.

Technical communication differs from other forms of communication in several ways, including audience, purpose, and style. The chapters of this book introduce many of the concepts that make technical communication distinctive and effective for writers, readers, speakers, and listeners in the workplace. Most of the concepts apply primarily to written communication, but you can apply them equally to spoken communication.

The audience for this book is the student who is preparing for a technical career in which communication will be part (but not all) of the job.

What Is Technical Communication?

Technical communication is the process of transmitting facts and information to a defined audience for a specific purpose. More simply, technical communication is writing for understanding. Technical writers present information in science, electronics, or other technical areas on a professional level, backed up by data and facts, so that the information is complete and accurate.

Most companies expect employees to communicate regularly on a business level through e-mails, letters, and memos. Some positions, usually higher-level, require reports, proposals, instructions, and presentations. Employees who can produce focused, clear documents often have a competitive edge over others who are less skilled with written communication.

The History of Technical Communication

Technical writing dates back to the earliest recorded language. Ancient civilizations drew pictographs on cave walls to describe how they hunted and where they journeyed. In more recent centuries, leaders in science and mathematics developed a language and style to record their theories, inventions, and discoveries by documenting their procedures and proofs for others to understand or replicate.

For the western origins of technical writing, we need to go back only to the 17th century, when the Royal Society in London produced the first written attempts to describe and categorize the physical world systematically. Today, in the 21st century, scientists use a time-honored methodology for conducting and reporting discoveries

and technical developments. As a professional in a technical field, your peers will expect you to conform to the writing conventions in your industry.

We have available to us today a wide variety of equipment and computer software to make our writing, research, and distribution efficient. We use word processors, desktop publishers, graphics programs, and multimedia presentation software, rich with features and versatility. We have innovative equipment such as modems, digital cameras, and scanners that transmit or process information into usable forms. The Internet offers millions of Web sites on which we can shop, research, and communicate with others. We have entire encyclopedias on the Internet, and multiple Internet search engines to research topics and locate data from thousands of sources online. We have laptop and hand-held computers, portable digital assistants (PDAs), and portable digital notepads (PDNs) to take with us from place to place. We even have computer software that types what we dictate (called speech-recognition software) and reads our computer documents and e-mails out loud using a computer-generated voice (called text-to-speech software). In recent years, the variety, power, and speed of communication tools have given us access to a staggering amount of data that we must then convert into information.

We are surrounded by technical information. As an example, look at the back of a common frozen-food carton. It is a marvel of verbal efficiency and visual layout, containing everything you need to know about the product. It contains preparation instructions, an ingredients list, nutrition facts, and pictures showing the prepared product. In addition, packages include serving suggestions, marketing information, a bar code for grocery store scanners, safety warnings, and recycling information, all carefully worded and laid out so they are succinct, accurate, visible, and understandable. This example might seem like a simple writing project, but in fact, it is a challenge undertaken only by teams of experienced technical writers, designers, and marketers. Every carton design has probably gone through extensive reviews, edits, redesigns, marketing feedback, and usability testing. We'll leave that kind of assignment to the pros.

Factors to Consider in Technical Communication

For most of us, the ability to write a meaningful report or a hard-hitting memo develops slowly. As with any skill, practice and experience are essential. To get started, look at the factors technical writers must consider before beginning any type of technical communication. These are audience, purpose, format, and style.

Audience

The audience could consist of managers, coworkers, customers and clients, the general public, or any combination. They will have different levels of understanding and different information needs that require specific formats and styles of communication.

Purpose

The purpose of a technical document could be to inform, explain, describe, persuade, or record your actions. Some documents have multiple purposes, and some purposes overlap, such as a request for equipment (to persuade) that includes a technical description of the equipment (to describe).

Format

Technical communication can be written in the following formats:

- Reports or documents, such as proposals, lab reports, product specifications, or quality-test results.
- Record-keeping forms, such as service reports, travel and expense forms, or troubleshooting logs.
- Instructions, such as user guides, online help, and training manuals.
- Correspondence, such as letters, memos, and e-mails.
- Presentations, such as interviews, marketing calls, or training seminars.

Some types of communication employ a combination of formats, such as a letter or e-mail that includes a message and a backup report as an attachment.

Style

Writers base the style (language, organization, and layout) of the document on the audience, purpose, and format. The language can include many technical terms, called *jargon,* or it can include general terms and definitions of technical terms. The document might need a visible structure of headings and subheadings or even chapters to identify the flow of information, such as a product specification or manual. Or the document might not need visible clues for structure, such as a casual e-mail or memo focused on only one topic, comprised of only a few paragraphs.

The layout can consist of condensed paragraphs that fill the pages of the document, or it can provide lots of white space, with examples, charts, or graphics to illustrate points and bulleted or numbered lists to highlight main points. These types of visual aids allow for a quick or scanned reading.

Preferences of Technical Readers

Generally people who read technical information prefer sentences that get straight to the point. They prefer words that are functional, exact, and clear. They prefer paragraphs that are short, with each paragraph focused on only one idea. And they prefer a visible organization with headings, bulleted lists, and numbered steps, and graphics and examples that illustrate the details of the subject.

Style Guides

A style guide is a reference book for writers. It offers guidelines on the finer points of word usage, punctuation, and mechanics for standard communication, beyond the basic rules of grammar and punctuation. The differences between style guides might appear to be slight, but they provide for consistency in such things as formatting headings, citations, and quotations.

Some style guides develop within an industry. For example, English departments in colleges and universities often prefer the *MLA Handbook for Writers of Research Papers,* written by Joseph Gibaldi and published by the Modern Language Association. The MLA style guide uses the following format for citing a book in a bibliography:

> Paynter, Robert T. *Introductory Electronic Devices and Circuits.* Upper Saddle River, NJ: Prentice Hall, 1991.

Newspapers and book publishers often prefer *The Chicago Manual of Style,* published by the University of Chicago Press. It uses the following format for citing the same book:

> Paynter, Robert T. 1991. *Introductory Electronic Devices and Circuits.* Upper Saddle River, NJ: Prentice Hall.

Companies usually adopt a style guide, which the writers in the company follow. In addition, some companies develop their own conventions and styles for certain documents. If your company makes such recommendations, familiarize yourself with the approved style and adjust your writing accordingly.

Getting Started: Focus on the Audience

Does this overview make technical writing sound overwhelming? It is a challenging task to most people, even those who have been writing professionally for many years. So where do you start?

Focus On the Audience

The first step in writing technical communication is to focus on the audience. While you are in school, you know that your audience is your professor or possibly your classmates, and you have a good idea of your audience's technical background and expectations. In the workplace, however, you must analyze your audience more carefully by asking the three questions that introduce the following sections.

1. What does the audience already know about the subject?

In communication, the "audience" is the person or group of people whom you expect to read your information. Even though writers do not know exactly who will read their documents, they can usually define an intended audience as either technical, semi-technical, or nontechnical.

Technical Audience

The technical audience includes practitioners in your field: those with technical experience and training, such as technicians and engineers. A technical audience understands fundamental concepts and jargon without definitions or background information. Readers expect the writer to use technical language efficiently and appropriately. For this audience, writers use technical terms and precise data to convey information. The following example is a paramedic's report written for a technical audience: the medical staff at an emergency room.

If you do not have medical training, the example above probably made little sense

> **REPORT FOR A TECHNICAL AUDIENCE**
>
> **Subjective:** Patient C/O of SOB secondary to MVA.
>
> **Objective:** —40 y.o. W/M found conscious, sitting behind bent steering wheel of passenger car with extensive front-end damage.

HEAD: ARWY patent; EYES: attntv; PERL; SKIN: W&D; c̄ some cyanosis present. NECK: No pain upon palp. No J.V.D., resp., retract., or trach. dev. present. CHEST: Chest movement asym. Flail segment @ right chest, ribs 3-7. NEURO: Patient CAO X 3. GCS = 15. *And so on . . .*

Assessment: Flail chest.

Plan: P/E, C-spine immob. (KED), Vx, SaO₂, O₂, IV (L.R., 14GA., ® a.c., #16541), EKG. Patient removed from vehicle to LSB. Telemetry: A.L.S. transport to Chicago Hope, #1236 attending.

to you. It includes many abbreviations common to emergency medical services (EMS), and it communicates a patient's condition efficiently to others in EMS.

Semi-technical Audience

This type of audience has some technical training or works in the industry, but not directly in the field, such as those working in related departments or those with training in related technical areas. This might even include personnel in marketing, finance, or administration of a technical company. The semi-technical audience needs some explanation of concepts, abbreviations, and jargon. Writers use technical terms only if they are common in the company or industry. For this audience, you might provide an orientation to the subject and explain or interpret the terms and information. The following example is a version of the first example revised for students in an emergency medical technician program or a first-aid class:

REPORT FOR A SEMI-TECHNICAL AUDIENCE

Subjective: Patient complains of shortness of breath, secondary to motor vehicle accident.

Objective: —40-year-old, white male found conscious, sitting behind bent steering wheel

---

of passenger car with extensive front-end damage. *HEAD:* airway patent (breathing freely). *EYES:* attentive, pupils equally reactive to light. *SKIN:* warm and dry with some cyanosis present. *NECK:* No pain upon palpation. No jugular venous distention, respiratory retraction, or tracheal deviation present. *CHEST:* Chest movement asymmetrical. Inspection revealed flail segment on right chest at ribs 3 through 7. *NEURO:* Patient cooperative, alert, and oriented and able to respond to questions about his name, date, and location correctly. Glasgow coma scale of 15 (normal). *And so on . . .*

Assessment: Flail chest.

Plan: Administer physical exam, stabilize flail segment (chest wrap), immobilize C-spine using a KED (Kendrick Extrication Device), monitor vital signs, monitor percentage of oxygen in blood using a pulse oximeter, administer oxygen and IV line (lactated ringers, 14-guage, right antecubital, by paramedic #16541), monitor with electrocardiogram. Remove patient from vehicle to long spine board. Telemetry: radio consult while transporting to Chicago Hope, paramedic #1236 attending.

Did you understand more of the report this time? Without all the abbreviations and acronyms, a wider audience can understand the report. But, without at least some training, you still will not have a clear picture of the extent of the patient's injuries or his condition.

Nontechnical Audience

The last type of audience is the general public, an unknown audience, or any combination of technical, semi-, and nontechnical readers, including customers, clients, and patients. It might also include upper management—a group that is uninvolved with technical activities, but that must have enough information to make decisions for the company. This audience expects a clear organization that progresses from the background to the new information, with examples or illustrations to explain points that may be confusing.

For this audience, writers provide the most comprehensive treatment of the subject, such as common terminology, simple language free of jargon and technical data, a full background and orientation to the subject, and a complete discussion of the main points. To simplify difficult concepts, writers often compare technical processes to more familiar ones through analogies and metaphors. The following revision of the prior examples is addressed to a family member of the victim.

REPORT FOR A NONTECHNICAL AUDIENCE

Your husband was involved in a car accident. He's alert, cooperative, and oriented. We're treating him for five broken ribs on his right side, which are each broken in several places, called flail chest. This can cause breathing difficulties and even lung damage. When Emergency Medical Services arrived, he complained only of shortness of breath and seemed a little blue from oxygen deprivation. They removed him from the car, placed him on a spine board, gave him additional support for his chest, and gave him supplemental oxygen and saline. His initial examination revealed the broken ribs, but no other neurological or respiratory problems. He is presently receiving oxygen and saline, and we'll continue to monitor his heart and other vital signs.

---

2. What does the audience want to know?

The audience, whether technical or general, might want only the highlights of the information. For example, a manager might want bottom-line information, such as total cost, time frame, or budget impact.

Or the audience might want detailed information, including all the background, procedures used, visual aids, data tables, and your conclusions. For example, customers will want estimates and explanations for repairs, especially if it's bad news, or troubleshooting information to solve or prevent a problem. Or coworkers might want you to provide exact procedures for a process.

3. What does the audience intend to do with the information?

This is the critical question. People read technical information for a purpose. Sometimes that purpose is simply for general interest. If so, you can make the subject more interesting for this audience by providing graphics, examples, and colorful details. Journalists and science writers address this audience, as you will see in a few of the reading articles in this book.

Other times, the audience wants to follow a procedure, solve a problem, or make a decision. Writers must anticipate questions and provide the organization and details this audience needs. For example, a manager might want the information needed to complete a projected budget for next year. A colleague might want to replicate a lab procedure. A customer might pay a bill (or refuse to pay it) based on an explanation of your service.

4. False assumptions about audiences

Unfortunately, writers sometimes make false assumptions about their audiences.

Assumption: My audience speaks and reads English.

Fact: About 10% of the people living in the United States were born in another country. Avoid those prize-winning vocabulary words—stick to the simplest appropriate terms you can find. And avoid words without a precise meaning, such as *really, very,* and *nice.*

Assumption: My audience will read the complete report or manual.

Fact: People normally don't read long documents or manuals from cover to cover. When's the last time *you* read an entire user manual? Include an executive summary, table of contents, and headings and subheadings to help your readers locate specific information when they might not have time to read the entire report.

Assumption: My audience will remember what I tell them.

Fact: Studies show that people forget up to 50% of what they hear within 10 minutes, and memory declines even more after that. They tend to remember more of what they read, and even more if they take notes or apply what they learn right away. For presentations or customer calls, experienced marketers provide something for the audience to read and take notes on. They use repetition and visual aids to help the audience remember. For example, they might create transparencies to project during the presentation and provide an introduction, such as a bulleted list of main points that will be covered, and a summary with a conclusion that reviews the main points.

Assumption: When listeners or readers don't understand, they will ask questions.

Fact: Formulating a question requires some degree of understanding. If the subject is too difficult or too new, the listener or reader might not be able to put a question into words. You can help by anticipating typical questions or trouble spots. Also, ask for feedback. Ask one person in your audience to recap the message in his or her own terms. Sometimes you can clear up misunderstandings or clarify points by listening to others paraphrase you.

Reading Comprehension Questions

1. What is the purpose of technical communication? _____

2. Define "audience" as it applies to technical communication.

   _____

3. What are the three questions you can use to analyze your audience?

   _____

   _____

4. Describe the types of technical communication you have written so far in your training or work experience, and the type of audience for whom they were written (technical, semi-technical, or nontechnical). _____

   _____

5. What types of technical writing do you expect to do in your career field? For what type of audience?

   _____

6. Describe one of the false assumptions about audiences and what you can do to prevent misunderstandings.

   _____

   _____

WRITING: Analyzing Your Audience

Writing is composed of many skills: thinking, researching, organizing, choosing words, and editing for spelling and grammar. It is easy to see why many people avoid writing altogether—it can seem overwhelming. However, as a technical student, you already have an advantage over other students: you are accustomed to learning individual skills that you can apply as needed to complex situations. Similarly, writing can be broken down into individual skills that you apply as needed for different assignments and purposes.

The writing sections of this text present individual skills used in technical writing, accompanied by examples and exercises to help you develop those skills. In Part I, you will practice many fundamental skills of writers: analyzing the audience, adjusting language and style, and organizing information. In Part II, you will practice many short types of writing, called the elements of technical writing, such as technical definitions and descriptions. Finally, in Part III, you will apply these elements to produce reports, business communications, presentations, and job-search materials. Skills are organized from simple to complex, with guided practice in writing, as well as in related skills such as spelling, vocabulary, grammar, and mechanics.

TIPS   Technical communication begins with analyzing the intended readers, called the audience. Technical writers answer the following questions to analyze their audience before they begin writing:

1. *What does the audience already know about the subject?*
2. *What does the audience want to know?*
3. *What does the audience intend to do with the information?*

This section provides practice in identifying how writers edit their passages for different audiences.

Exercise 1.1   *The following examples demonstrate how information changes for different audiences. Read the examples, keeping the intended audience in mind.*

The first passage was written for people who are familiar with fly-fishing. The author uses several terms and concepts that are unfamiliar to readers who do not fish. Although it is efficient and clear to fishing enthusiasts (a technical audience), it might be difficult for a nontechnical audience. Underline the words or phrases that seem to be specialized for the subject of fly-fishing.

PASSAGE 1

I sat on the bank studying the riseforms for several minutes. They were taking something too small to see. Finally, I selected a size 16 Griffith's Gnat and attached it to my tippet. It looked like the Trico hatch was on, and I wanted to get in on the action!

Passage 2 is revised for a general audience, such as people who enjoy regular fishing or the outdoors. The author explains each term or uses simpler terms than those in the first passage. And each fact is stated individually, with a logical progression of information. Underline the added information.

PASSAGE 2

I sat on the bank studying the surface of the water for several minutes. The trout were feeding on something too small to see, leaving spreading dimples (riseforms) as they slurped the invisible insects floating on the surface of the stream. Judging from the time of day, the season, and the minuteness of the insects, I reasoned they were feeding on Tricos, a small mayfly of the family *Tricorythodes*. I took out my flybox and selected the smallest artificial fly that I had with me, a black Griffith's Gnat. Though larger than the natural Tricos, the Griffith's Gnat will often fool trout when Tricos are abundant. I tied it onto the end of the fly line and said a short prayer.

Passage 3 was written for people learning about rigging a sailboat. The author uses several terms and concepts that are unfamiliar to nonsailors. While it is efficient and clear to professionals or even to advanced sailing students (a technical audience), it is too difficult for a nontechnical audience. Underline the words or phrases in Passage 3 that are specialized for sailing.

PASSAGE 3

Rigging is broken down into two major categories: running rigging and standing rigging. Running rigging consists of all the lines on a boat that are easily adjusted, including halyards and sheets. Standing rigging consists of the wires that hold up the masts of a sailboat, including stays and shrouds.

Passage 4 is revised for a general audience, such as people who enjoy watching sailboats or those wanting to become sailors. The author explains each term or uses simpler terms than those in the original passage. And each fact is stated individually, with a logical progression of information. Sailors will find it slow, but a more general audience needs the extra details. Underline the words or phrases that the author added to make the fourth passage appropriate for a general audience.

PASSAGE 4

Sailboat rigging consists of all the wire and lines (ropes are called "lines" on a sailboat) that are attached to the mast and sails. It is broken down into two major categories: running rigging and standing rigging.

Running rigging consists of the lines that are easily adjusted, including halyards and sheets. Halyards are the lines that raise and lower the sails, and sheets are lines that adjust them in and out laterally. The running rigging controls the sails and keeps the sailboat running (moving) with the wind.

Standing rigging consists of the wires that hold up the masts of a sailboat, including stays and shrouds. Stays are the wires that keep the mast from falling over the bow (front) or stern (back) of the sailboat. Shrouds keep the mast from falling athwartship (over either side of the boat). The standing rigging keeps the mast standing upright.

Exercise 1.2   *Read the following passage, which was written for a general audience interested in rock climbing. Then complete the exercises that follow.*

PASSAGE 5

All beginning climbers will have the protection of a belay—a top rope to check their falls—secured by the climbing leader. The first task of all students is to learn to tie the bowline (pronounced "bō'-lynn"), the knot used in belaying. Climbers are also customarily taught a set of verbal signals, or words and phrases that communicate important safeguards. The climb leader ascends the first leg of the route and finds a ledge from which to begin the belay. The first climber watches until the leader stops climbing. The climber secures the belay rope around her waist and calls, "On belay?" The belayer (climb leader) assumes brace stance above the climber, and calls back, "Belay on." The climber then calls, "Climbing." The belayer responds with "Climb" and begins to pull up any slack in the rope as the climber slowly ascends.

During the climb, the climber might yell "Slack!" which tells the belayer to pay out more rope for the climber to maneuver before resuming pulling up the slack rope. The "Rock!" signal, yelled downward loudly, is obligatory because it signals loose, falling rock (or equipment) to climbers below. When the climber reaches the belayer's position, the belayer yells "Belay off!" which informs the next climber to prepare to climb.

1. Underline the technical terms defined in the article.
2. Rewrite the passage for a technical audience composed of experienced rock climbers in training to become climbing leaders. Delete the information that the professionals will not need.

_____

_____

_____

_____

_____

_____

Exercise 1.3   *Read the following passage, which was written for a general audience attending a first-aid class on caring for an accident victim suffering from shock. Then complete the exercises that follow.*

PASSAGE 6

Shock is the result of hypoperfusion (inadequate or low perfusion) of the body's cells and tissues caused by insufficient blood flow through the capillaries. Before treating any patient, put on protective gloves. The next step of shock management is to maintain an open airway using the standard method of tilting the head. Administer a high concentration of oxygen using a nonrebreather mask, a device that fits over the patient's mouth and nose for a controlled flow of oxygen.

Check the patient's heartbeat, and perform cardiopulmonary resuscitation (CPR), if necessary. Check the patient for external bleeding, and if any exists, control it using pressure bandages. If the patient has no serious injury, raise the victim's feet 8–12 inches, called the Trendelenberg position. Cover the patient with a blanket to stop heat loss. Transport the patient to an emergency room immediately, and treat any secondary injuries, such as broken bones or joints, along the way.

---

1. Underline the technical terms defined in the article.
2. Rewrite the article for an audience of paramedics and ambulance workers, who need just a quick reminder of the steps to follow. Rewrite the passage into *no more than ten brief steps.*

Step 1: _____

Step 2: _____

Step 3: _____

Step 4: _____

Step 5: _____

Step 6: _____

Step 7: _____

Step 8: _____

Step 9: _____

Step 10: _____

Exercise 1.4   *Copy a passage from a textbook or magazine, similar in length to those in the examples, that uses technical terms and jargon in your career field. Then rewrite the passage for a general audience. Define technical terms, and add the background necessary to understand at least one concept in the passage.*

SPELLING: Using the Spelling Check

Each chapter in this text contains a spelling unit that focuses on one spelling pattern. Appendix 4 contains a longer list of frequently misspelled words.

Correct spelling is a key ingredient in effective written communication. If you misspell common—or even uncommon—words in your technical writing, others perceive you as either careless or unprofessional. Spelling can be compared to hygiene: people notice bad hygiene, but they usually don't notice good hygiene. The same is true of spelling. When you read "good" spelling—words spelled correctly—you are probably not aware of spelling. But when you spot a misspelled word, you stop reading, stare at the misspelled word, lose your train of thought, and possibly start to lose faith in what you are reading.

You might ask why spelling is included in this text, especially since word processing programs provide handy spelling-check tools that highlight misspelled words. Indeed, spelling checks are wonderful aids for writers. Most spelling checks in word processing programs operate the same way. They have a standard dictionary of 50,000 or more common words. As you type, or when you start the spelling check, the program matches each word in your document with the words in its dictionary. If it doesn't find your word in its dictionary, the program highlights it and searches for possible alternative words. When used regularly, spelling checks can catch and correct simple mistakes, saving us time and embarrassment. However, you cannot rely on spelling checks to do the entire job for several reasons.

1. Automatic spelling checks are not always available, such as when writing on eraser boards, forms, and even some e-mail programs.

2. The spelling tool cannot identify misused words. It blindly matches words with its dictionary, not recognizing how the word is used. Consider the following sentence:

   I was *two* hungry *too* sleep.

   The sentence above contains the homonyms *too* and *two*, which are spelled correctly, but used incorrectly. The spelling check will not highlight them. Only careful proofreading will find the misused words. (The **Word Watch** sections of each chapter will alert you to sets of misused words.)

3. Spelling check dictionaries do not include the most recent technical language. Spelling checks will highlight not only the latest technical words, but also specialized words, acronyms, names, and foreign words in your documents, even though you spelled them correctly. One solution for this is to create your own customized dictionary in your word processor. Then you can add technical words so that your spelling check will be faster.

4. Even when the spelling check highlights a word, you still have to either select or type the correct spelling—a risky task. Sometimes words are misspelled so badly that the program cannot come up with alternatives, or at least none that resemble your intended word.

5. Riskier still, some word processors let you "add" words to the dictionary. When you choose to add a word, be extra careful of the spelling. If you are in a hurry and accidentally tell the program to add a misspelled word to the dictionary, the program

will always accept that misspelled word in your documents. Most spelling checks have no way to remove words once you add them to the dictionary. If available, choose an option that allows the word to remain, such as "Ignore," "Skip," or "Allow in document," until you confirm the spelling.

Exercise 1.5   *Open your word processor, and describe the options available to you when the spelling check highlights a word.*

_____

_____

_____

_____

_____

Exercise 1.6   *Use the online help or manual to find how to add a customized dictionary for technical words, and once added, how to open or close the dictionary.*

_____

_____

_____

_____

Exercise 1.7   *Use the online help or manual to find information on the "Autocorrect" feature, if available in your word processor. Describe how to add and remove words that are replaced using "Autocorrect."*

_____

_____

_____

_____

_____

VOCABULARY: Prefixes *pseudo* and *quasi*

The more words we know, the more precisely we can understand the subtle differences in words and articulate our ideas when writing and speaking. For this reason, each chapter in this text contains a section to help build your vocabulary. Instead of your memorizing a list of words (which is usually ineffective anyway), the sections introduce sets of prefixes, suffixes, or root words that are the foundation of our language. The first set of prefixes is *pseudo* and *quasi*.

> *Pseudo* is a prefix that means "false" or a "fake."
>> pseudo-intellectual: pretending to be a thinker or educated
>> pseudonym: false name used by artists or authors
> *Quasi* is a prefix that means "approximately" or "in some sense or degree."
>> quasi-intellectual: some degree of education
>> quasi-official: partially official

Exercise 1.8 *Write the technical meaning of each word. Relate the meaning to the prefix.*

1. quasi + judicial _____
2. quasi + public _____
3. quasi + binding contract _____
4. quasi + serious _____
5. quasi + legitimate _____
6. pseudo + athletic _____
7. pseudo + classic _____
8. pseudo + sophisticated _____
9. pseudo + symptoms _____
10. pseudo + psychology _____

Exercise 1.9 *Describe the difference in expertise between persons labeled as the following:*

pseudotechnical _____

quasitechnical _____

WORD WATCH: *its* and *it's*

Each chapter of this text contains a Word Watch section that examines sets of commonly misused words—words that are similar in spelling or pronunciation, but different in meaning and usage. These words are especially troublesome because spelling checks ignore the context surrounding words, which is what determines correct word usage. Appendix 4 contains a longer list. The first set of misused words is *its* and *it's*.

Unlike some pronouns, the possessive form of *it* does *not* have an apostrophe:

> The document and *its* conclusion were the subject of the memo.
> The writer studied the group and analyzed *its* information needs.

Use the apostrophe when forming the contraction of *it is,* spelled *it's.*

> *It's* not easy to anticipate the audience's questions.
> I looked at the document, and *it's* clear that the writer knew the audience.

Hint: If you can substitute *it is,* add the apostrophe for the contraction. If not, leave it out.

Exercise 1.10 *Complete the sentences with the correct form of* its/it's.

1. I analyzed the audience and _____ needs on the subject.
2. _____ not clear whether upper management is included in the audience.
3. We decided to include the term and _____ meaning.
4. _____ unwise to categorize the marketing department and _____ manager as a "technical" audience.

5. While the marketers know who purchases the device, they are not familiar with _____ components.

6. When you describe a component, include _____ location in the device.

7. Everyone involved agreed that _____ time to distribute the document for review.

8. After the document was reviewed, I decided to change _____ title.

9. My first reviewer thinks that _____ not necessary to add graphics.

10. My second reviewer suggested that I add a graphic and make sure _____ the latest model.

CHAPTER 2

Language and Style

- Write sentences using the active and passive voice.
- Adjust sentence length.
- Eliminate single and double negatives.
- Write using a neutral tone.
- Double the final consonant correctly when adding a suffix.
- Use negative prefixes correctly.
- Use *a/an/and* correctly.

READING: Strong Writing Skills Essential for Success, Even in Information Technology

by Paula Jacobs

> ## Hot Job
> **Electronics Foreman:** FT, top salary, growth potential, good benefits. AS degree preferred. Minimum 5-yrs' experience with broad background in electrical repairs. Strong computer and communication skills a plus. Writing ability required. Proficiency in Word/Excel required.

No matter how good your technical skills are, you probably won't move up the Information Technology (IT) career ladder unless your writing measures up. "One of the most surprising features of the information revolution is that the momentum has turned back to the written word," says Hoyt Hudson, vice president of IS at InterAccess, an Internet service provider in Chicago. "Someone who can come up with precise communication has a real advantage in today's environment." Whether you are pitching a business case or justifying a budget, the quality of your writing can determine success or failure.

InfoWorld, Framingham, July 6, 1998

Writing ability is especially important in customer communication. Business proposals, status reports, customer documentation, technical support, and even e-mail replies all depend on clear, written communication.

Alan Cunningham, a manager at Computer Sciences Corp. who is working on a project at NASA's Marshall Space Flight Center, in Huntsville, Ala., says many failed partnerships between business personnel and their IT counterparts can be directly attributed to lack of communication between the parties.

"Without good communication skills, IT professionals are little good to business people because there is no common platform," Cunningham says. "Just like all IT

How good writing can help you advance

▶ Increases customer satisfaction

▶ Saves time

▶ Improves communication across the organization

▶ Enhances your professional image

▶ contributes to business success

▶ Raises your professional status

professionals should have to take some elementary finance and accounting courses to better understand business processes and methods, every IT professional should be able to write cogently and explain technical elements in readable English."

"Knowledge may be power, but communication skills are the primary raw materials of good client relationships," Cunningham adds. Every job description for a new position on his staff includes the following line (which would include other languages if the business were international): "Required: effective organization and mastery of the English language in written and oral forms." Clear communication can enhance your reputation as an IT professional, says Kevin Jetton, executive vice president of the Association of Information Technology Professionals (AITP) and president of GeniSys Consulting Services, in San Antonio. It is especially important to communicate in plain English and not technical jargon when you are talking to a non-IT business executive.

"You can have the greatest technical skills in the world, but without solid communication skills, who will know and can understand?" Jetton says.

Even if you have limited customer contact, writing skills are essential. Larry McConnell, deputy registrar for information services at the Massachusetts Registry of Motor Vehicles, in Boston, says that unless you can communicate, your career will level off.

Your job efficiency may depend on how well others communicate, as well. Joe Thompson, product support lead at Kesmai, an online games developer in Charlottesville, VA, says his daily work often depends on somebody's writing skills. Whether he's communicating with the test department or with a customer, Thompson sees writing as the key to effective two-way communication.

Even if writing is not your forte, you can improve your skills. Many companies offer on-site writing courses or send their staff to business writing workshops such as those offered by the American Management Association (http://www. amanet.org) and other training organizations.

Pete McGarahan, executive director of the Help Desk Institute, in San Francisco, says one of the best investments of his career was hiring a trainer to teach business writing for IT professionals.

Check out writing courses at colleges and community education programs, as well. "College-level courses in English composition and creative writing help broaden skills beyond the technical 'myopia' common to many IT professionals, enabling them to establish rapport and truly communicate with their clients," Cunningham says.

Good writing requires practice. AITP's Jetton suggests becoming involved in community volunteer opportunities or professional societies, where you can work on newsletters or write committee reports. "Communication skills are an ever-evolving skill set," Jetton says. "You never have enough practice."

Reading Comprehension Questions

1. What does Mr. Hudson say is the biggest surprise of the information revolution?

2. In the sentence, "Even if writing is not your forte, you can improve your skills," explain the meaning of the word *forte*. Write the phonetic pronunciation.

3. According to the contributors, what are some of the things that happen when an employee does not have good communication skills?

4. List three suggestions made by contributors to improve your writing ability.

5. Suppose, as you get close to graduation, you see a job description that states: "Required: effective organization and mastery of the English language in written and oral forms." Discuss potential ways that you can demonstrate these skills to a prospective employer.

WRITING: Language and Style

The language and style writers use for technical communication depends on the audience—the background, need, and purpose of the people who will read the information. Writers choose the most effective terms, writing style, and organization to make the subject understandable to their audience. This section focuses on four factors: voice, sentence length, negatives, and tone.

Active and Passive Voice

In speaking, you can vary your voice and tone to communicate a message more effectively. For example, speaking slowly and clearly with a strong voice adds authority to a message. Speaking in a lively voice with changes in pitch and inflection can make the message more interesting. Good speakers learn how to use their voices to inform, persuade, and entertain an audience.

In writing, we also have different voices. The most common are called active and passive voice.

The **active** voice emphasizes the fact that the subject of the sentence does something. It directs attention to the subject:

The architect placed the blueprint on the table.

The emphasis is on the architect and what she did. There is no confusion about who put the blueprint on the table. Most technical writers today write predominantly in the active voice because it is more direct and easier to understand and follow than the passive voice.

The **passive** voice emphasizes the idea that the subject is acted upon. In doing so, it directs attention away from the person or thing doing the action, and focuses on the thing that was acted upon.

> The blueprint was placed on the table.

Now the emphasis is on the blueprint, not who moved it. Notice that the verb has a helping verb (*was*). This is a common signal for passive voice. Passive voice is effective if the acting agent (who put the blueprint on the table) is not important. However, most technical writers avoid the passive voice, particularly when writing procedures or instructions. In the pair of sentences that follow, which instruction is easiest to follow?

> The disk is inserted in the CD drive. (passive voice)
> Insert the disk in the CD drive. (active voice)

In the first sentence, the reader might ask whether she has to insert the CD or whether this is optional. The second sentence, written in the active voice with an implied "you," clearly tells the reader what to do.

Consider the following sentences, and decide which message sounds stronger:

> Communication skills can be improved by taking college writing classes.
> Take college writing classes to improve your communication skills.

The first sentence is weaker due to the indirect language and passive voice. The second sentence, using the active voice, has a clearer message with a stronger tone.

Exercise 2.1   *Underline the verbs in each sentence. For each set, write* **A** *next to the sentence written in the active voice.*

1. The planner included a new printer in the budget.
   A new printer was included in the budget.
2. We mailed the final proposal to the customer.
   The final proposal was mailed to the customer.
3. Push the Enter key to start the program.
   The program is started by pushing the Enter key.
4. The spacecraft continues to take pictures in space.
   Pictures are continually being taken in space.
5. The composition and structure of the moon are revealed in pictures.
   The pictures reveal the composition and structure of the moon.

Exercise 2.2   Rewrite the following sentences in the active voice.

Example

Passive voice: The gasoline engine has been re-engineered by the auto industry.
Active voice: The auto industry re-engineered the gasoline engine.

1. Old engine designs have been improved by engineers.
2. Fuel injection can be coordinated by engine management computers.
3. Metals have been replaced by plastics and ceramics.
4. Less power is consumed by plastic engine components.

5. No heat is transferred by ceramic.

6. The car is started by turning the key.

7. Chassis systems have not been neglected by manufacturers.

8. More power is needed by small engines.

9. Present innovation has been dominated by advancements in digital control.

10. The economy has benefited from the new innovations.

Exercise 2.3   *In the reading passage, identify examples of active and passive voice and determine the predominant style.*

Sentence Length

Technical communication can be full of ideas and facts. Shorter sentences are usually easier to understand than long, complicated sentences. Critics agree that sentences over 25 words are too long for most readers to understand. However, use this as a guideline, not a rule. A message consisting only of short sentences will sound choppy and artificial; sometimes the relationship between ideas gets lost when sentences are too short. Compare the following sentences:

> The bird flew into the yard. The cat was waiting. The cat was in the shadow. The tree had the shadow.

These sentences are too short and too much alike.

> When the bird flew into the yard, the cat was waiting in the tree's shadow.

In this sentence, the logic is simplified by combining ideas and using a signal word (*when*) to relate the ideas.

A document with sentences that are similar in length and structure sounds dull. Variety in sentence length and construction makes the message more interesting; however, sentences can quickly become complex and dense. Technical writers must pay close attention to sentence length for understandability and interest.

Examples

Too Long: As the brain cells grow, connections called synapses form between the billions of brain cells, or neurons, that process information and allow excitations to pass from one neuron to another.

This problem sentence can be revised by breaking it into two sentences:

Better: As the brain cells grow, connections called synapses form between the billions of brain cells, or neurons. Synapses allow excitations to pass from one neuron to another.

Too Long: The technique involves infecting some neurons in a slice of brain with a benign virus that causes cells to produce internally a fluorescent dye that can be scanned by an infrared laser.

This problem sentence can be revised by breaking it into two sentences:

Better: The technique involves infecting some neurons in a slice of brain with a benign virus. This process causes cells to produce internally a fluorescent dye that can be scanned by an infrared laser.

Too Choppy: The laser has energy. The energy was too weak. The laser did not excite the neurons.

These problem sentences can be revised by merging them into one sentence:

Better: The laser's energy was too weak to excite the neurons.

Exercise 2.4   *Rewrite the following paraphrased sentences from the reading passage into appropriate sentence lengths.*

1. College-level courses in English composition and creative writing help broaden skills beyond the technical 'myopia' common to many IT professionals, enabling them to establish rapport and truly communicate with their clients.

2. You could have limited customer contact. Writing skills are still essential. You must have communication skills. Without them, your career will level off.

3. Knowledge may be power, but communication skills are the primary raw materials of good client relationships, and every job description for a new position on the staff includes the following line (which would include other languages if the business were international): "Required: effective organization and mastery of the English language in written and oral forms."

4. Your job efficiency may depend on how well others communicate, as well, because your daily work often depends on somebody's writing skills to communicate with the test department or with a customer.

5. Writing ability is especially important. It is important in customer communication. Business proposals, status reports, and customer documentation all depend on clear, written communication.

Negatives

Understanding negative sentences is difficult. People often misread or fail to see the negative word, such as *not,* or a negative prefix, such as *non* or *un.* Understanding a negative sentence is especially difficult when the sentence contains two (or more) negative terms.

Consider this quotation from author Ernest Hemingway:

"Never think that war, no matter how necessary nor how justified, is not a crime."

It takes some re-reading to determine that his opinion of war is that all war *is* a crime. Mentally, the reader must shift the negative to a positive to understand the author's opinion of war. In the quotation, the sentence contains the two negative terms: *never* and *not.* As in mathematics, the two negatives "cancel" each other, and we can remove both negatives and reword it into an active, positive form. However, this takes time and effort, and it is risky. Fast readers might miss one of the negatives and arrive at the opposite conclusion from what the author intended.

Consider another example:

It is not unlike Jim to wash his car on Saturday.

This passive sentence with two negatives (*not* and *unlike*) leaves the reader wondering whether Jim washes his car on Saturday or not. The reader must mentally reword the sentence to get the correct meaning:

Jim usually washes his car on Saturday.

Consider the following sentence:

> He has little to gain by not finishing the class.

The negative terms, *little to gain* and *not finishing,* make a positive statement. The sentence can be reworded into the positive form:

> He has much to gain by finishing the class.

Even single negatives in sentences are risky. Consider this direction:

> Do not remove the battery before turning off the device.

Many readers overlook *not* in sentences. If possible, reword the sentence to avoid negative instructions.

> Turn off the device before removing the battery.

If using the word *not* is unavoidable, underline or capitalize it to call attention to it:

> Do NOT touch the casing while the machine is running.

Exercise 2.5   *Reword the following sentences to eliminate the negatives.*

1. Do not misread the directions.

_____

2. It is not uncommon for a manager to write many types of documents.

_____

3. Do not start your career without taking at least one writing class.

_____

4. If you cannot communicate well, you will not climb the career ladder.

_____

5. You cannot establish rapport with customers if you do not have good communication skills.

_____

Tone

Look in any thesaurus and you will find many words to express an idea. For example, if you look up *thin* or *narrow* in a thesaurus, you will see pages of terms to substitute or clarify the idea of thinness or narrowness. These words are not interchangeable. Although they have similar basic meanings, they might have a specific tone, which writers refer to as positive, negative, or neutral. Tone is the attitude, or emotion, of the writer, expressed in word choice. Tone can be casual or formal. It can be positive, negative, or neutral. Writers set the tone of their documents by the words they use.

- Formal English creates a formal tone. (Business people consider this stuffy.)
  "The purpose of this interview is to determine your predilection toward written communication."
- Informal English, using slang or abbreviated words, sets a casual tone. (Business people consider this unprofessional.)
  "Interested in writing? Check this out."

- Flowery words set an insincere tone. (Business people consider this gushy.)
"We thank you very much for your exceptionally enthusiastic interest in our opening."
- Sarcastic or angry words set a negative tone. (Business people consider this unproductive or mean.)
"We need a writer, and with some recycling, you might fit."
- Neutral words set a businesslike tone, free of emotion or manipulation. (This is the preferred tone in business writing.)
"We have an opening in our technical writing department. Would you like to hear about it?"

Read the following pairs of synonyms, and determine which column has a more positive tone, and which a more negative tone.

Synonyms with different tones:

suggest	insinuate
rumor	gossip
assertive	aggressive
pause	hesitate
eager	brash
enthusiastic	fanatical
knowledgeable	know-it-all

The words in the first column are considered complimentary. The words in the second column, although they have the same general meaning, are considered more negative in tone.

Some words or phrases are considered "loaded," which means they have strong emotional meanings. Examples of loaded words are listed below.

Loaded words:

red-neck	hippie	yuppie
hot-head	sexy	leading edge
kook	excellent	state-of-the-art
lean and mean	macho	feminist

Other words with loaded meanings include nicknames for ethnic, religious, or political groups; political sayings; and sexual terms. If a loaded word or phrase is insulting, it is not appropriate in any type of professional writing. Some loaded words are harmless or possibly useful. For example, a "lean and mean" department currently implies a small, highly trained staff capable of working intensively on projects. However, some expressions have become overused and have lost a clear meaning, such as "a nominal fee." Overused or trite expressions should be avoided.

Some words and expressions are considered formal, neutral, or slang. In the following list, compare the synonyms that set a formal, neutral, or slang tone.

Formal Words	Neutral	Slang
fortuitous	fortunate	lucky
contemplate	consider	chew on
copious	many	gobs
reiterate	repeat	ditto
elucidate	explain	draw you a picture
dialogue	conversation	rap
recalcitrant	stubborn	muleheaded
disconcerting	upsetting	a downer

Try to keep a neutral tone in most business writing. Avoid abrupt shifts in tone. Shifting from one tone to another is confusing to readers. They may feel manipulated or misled. This problem in writing is the hardest to identify and correct because it does not consist of grammatical errors. It consists of word choice and phrasing. Look for the tone shifts in the example below.

> I am interested in applying to your company for the position of lab technician. I have this brainy friend who works for you and he thinks your company is proliferating and in tip-top shape.

The revised paragraph eliminates the loaded words (*brainy*), slang (*tip-top*), and overly formal word (*proliferating*).

> I am interested in applying to your company for the position of lab technician. A friend who works for you has described your company as productive and well managed.

Exercise 2.6 *Decode the following common expressions which have been garbled with overly formal words. Write the common expression.*

Example

Formal: Dispatch remuneration expeditiously.
Decoded: Send money soon.

1. Nothing flourishes like prosperity. _____

_____

2. If you can't subjugate them, enroll. _____

_____

3. Bestir yourself and detect the redolence of java. _____

_____

4. Rapidity of motion makes refuse and debris. _____

_____

5. Dual craniums are superior to sole examples. _____

_____

Exercise 2.7 *Revise the following messages to eliminate shifts and set a consistent, businesslike tone. Answers may vary.*

Example

Example: I'm sending back this piece of junk.
Revised: *I am returning this product.*

1. What's happenin'? I caught a jet to drop in on you and found you had cut out. I'd like to get a line on that new company you slave over.

_____

_____

2. I understand you have a prodigious assortment of topical technical periodicals. Your appraisal would be sincerely appreciated. Give me a jingle.

_____

_____

3. I won't tolerate any shilly-shallying on this undertaking. Vacillation will only retard the resolution.

_____

_____

4. We would like to hang out while you put on the feedbag and deliberate our druthers.

_____

_____

5. What do you envision as our game plan? Let's get the prerequisite go-aheads, like yesterday.

_____

_____

Review of Preferred Language and Style for Technical Writing

TIPS   Follow these guidelines for technical writing:
- *Use active voice, which emphasizes that the subject does an action.*
- *Keep sentences shorter than 25 words in length.*
- *Write statements in the positive form. Avoid negatives (single or double).*
- *Use a neutral, businesslike tone.*

SPELLING: Doubling the Final Consonant

The problem of doubling or not doubling the final consonant when adding an ending, or suffix, is one of the easiest and most consistent spelling rules. The doubling rule is used only if the *suffix,* or new ending, begins with a vowel (ING, ED, ANCE).

First, remember that the five **vowels** are A, E, I, O, and U (we do not consider Y for this rule). All the rest of the letters in the alphabet are called **consonants.** Now, let's look at **one-syllable** words.

Rule: If the last two letters in a word are a single vowel followed by a single consonant, double the final consonant before adding the ending.

We'll call this the **one and one rule.** Notice that each of the following words ends in one single vowel followed by one single consonant.

> *Examples:*   r *un* + ing = running
> h *op* + ed   = hopped
> pl *od* + ing = plodding

Notice that the following words do not follow the *one and one* rule.

> *Example:*   seat + ed   = seated
> test + ing = testing

Note: Certain letters are never doubled, even if they follow the rule. They are W, X, and Y.

> *Example:* draw + ing = drawing
> say + ing = saying
> box + ed = boxed

Exercise 2.8 *Add the following endings to the words.*

1. jam + ed _____

2. band + ing _____

3. hum + ing _____

4. link + age _____

5. trim + er _____

6. drop + ed _____

7. fix + ed _____

8. loop + ing _____

9. trip + ed _____

10. ring + ing _____

There is one more part to this rule. It concerns words with **more than one syllable.** When a word has more than one syllable, one of them will be **stressed**—pronounced louder or with more emphasis than the others. Dictionaries have an accent mark after the stressed syllable (in the phonetic spelling).

Rule: Double the final consonant of a word with two or more syllables if it follows the previous rule (one and one), and if the stress is on the last syllable.

The following words conform to the rule above, called **one and one and last rule.**

> *Example:* refer + al = referral
> submit + ing = submitting

The following words do not conform to the *one and one and last* rule.

> *Example:* resist + or = resistor
> system + atic = systematic
> relax + ing = relaxing

Exercise 2.9 *Add the following endings to the words.*

1. transfer + ence _____

2. decay + ed _____

3. fit + ing _____

4. travel + ing _____

5. display + ed _____

6. control + ed _____

7. occur + ence _____

8. transmit + er _____

9. gather + ed _____

10. admit + ance _____

Exercise 2.10 *Now check your memory by writing the rule for doubling the final consonant.*

VOCABULARY: Negative Prefixes

There are several negative prefixes that we use to change a word to mean the opposite or to signify *not*.

Negative Prefix		Root Word		New Word
a	+	symmetric	=	asymmetric
de	+	activate	=	deactivate
dis	+	appear	=	disappear
il	+	legal	=	illegal
ir	+	relevant	=	irrelevant
mis	+	placed	=	misplaced
non	+	sense	=	nonsense
un	+	known	=	unknown
counter	+	act	=	counteract

Note: Do not change the spelling of the root word when adding a prefix.

Exercise 2.11   *Choose the correct negative prefix. Use* mis *or* dis.

1. Theo used his calculator so that he would not _____calculate.
2. We stared at the spectacle in _____belief.
3. The _____orderly appearance of the lab was corrected.
4. The technician corrected the _____spelled words.
5. Susanne knew that leaving work early would _____please her manager.

Use ir *or* il. *(Notice that the beginning letter of the root word sometimes gives a clue to the prefix.)*

6. Some of the damaged equipment was nearly _____replaceable.
7. The results of the experiment seemed _____logical and _____ rational.
8. The assignment was _____ legible because the pencil lead smeared.
9. The _____regular towels were unusual shapes and sizes.
10. The _____ literate man could not read the documentation.

Note: The prefix *il* (usuallyhyphenated *ill-*) means "bad".

   Examples: an *ill-fated* experiment
         an *ill-natured* boss
         an *ill-advised* procedure

Use un *or* non. *(Often* un- *and* non- *are used interchangeably, but one will be used more commonly.)*

11. This section of the restaurant is reserved for _____smokers.
12. My supervisor was _____willing to let us miss our deadline.
13. The happy-go-lucky worker seemed _____troubled and relaxed.

14. The American Cancer Society is a _____profit organization.

15. Michael finally realized that his _____professional behavior cost him a promotion.

Use a, de, *or* counter.

16. Right triangles and equilateral triangles are _____symmetrical.

17. The radio receiver had to _____code the message.

18. Lynn turned the dial _____clockwise.

19. Since the clocks were _____synchronous, the machines could not communicate with each other.

20. The carpenter had to _____sink the drilled hole.

Note: These prefixes *do not always* mean "the opposite of," so be careful when using them.

> *Examples:* He read the passage *aloud* (out loud).
> The American dollar has been *devalued* (reduced in value).

WORD WATCH: *a, an* & *and*

Using *a, an, and*

The words *a, an,* and *and* are often misused. Sometimes this happens because the writer is not sure when to use *a* or *an.* Other times, the writer means to say *and,* but forgets to add the final *d.* These three words are *not interchangeable.* They each have a specific function, or job, in sentences. Using the wrong word will make you appear careless.

A / an (as well as *the*) are used in the same situations (in front of single nouns), but one *or* the other is used.

a car	an automobile
a circuit	an integrated circuit
a direct current	an alternating current
a slope	an elevation
a uniform	an hour
(uniform begins with a hard *u* or *y* sound)	(hour begins with an *o* sound)

Do you notice a pattern? Look at the first *sound* following *a / an.* Notice that sometimes the first *sound* of the word is different from the first *letter.*

> Use *a* before words beginning with a consonant sound.
> Use *an* before words beginning with a vowel sound.

Remember that **vowels** are the letters *a, e, i, o, u* (and sometimes *y*). All other letters are **consonants.** Notice that the first *sound* of a word is not always the first *letter.*

And is a conjunction used between words. Sometimes people do not pronounce the final *d* or fail to add it when they write. Remember that *an* is not a substitute for *and.*

Rules: Use *a* in front of a word beginning with a consonant sound.
Use *an* in front of a word beginning with a vowel sound.
Use *and* to join words.

Exercise 2.12   *Fill in each blank with a, an, or and.*

Time management is _____ skill that students must learn if they hope to handle all their responsibilities. Reducing traveling time saves _____ hour here _____ there. Bringing _____ bag lunch means being able to eat without hunting for _____ inexpensive cafe. These things are necessary for people trying to go to school, keep _____ part-time job, study, _____ most importantly, find some time to relax. Many students find _____ hobby that is inexpensive _____ active. Jogging, calisthenics, _____ dancing are _____ few activities that provide _____ healthy outlet for stress that builds from keeping _____ demanding schedule. With good pacing and planning, students can make the most efficient use of their time _____ keep _____ enthusiastic, energetic attitude.

CHAPTER 3

Organization

- Classify words by category.
- Write topic sentences with colons and without colons.
- Identify main ideas and supporting details in paragraphs.
- Write a paragraph with a main idea and supporting details.
- Write an outline of a paragraph.
- Take notes on a lecture.
- Spell words ending in *ise, ize,* and *yze* correctly.
- Use *mono, bi, poly,* and *semi* correctly.
- Use *a lot* and *allot* correctly.

READING: Taking the Noise Out of Technical Writing

by John W. McDonald

In any communication system, the accuracy and intelligibility of the message is limited by noise. Noise enters written communication as "linguistic noise" (misfit words, confused syntax, and faulty organization) and "typographic noise" (printing errors, poor graphic design, and faulty reproduction). The concepts of noise filters, modulation, and signal-to-noise ratio are applied to technical writing in terms of careful writing, editorial monitoring, and relevance of the message to a particular audience.

Introduction

Noise has recently become a topic of great interest. Unwanted and excessive noise pervades our society, with effects ranging from the distractive to the destructive. Legislative bodies have responded to public awareness of the detrimental effects of noise by imposing standards for industry, construction, and transport. One wonders whether the sonic boom of the SST will ever be as loud and disturbing as the political noise it has already generated.

All communication, with the possible exception of divine revelation, takes place in the presence of noise. In all person-to-person communication, we must rely on our truly remarkable ability to detect useful signals in an environment of ubiquitous noise. And hope that we don't make matters worse by irritable reaction to the task of sorting out what is useful in a badly constructed message that is garbled in transmission, distorted by the carrier or medium, and filtered through our own sieve of preconceptions and prejudices.

Before we talk about how to reduce noise in communication, let's discuss what noise is, and how it gets into the communication system.

"Taking the Noise Out of Technical Writing," by John W. McDonald from *Technical Editing: Principles and Practices,* May 1975, pp. 28–32. Reprinted by permission of The Society of Technical Communication.

Sources of Noise

The basic limitation on the speed and accuracy of any mode of communication—whether oral or written—is due to noise that distorts the message, with consequent confusion for the receiver. For this talk, we will take the usual physics textbook definition: noise consists of unwanted disturbances, especially those that are random and persistent, that obscure or reduce the clarity and quality of a signal.

A simplified communication model consists of (1) the source, (2) the message, (3) the channel, and (4) the receiver, all superimposed on a background of ambient noise. A fifth element, that of feedback, establishes a link between the source and the receiver so that we can instantly evaluate how well the receiver is getting and understanding the message and so that the sender can modify or reinforce his message by reducing noise or by modulating the channel. Here we take modulation to mean any process that varies the signal for more efficient transmission, with the intent of superimposing intelligence on, or increasing the information content of, a carrier wave. The concept of modulation as applied to the editorial process may give rise to a quite different sort of noise if an author thinks his "style" is being violated.

In face-to-face conversation, noise comes from mispronounced words, distracting mannerisms, esoteric vocabulary, inattention, and such environmental effects as traffic, air-conditioner hum, and the like. To partly overcome interference, man's language is highly redundant so that many words and sounds in speech can be lost without serious effect on the message being sent. Redundancy allows man to detect signals in noise, and intelligibility can be achieved even when the noise intensity is very much larger than the message sound. This is described as the "cocktail party effect," as many of you will remember from the Wednesday night reception, whereby conversations can be carried on in an extremely noisy room. But just barely.

In conversation, we can use the instant feedback of a question, a raised eyebrow, a puzzled grimace, a rebuttal, or a glazed eye to alert the sender that the message isn't coming through.

All person-to-person communication suffers from noise, some more than others, but none so much as writing. The elements of the written communication system are (1) the writer, (2) the report, and (3) the reader. The system lacks an instant feedback channel that links the reader with the writer, and the first transmission has to be clear—and clearly understood. The reader has no way to influence the quality of the message except by angry "letters to the editor" long after the fact, and the writer cannot be certain that he is not creating noise.

In written communication, noise can occur at each of the three elements.

The writer can contribute linguistic noise at his semantic encoder. He may become the victim of his own semantic inadequacy that allows misfit words, mangled metaphors, and deadwood to creep in. The report itself may contribute typographic noise by printing errors, faulty reproduction, cluttered illustrations, and poor graphic design. Finally, the reader may be the victim of psychological noise, the source of which may be the message itself, semantic or typographic noise in the report, or some external or subjective effect.

Noise Abatement at the Source

Technical reports, by their nature, resist being measured by a common standard. But if there is any common measure for judging technical writing, it is organization—and faulty organization is most readily revealed by comparison with an outline.

Outlines. Writers fall into two classes: those who find outlines useful or indispensable, and those who consider them a nuisance. If you're in the first category, use an outline as a starting point and in-process guide; if you're in the second class, use an outline anyway but prepare it after the report is completed as a *post hoc* verifier.

Outlines have no standard form; they must be variable and flexible to suit the subject and scope of coverage. Nevertheless, an outline is as essential to solid writing as blueprints are to construction. The final product should not reveal the skeleton; the writing should be strong enough to hold the report together and solid enough so that no bare bones poke through. Minimum requirements for an outline are:

- State the epitome of the subject and your attitude about it.
- Give evidence to support your thesis.
- Summarize what you want the reader to remember, with sufficient support from evidence and sufficient clarity of expression to keep him from asking "So what?"

Another noise filter applied at the source is to remember that it is impossible to "tell all." The author must make a selection of data and arguments, but the selection must fairly correspond to the mass of evidence and it must offer a graspable design to the beholder.

Carelessness. The commonest source of noise is not the willful distortion of intent, but innocent lapses of attention that jerk us to wakefulness. It requires the unnodding vigilance of writers and editors alike to keep us from taking sweeping statements with "a dose of salts" rather than "a grain of salt."

Our daily work abounds with good examples of bad habits, all sources of noise. We have to know what correct usage is, which depends in part on knowing the difference between what a word denotes and what it connotes. We must draw fine shades of difference between *accuracy* and *precision,* between *activate* and *actuate,* between *alternate* and *alternative.* The list is long.

We must also watch for mangled metaphors that result from trying so hard to reach the descriptive high notes that the singer dies of syntactical strangulation. We all have lists of boners gleaned from our editorial labors. Some of my favorites are:

- "A virgin field pregnant with possibilities."
- "The need is evident for a list of physicists broken down by specialization."

- "No authenticated case is known in which sterile parents transmitted this quality to their offspring."
- "This publication fills a much-needed gap."

Our attention is demanded in culling out tautological repetition of the same sense in different words, as in "general consensus of opinion." And we must prune deadwood—it's as burdensome to prose as it is to a tree.

Someone may object that the distinctions are too finely drawn. Not so. Strictly speaking, technical communication is a matter of speaking strictly.

Spelling. A word is more than the sound it makes. It is, among other things, the way it looks on a page—and it looks best when it's spelled right. An author who writes with no misspelled words has already dismissed the first grounds of suspicion of the limits of his scholarship. He might remember that the fancy way of saying "correct spelling" is "orthography" (from the Greek roots *orthos* meaning *straight, correct, normal;* and *graphein* meaning *to draw, to inscribe*).

A good dictionary is a communicator's most useful single tool. An author who frequently consults a dictionary will usually learn something to his—and his reader's—advantage.

Proofreading and Goof-proofing. Examine the finished product to be sure that the message cannot be misinterpreted. Substantiate your claims, check your data, and ask colleagues for their opinions. If you are uneasy about your data or insecure in your interpretation, your writing will show it. And proofread! Proofreading is equivalent to goof-proofing. Look up words whose meanings have been dulled by familiarity, get the typos out of the final draft, and watch out for grammatical lapses. Don't be hasty to rush your literary brainchild to a baptism of printer's ink.

Editorial Modulation

Only rarely in creative activity, whether in research, in writing, or whatever, can we achieve a desired result at the first try. Rarer still is a draft manuscript that goes from writer to editor to printer without retracing part of its path for correction or improvement.

The act of creating any written piece starts a process that resembles the oscillation of a vibrating spring; this concept enables us to follow the path of a manuscript on successive trips between author and editor (see Figure 3–1). Imagine that the initial displacement from the solution axis is proportional to the degree of uncertainty, and that the restoring force of a spring represents the process of evaluation and generalization. At the maximum amplitude of each oscillation we apply a "damping" factor that consists of the act of defining and refining, and thus limit the next oscillation. Successive definitions at peak amplitudes, as the draft goes from author to editor and back again, contribute to decreases in the degree of ignorance. The envelope formed by decreasing displacements of these quantities is the convergence of understanding. It approaches, at infinity, the truth.

At the Los Alamos Scientific Laboratory, we construe editorial modulation to mean editing for consistency, organization, and language correction, with the specific intent of retaining the author's style and the accuracy of technical content. Some writers complain that an editor's efforts have changed the intended meaning. When that occurs, the writer was probably not clear in the first place. A writer who is concerned

lest editorial modulation violate his "style" can best preserve it by

- Writing clearly and directly.
- Making sure of his facts and marshaling them in logical sequence—with the "news" of his findings or a statement of the problem first, followed by details and supporting arguments.
- Imparting the significance of his work to readers outside his field.
- Providing enough details of method and equipment so that others can repeat his work.

He should apply a noise filter at the source by avoiding overspecialized jargon and acronyms, by using the active voice in preference to the passive, and by using personal pronouns (I, we) where appropriate. "The present author" went out with "gentle reader," so say "I" naturally and honestly when it fits. "We," unless the work is reported by more than one author, is used only by royalty, editors, and people with a tapeworm. Finally, it is no sin to occasionally split an infinitive or to use a preposition to end a sentence with.

Noise Abatement in the Report

Noise filters applied to the reproduction of a report are easier to use and control than the human factors of source and receiver. We can, by consistent and conscious effort, filter out printing errors, faulty re-

FIGURE 3–1

The creative process as a damped oscillation.

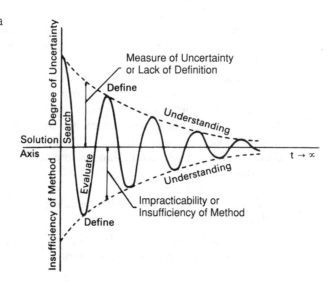

production, cluttered illustrations, over-crowded text, inadequate margins, stiff bindings, misplaced references, and poor display of mathematical expressions. The filter of language correction will take care of most lapses in punctuation, orthography, transitions, unity, consistency, and conformance to "house" style.

It is another matter, however, to nurture a happy editor-author symbiosis wherein writers accept the suggestions of editors who understand the limitations and requirements of printshop and audience, wherein editors learn the subject matter from authors in a mutual search for clearer writing, and neither insists on having his own way.

Noise Abatement at the Receiver

When we're at the receiving end of the communication chain, psychological noise may arise from misinterpreting the sender's intent, or from listening only to what we want to believe. Aside from this irreducible residuum of human cussedness, we know that the linguistic and typographic noise of a report, added to noise from the environment or from the recesses of our subconscious, can become psychological if it's bad enough. Our impatience with an author who seems not to be meeting his obligation of clear exposition can contribute to psychological noise. Noise filters at the receiver must, by and large, be applied by the receiver himself if he's willing to understand and takes the effort to seek and tune to the author's frequency.

Optimizing Signal-to-Noise Ratio

I guess it's possible for one person's noise to be another person's signal, but it is the relationship of what is relevant to what is extraneous that must be our constant concern. We, as communicators, are obliged to be source, message, channel, receiver, and feedback at one time or another—and sometimes we are asked to be all of them at once. To my mind, then, we must

- Have something to say (optimize the signal).
- Say it concisely and simply; avoid adjectival overkill (apply noise filters at the source).

- Say it in terms of interest to the reader (minimize noise at the receiver).
- Write correctly so that editors cannot distort the message or suppress the style (optimize editorial modulation; reduce noise in the report).
- Adjust the emphasis or choice of words as necessary to make sure that the message is getting through (pay attention to the feedback signal).
- Try to inform, not impress. Write as we would be written to.

Writing for Understanding

It has become a commonplace to say that scientists can't write. This is not true. Many of them write creditable prose every day, and some of them write very well indeed. To say that scientists can't write is an insult to those who do, and discourages those who should. Nevertheless, the quality of many manuscripts submitted to our Department for publication indicates that research scientists regard the recording and reporting of their work as an unpleasant addition to their already heavy burdens. Nothing of the sort. Writing, rather than impeding research, may properly be regarded as the mechanism by which the work is accomplished.

Michael Faraday, the great British physicist and lucid expositor of his own work, is reported to have defined research as consisting of three steps: Start, finish, publish. Research, in this view, starts on the basis of well-made plans that define a specific field of interest and finishes when results are obtained that are of a degree of accuracy compatible with the experimental design necessary to prove the hypothesis. Results are then written and distributed in a clear exposition of observations made and results obtained. I would encourage the writing of laboratory notes in daily increments as a stimulus to the scientist to think about his work in a more organized way and to use the intellectual discipline of writing as daily feedback to the actual progress and shape of the work.

Some complicated concepts are difficult to explain in an uncomplicated way, but the intellectual challenge requires an attempt to solve puzzles, not create them. The challenge lies not only in achieving scientific and technical advances, but also in meeting the demand for conveying involved concepts clearly and concisely; additional effort is required to present good ideas effectively. Scientists from different disciplines, in their attempts to explain something to a third party, may become comprehensible to each other. It is not enough to be understood; we must make sure that we are not misunderstood.

Reading Comprehension Questions

1. What is "noise" in writing? _____

_____

2. a. How does the writer contribute noise?_____

_____

b. How does the report contribute noise? _____

_____

c. How does the reader contribute noise? _____

_____

3. What are some methods for eliminating noise at the source? _____

_____

4. What are some methods for eliminating noise in the report? _____

_____

5. What are some methods for eliminating noise at the receiver?_____

_____

6. Why do you think it is important for technicians to write for understanding?

_____

WRITING: Organization

Analogy
Comparison of a new idea to a familiar idea

In the reading article, the author uses a technique called an "analogy" to impress upon an audience of scientists the value of clear writing. Analogies are comparisons of something familiar to something unfamiliar. An analogy is effective when the audience can draw a parallel between the elements of the familiar concept to explain similar or related elements in the new concept.

The analogy in the article draws on the physics concept of noise reduction to explain why and how science writers should reduce "linguistic noise" in their reports or presentations. Any scientist who is familiar with communication systems will, presumably, understand the technical definitions of *noise, modulation, abatement, source, receiver, channel, signal-to-noise ratio,* and *damped oscillation.* Even if you are not familiar with those terms, you could still understand the comparisons and the point of the article: that careful organization and editing are necessary to make technical writing clear and understandable.

Organizational Aids

In Chapters 1 and 2, you practiced some of the techniques mentioned by the author for writing to a specific audience and using appropriate language and style. This chapter focuses on the organization of paragraphs, which is also the main skill in writing outlines and taking notes.

TIP   The author uses several techniques that you can use as organizational aids:

- **An introduction** that sets the stage for the rest of the article
- **Headings** in large font to identify transitions to new ideas
- **Bulleted lists** to draw your eye to material, making it easy to read
- **A figure** that illustrates the comparison of verbal communication to the electronic analogy
- **A summary** ("Writing for Understanding") that restates the main idea

Writers of all types of technical documents, from e-mail to formal reports and proposals, can increase the clarity of their information by providing organizational cues such as introductions, topic headings, bulleted or numbered lists, transition words, graphics, and summaries. Some of these topics are included in later chapters in this book. This chapter covers paragraph development, starting with main ideas and topic sentences. But first, we must take a look at the overall process of organizing information into categories.

Classification

Before starting a paragraph, writers must decide what the topic of the paragraph will be. Usually a paragraph focuses on one main idea and the supporting ideas for that main idea. Sometimes writers start with an outline of the topics they want to include in the entire document, and then start to organize the ideas into logical paragraphs and sections.

Well-written technical articles present information in an order that eliminates the need for repetition. Once a term is defined, it can be used freely. Each section of the article discusses only one idea. Key ideas are underlined or titled. Ideas are presented in an order that prepares the reader for the next idea. It is this ordering that keeps technical writing "to the point."

Technical writers plan the logical ordering, or organization, of information before they start writing. They often use a strategy called **outlining,** which is a numbered and lettered sketch of the contents. They may then use the outline headings and subheadings as labels within the article or report.

Grouping or classifying information is also a method that some people use for memorizing information. For example, try to memorize the following list of words.

amplifier	fuse	Ohm
Boyle	giga	Pascal
capacitor	kilo	resistor
Curie	mega	tera

Now try to memorize the same words in a different order.

Metric units	Devices	Scientists
giga	amplifier	Boyle
kilo	capacitor	Curie
mega	fuse	Ohm
tera	resistor	Pascal

The first list is in alphabetical order, and the second list is in categorical order. It is often easier to remember items grouped by categories. When we organize, or group information in categories, it is also easier to discuss or write about the information.

Exercise 3.1   *Rearrange the following words into categories (four words per category) and label each category.*

antifreeze	gasoline	oil	sphere
cone	horitzontal	parallel	vertical
cube	manometer	perpendicular	voltmeter
cylinder	multimeter	speedometer	water

Categories: _____   _____   _____   _____

1.	1.	1.	1.
2.	2.	2.	2.
3.	3.	3.	3.
4.	4.	4.	4.

A paragraph, or a whole report, will usually have one main idea, such as a *category,* and several supporting details, such as the items within a category. Sometimes an author will state the category and list the details within the category, as in the following sentences.

> Kirchhoff's law is used in branch current analysis, mesh current analysis, and node voltage analysis.

> There are three methods of analysis that use Kirchhoff's law: branch current analysis, mesh current analysis, and node voltage analysis.

Writing Lists

A list at the beginning of a paragraph, section, or report is an effective way of telling the reader how you have organized your ideas. The second sentence in the example above states the list in an obvious way by using a **colon** (:) before the items. Although either way of stating the list is correct, the colon is a visual cue and thus is more direct.

The colon is a punctuation mark that means "as follows." It is a signal to the reader that a list is being given. Other signals for lists are words and phrases such as the following:

There are three kinds: . . . can be divided into three areas:

types:	classified
groups:	grouped
classes:	classed
categories:	categorized

Example: Resistors can be divided into two groups: fixed and variable.
There are two types of current: alternating and direct.

A colon is usually not used following a verb. After a verb, simply write the list.

Example: The two main categories of resistors *are* fixed and variable.

The two types of current *are* alternating and direct.

If your list includes more than two items, put a comma between the items.

Example: The company has three regions: the northeast, southeast, and west.

If commas are needed within items, separate the items with a **semicolon** (;).

Example

The article can be divided into two sections: the specialized and multipurpose information utilities; and the requirements of the hardware, computer-communications network, and software.

For typing rules about these punctuation marks, see Appendix 2: Tips for Word Processing.

Exercise 3.2   *For each of the following exercises, write a sentence that names the category, signals a list, and lists the items in the category. You are asked to write three sentences using a colon, and two sentences without using a colon.*

Example

A. Resistors

    1. fixed

    2. variable

    Sentence: There are two categories of resistors: fixed and variable.

1. Engineering

    a. electrical

    b. mechanical

    c. civil

    Sentence (use a colon)_____

_____

2. Girder types

    a. steel I-beams

    b. built-up wood girders

    c. concrete reinforced beams

    Sentence (use a colon)_____

_____

3. Rectangular Coordinates

    a. origin

    b. x-axis

    c. y-axis

    Sentence (use a colon)_____

_____

4. Rays produced by welding

    a. visible light rays

    b. infrared rays

    c. ultraviolet rays

    Sentence (do not use a colon) _____

_____

5. Four-stroke piston cycle

    a. intake

    b. compression

    c. power

    d. exhaust

    Sentence (do not use a colon) _____

_____

Topic Sentences

Topic sentence
Main idea of a
paragraph

Each paragraph should have a main idea. If it is stated in a sentence, it is called the *topic sentence*. The topic sentence sets the limits of the paragraph. It can appear at the beginning, middle, or end of a paragraph—or not at all (the inferred topic sentence). Good technical writers usually start each paragraph with a clear topic sentence to ensure readability and logic.

Consider the topic sentence, "Resistors are divided into two groups: fixed and variable." You can easily predict that the content of the paragraph will be limited to supporting details about the two types. Perhaps the paragraph will describe one type, the first one listed, and a second paragraph, beginning with "The second type. . . ," will complete the idea. Identifying the topic sentence is an important task for the reader.

Thesis statement
Central idea of a
report or docu-
ment

If your topic sentence becomes the central idea of a formal report or document, it is called a *thesis statement*. Formal reports begin with a thesis statement, followed by main ideas, each supported by details, as in the structure below.

Thesis Statement (overall topic)

Topic Sentence (main idea)
Supporting detail
Supporting detail

Topic Sentence (main idea)
Supporting detail
Supporting detail

If your topic is the central idea of an informal report or memo, it is called a *topic statement* or *subject statement*.

Exercise 3.3   *Outline the supporting details from the following paragraph.*

LINGUISTIC NOISE. The writer can contribute noise to the report. He may become the victim of his own semantic inadequacy that allows misfit words, mangled metaphors, and deadwood to creep in. The writer might skip important logical steps in the development of ideas, leaving readers confused. The report itself may contribute typographic noise by printing errors, faulty reproduction, cluttered illustrations, and poor graphic design.

Topic Sentence: Writers contribute noise to reports.

Supporting Details:

1. _____

2. _____

3. _____

Exercise 3.4   *Outline the following paragraphs, listing the topic sentences and supporting details.*

Sources of noise   In face-to-face communication, noise can come from many sources. Speakers who mispronounce words cause listeners to stop listening briefly and mentally repronounce the words correctly, possibly missing part of the message. Distracting mannerisms can interfere or conflict with the message. Inexperienced speakers might use esoteric vocabulary, those vague, multisyllabic words, trying to impress listeners. Sometimes interference comes from environmental sounds, such as traffic, machines, ringing phones, or nearby conversations.

Topic Sentence: _____

Supporting Details:

1. _____
2. _____
3. _____

Noise abatement Writers must filter out noise in technical reports. The most effective method to filter out noise is by writing an outline that clarifies the organization report. Writers must carefully edit their writing, evaluating the word choice, eliminating redundant terms and mangled metaphors, and pruning unneeded words. Writers should check the spelling of words because even one misspelled word can cause readers to mistrust the rest of the information. And the final draft needs special proofreading to catch mistakes and verify that the message is clear, substantiated, and grammatically correct.

Topic Sentence: _____

Supporting Details:

1. _____
2. _____
3. _____
4. _____

Exercise 3.5   *Write a one-paragraph description of ways that you can improve your writing. Begin with a topic sentence. In the rest of the paragraph, add the details that support your topic sentence.*

_____
_____
_____
_____
_____
_____
_____
_____
_____
_____
_____

Outlines

When you set out to write a document, it is sometimes difficult to get started. What and how many ideas will you include? In what order will you write them? Most word processors provide the tools needed to reformat and restructure quickly. However, the word processor cannot provide clues for logical development of ideas. This is still up to the writer.

One of the simplest steps to resolve the uncertainty is to write a quick outline of the ideas you want to convey. Writing an outline forces you to consider all the information you should include for your audience.

Outlines also save time. Once you have the main ideas or topics written down, you can then focus on one topic at a time and develop it fully. You can devote your concentration and research to each topic in any order. Each main topic becomes a main heading in your document.

Topics can require subtopics. If a topic gets long and complex, write a list of ideas just for that topic, and you have your list of subheadings.

Formal outlines can become elaborate, with Roman numerals, letters, and Arabic numbers, all indented to show a parallel structure.

The following outline represents the organization of the article in the reading section:

Thesis statement: Technical writers must take care to eliminate "noise" from their documents.

I. Introduction

    A. Lead-in with examples of excessive noise in everyday life.

    B. Thesis: "All communication takes place in the presence of noise."

    C. Organizational preview: "Before we talk about how to reduce noise, let's discuss what noise is, and how it gets into the communication system."

II. Sources of Noise (how noise gets into the communication system)

    A. From the source (writer or speaker)

    B. From the message (subject)

    C. From the channel (report)

    D. From the receiver (reader or listener)

    E. Feedback (speaker modifies message by reducing noise)

III. Noise Abatement at the Source (how the writer removes noise)

    A. Write an outline to improve organization

    B. Prevent carelessness

        1. Choose words carefully

        2. Eliminate mangled metaphors

        3. Eliminate redundancy and deadwood

    C. Check spelling

    D. Proofread

IV. Editorial Modulation (continual revisions of drafts)

    A. Reduce uncertainty (confusion or ignorance)

    B. Evaluate continually (edit for consistency, organization, language, style, and technical accuracy)

    C. Preserve own style

        1. Write clearly and directly

        2. Verify facts

        3. Convey significance of report

        4. Provide details so others can repeat the work.

    D. Use good technical style (avoid unnecessary jargon and acronyms, use active voice and personal pronouns)

V. Noise Abatement in the Report (filter out printing errors)

VI. Noise Abatement in the Receiver (listener or reader filters out noise)

VII. Optimizing Signal-to-Noise Ratio

    A. Have something to say

    B. Say it concisely and simply

    C. Use terms of interest to reader

    D. Write correctly

    E. Adjust language to ensure clarity

    F. Try to inform, not impress

VIII. Writing for Understanding

    A. Why scientists don't like to write

    B. Writing as the end-product of research

    C. Challenge to explain complicated concepts

Outlines can also be simple lists of ideas. Experienced writers learn to develop full documents based on simple lists. The following example is a modified outline of the formal outline above.

Thesis statement: Technical writers must take care to eliminate "noise" from their documents.

- Introduction: noise is present in all communication, linguistic or electronic
- Sources of Noise: how noise gets into the communication system
- Noise Abatement: how to get noise out of the communication system
- Optimizing Signal-to-Noise Ratio: how to edit and revise to prevent noise
- Conclusion: challenge to write for understanding

Note-Taking Organization

When you take notes on a reading or a lecture, use an outline to simplify complicated subjects. Notes not only force you to organize spoken or written information (which is necessary for memorizing), but they become a valuable tool when reviewing for tests.

To write an outline, write the main ideas and important details in condensed phrases. Writing an outline is a way of selecting the main ideas and important supporting details. Practice and careful reading or listening will improve this skill.

Exercise 3.6   *In each of the following paragraphs, a category will be given, followed by at least two details. Write the outline of each paragraph. Remember to be brief, accurate, and complete.*

Example

Switches are commonly used to open or close a circuit. Closed is the ON, or make, position; open is the OFF, or break, position. The switch is in series with the voltage source and its load. In the ON position, the closed switch has very little resistance. Then maximum current can flow in the load, with practically no voltage drop across the switch. Open, the switch has infinite resistance, and no current flows in the circuit.

Main Idea: Uses of switches

Details: a. Closed, ON, or make position—low resistance

                 b. Open, OFF, or break position—high resistance

1. A battery is a combination of cells. After discharge, a primary-cell battery cannot be recharged because the internal chemical reaction cannot be restored. After it has delivered its rated capacity, the primary cell must be discarded. A secondary-cell battery can be recharged because the chemical action is reversible. Because it can be recharged, it is also called a storage cell.

   Main Idea: _____

   Details: _____

   _____

2. The first law of thermodynamics states that energy can neither be created nor destroyed—it can only be converted from one form to another. Energy can have several forms, such as mechanical, electrical, chemical, and heat. All forms of energy generally end up as heat.

   Main Idea: _____

   Details: _____

   _____

3. The second law of thermodynamics states that to cause heat energy to travel, a temperature difference must be maintained. Heat energy travels downward on the temperature scale. Heat from a warmer material will travel to a cooler material, and this process will continue as long as a temperature difference exits. The rate of travel varies directly with the temperature difference.

   Main Idea: _____

   Details: _____

   _____

4. All known matter exists in three physical forms: solid, liquid, or gas. Solid matter will retain its shape and physical dimensions. Solids of sufficient density will retain their size and weight. A matter in liquid form will retain its quantity and weight but will take the shape of the vessel surrounding it. The dimensions of a liquid will change with the size of the vessel. Matter in a gaseous state, or vapor, will not hold its dimension or density unless contained in a vessel, and if contained, will expand or contract to occupy the volume of the vessel.

   Main Idea: _____

   Details: _____

   _____

5. Two basic designs make up the majority of roofs in the United States. These are the gable roof and the hip roof. The gable roof slants toward the long sides of the house, and the flat, triangular areas at the ends of the roofline are called the gables. The hip roof slants toward all the exterior walls. The ends taper toward the inside of the building. Other styles simply combine the slants into different profiles.

   Main Idea: _____

   Details: _____

   _____

Note Taking

Taking notes during classes is a challenge for people who are not comfortable writing at all. Although paying attention and listening well help, most people do not remember enough from listening. You don't have to be a speed writer. Writing just the main ideas in a notebook can be a valuable memory aid and review tool for tests.

TIPS   Try these strategies for taking better notes:

- **Keep a separate notebook for each class.**
- **Read the assignment before the class.** Yes, this sounds obvious, but some students think instructors cover all the material in class and get lazy about reading. This means most of their attention will be on listening, not taking notes. If you read the lesson first, the lecture will make more sense and so will your notes. And you'll remember the material better when hearing it a second time.
- **Start notes for each class on a new page, dated, and with any reminders,** such as assignments or upcoming tests, at the top of the page—it's much easier to find important information at the top of a page. If you are absent for a class, mark the date that you missed the class. Then try to find someone else's notes for that day.
- **Copy the objectives of the class** if the instructor writes the objectives for the class on the board. Use it as a pre-organizer for the rest of your notes.
- **Develop your own shorthand.** For example, use symbols or consistent abbreviations for common words in your area of study. Use a question mark for missing information or things you don't understand. Use a check mark to indicate something that requires follow-up or more information.
- **Copy all diagrams or notes that the instructor writes on the board.** Usually you can keep up with the instructor as she writes.
- **Write the main ideas as the instructor talks.** Do not try to write every word, and do not try to write complete sentences. Use key words that will jog your memory later.
- **If you miss material, leave a space and mark it with a question mark.** (Everyone daydreams or gets distracted sometimes.) After class, try to look at someone else's notes to fill in the information you missed.
- **If you find yourself getting restless or sleepy, take even MORE notes.** Sometimes the process of taking notes can keep you attentive.
- **Review your notes after class,** periodically during the course, and before tests. Fill in missing information as you review your notes, especially if your notes were wrong or incomplete when you originally took them.
- **Start a study group.** Find a few students who also take notes, and meet regularly to review class notes—don't wait until just before the final exam. Talking ideas through can clear up misunderstandings and fill in missing pieces.

Exercise 3.7   *Take notes of a lecture in a technical class. Compare notes with another student in the class.*

SPELLING: Suffixes *ise, ize* and *yze*

The three final syllables *ise, ize,* and *yze* all sound exactly alike–we hear them as "eyez." When we add one of these endings to an appropriate noun or adjective, a verb is formed. The new ending means "to make," so the word *standardize* means "to make standard."

To determine the correct spelling of the ending, there are a few commonsense rules. Use your visual memory for frequently used words, as it will help more than remembering the rules. Always check the dictionary when the word "just doesn't look right." Notice these endings when you read.

Rule 1:   Use a final *yze* for only a few technical words:

analyze   paralyze   electrolyze

Rule 2:   Use a final *ise* when it is part of a word, such as *wise, vise, rise,* and *guise.*

likewise   advise   sunrise   guise

otherwise   supervise   arise   disguise

Rule 3:   Use a final *ise* for words ending in *-mise, -prise* and some words ending in *-cise.*

surmise   surprise   exercise

compromise   comprise   incise

Rule 4:   Use a final *ize* for nearly all other words.

apologize   minimize   organize

alphabetize   mechanize   symbolize

emphasize   memorize   visualize

In most cases, when you add suffixes to the verb form of one of these spellings, just drop the final *e*, keep the *yz, is,* or *iz,* and add the suffix.

ANALYZE	ADVISE	ORGANIZE
*analysis	adviser	organizer
analyzing	advisement	organization
analyst	advisory	organizational
*analytical	advisable	organizing

*These words vary slightly from the rule.

Warning: Some writers attempt to sound formal or technical by adding *-ize* to form words that are not normally verbs. Although these words often indicate an action, try not to overdo them. Overusing *ize* can make the writer sound like a robot—artificial and distant. Notice the difference in tone in the following two sentences.

Mechanical: In an effort to minimize errors, we prioritized and analyzed our objectives and organized our procedures.

Human: To reduce errors, we first established our objectives and procedures.

The same overkill can happen when writers add *-wise* to a noun and use the new word as an adverb, as in *timewise* and *costwise.*

Mechanical: The procedure was more efficient timewise and costwise.

Human: The procedure's efficiency saved time and money.

Writing that sounds human is easier to read than mechanical-sounding writing.

Exercise 3.8   *Using yze, ise, and ize, add the correct ending to each word.*

1. Turn the dial clockw_____.

2. Use the oscilloscope to visual_____ the waveforms.

3. To anal_____ the results, correct measurements have to be taken.

4. It helped to priorit_____ the needs.

5. The manager took the situation under adv_____ment.

6. The cable was placed lengthw_____ on the bench.

7. She was thrilled when she was asked to superv_____ the project.

8. If the need ar_____s, more staff will be added.

9. No one was more supr_____d than he was to hear of his promotion.

10. In a schematic drawing, sawtooth marks symbol_____ fixed resistors.

Note: In publications using British English, you will notice slight differences in spelling compared to American English. One common difference is that American English words ending in *ize* are often spelled with *ise* in British English.

FIGURE 3.2
Comparison of some British English and American English words.

U.S. English	British English
apologize	apologise
civilize	civilise
organize	organise

VOCABULARY: Number Prefixes *bi, mono, poly,* and *semi*

Mono is a Greek prefix meaning "single" or "alone". It is combined with many scientific words.

monochromatic	=	one color
monovalent	=	having one valence

Bi is a Latin prefix meaning "having two."

bicycle	=	two-wheeled vehicle
biceps	=	muscle with two points of origin as in upper arm
binary	=	made up of two parts

Semi and its relatives, *demi* and *hemi*, all mean "half" or "partially."

semiconductor	=	a partial conductor
hemisphere	=	half the globe or sphere
demigod	=	a minor god

Poly is a prefix meaning "many" or "more than usual." The *y* is never changed to *i* when added to other words.

polygon	=	many-sided figure
polysyllabic	=	having several (four or more) syllables
polytechnic	=	providing instruction in many technical fields

Exercise 3.9   *Use the correct prefix to complete each word.*

1. An element with two poles is called _____polar.

2. A device with only one stable output state is _____stable.

3. A device with two stable output states is _____stable.

4. A _____ conductor has electrical conductivity between that of a conductor and an insulator.

5. A lie detector test involving the measurement of many types of rate changes uses a device called a _____ graph.

6. A small, half-sized cup for strong, black coffee after dinner is a _____ tasse.

7. A _____ tonous tone has only one pitch and can be tiresome to hear.

8. A person who knows two languages fluently is _____ lingual.

9. _____ styrene is a tough, colorless plastic material made by combining small molecules of styrene.

10. A job that requires only a little formal training is called _____ skilled labor.

WORD WATCH: *a lot* and *allot*

A lot and *allot* are homonyms: words that sound the same, but have different uses and spellings.

Allot is a verb meaning to divide up or assign as a portion or share. Other forms of this word are allotted, allotting, and allotment.

> Your manager will allot you 10 minutes in the presentation.

A lot is a two-word phrase meaning plenty of something. It is not precise in meaning, and therefore should be avoided in technical writing.

> We had a lot of time to complete the revision.
> Better: We had two weeks to complete the revision.

Note: The misspelling *alot* is not a word. It should be spelled correctly as either *a lot* (the most common use) or *allot*.

Exercise 3.10   *Fill in the blank using the correct form of* a lot *or* allot *in each sentence.*

1. Reducing noise is _____ harder than I thought.

2. My professor could not _____ more than one week to the subject.

3. I discovered I use the word "utilize" _____ more than I should.

4. We could not find _____ of examples of analogies.

5. He tried to _____ half an hour every week for peer editing of our reports.

6. I eliminated _____ of the redundancy by sticking to an outline.

7. Changing the report to the active voice made it _____ more interesting.

8. The lecture was over before I took _____ of notes.

9. She had to _____ several hours each night to writing the report.

10. I expect the report will contain _____ of complex ideas.

Writing Elements

CHAPTER 4

Technical Definitions

- Write formal technical definitions.
- Write an expanded definition.
- Spell plurals correctly.
- Use Latin roots correctly
- Use *to, too,* and *two* correctly.

READING: You Are Not Alone: Beware of What You Say on the Internet

by Rochelle Kaplan

Messages or information transmitted through e-mail over the Internet are not private, and this should be considered before one sends any type of message. People should not copy information from the Internet and claim it as their own.

Are there any legal restrictions on what I can say in e-mail or how I can use information I find on the Internet?

The Internet is fast becoming the communication tool of choice, even eclipsing the telephone. However, there is confusion over just what the Internet is and who owns the information on it. Is the Internet a mega-size warehouse where people deposit and gather information, or is it a giant electronic publication where we are all authors and publishers and must guard what we write or be careful of what we copy? Or is it a little of both? Consider the following situations.

- A former university student was indicted for hate crimes after he e-mailed threatening messages to minority students.
- A person was sued for defamation after he posted a message to a discussion group regarding the business conduct of a brokerage firm.

- A company asked a court to stop a private individual from using the company's logo on his WWW home page.
- An Internet provider was sued by a computer game manufacturer for copyright violation. A customer of the provider copied computer games and forwarded them to a chat group offered by the Internet provider.

Other, similar situations are likely to occur.

Before You Send That Message . . .

There are several things a computer user should consider when sending an e-mail message to another person or to a group, or when gathering information, pictures, and graphics from sites on the WWW. Questions to consider are:

- Would you send the same message on paper, knowing that it might be seen by

FIGURE 4–1

more people than just the intended recipient? Would you say those same words in a face-to-face situation, knowing that a crowd standing nearby can hear the conversation?

- Would you feel free to take the information and use it as your own if the information were printed in a book or newspaper?

Most of the information placed on the Internet can be read by millions of other people. This means you can't send a "confidential" message. You also can't take and distribute information because most of it is copyrighted or trademarked. Defamation, criminal and civil harassment, and copyright and trademark laws are presently being used to limit the types of communication on the Internet in the United States.

If You Won't Say It, Don't E-mail It

Defamation is the communication of untrue information about an individual or organization to a third person that harms the reputation of the person or organization and results in damages to that person or organization. Defamation can also include communication of harmful information with no regard to the truthfulness of the statements. An e-mail message about a person should be treated the same as a spoken statement or a message sent by mail or fax. Messages containing information harmful to a third person can be considered to be defamatory.

An e-mail discussion group (newsgroup, listserv, chat room, etc.) is nothing more than an electronic roundtable. As with any roundtable, the participants share information about a particular topic. A subscriber to the group may ask other group members for a reference about a product or person, e.g., "Has anyone used (fill in name of product or service)?" or "Has anyone worked with (fill in name of person or organization)?" Sometimes a subscriber "warns" or "advises" other subscribers about a product or vendor.

If you want to share information about a product, person, or organization as part of an e-mail discussion group, offer first-hand knowledge only. If your message is in response to a direct question, consider sending a "private" reply to the questioner rather than sending it to the entire group. It is acceptable to state, "This product did not work well in our office because . . ." and offer details. You may give an opinion. What you can't say is, "I heard that this product is defective" or "We used this product and the company that sells it is a bunch of rip-off artists because the product did not work in our office." If the facts change—the product performance improves or the issue between individuals is settled—then immediately put those new facts out to the group. When participating in online groups, keep in mind that your communication is instantly transmitted to hundreds or thousands of other individuals. If the communication contains untrue and harmful information about another person, product, or organization, the act of defamation has now been multiplied by the number of people that received the message.

Harassment is words or actions that are designed to threaten, intimidate, and/or make a person's workplace or educational environment unbearable and intolerable. E-mail can be harassing. If harassment occurs in the workplace and is directed toward employees of a certain race, ethnicity, age, disability, religion, or gender, then it is a violation of state and local EEO laws. If harassment is in an educational setting and is directed at students, then civil rights laws prohibiting discrimination in educational and/or public institutions are violated. Most schools and companies have strict policies prohibiting harassment. Hate crimes are the criminal extension of harassment. Most state criminal laws include harassment by electronic means. Until recently, that meant harassment by phone; now the definition includes e-mail. Moreover, the Federal Telecommunications Act makes it a federal offense to use the Internet to harass someone.

Don't be surprised if you are warned by your Internet provider or discussion group administrator that your privileges to participate in a group could be terminated if you post defamatory messages or copyrighted information. Internet providers do not want to be drawn into a defamation, harassment, copyright, or trademark case because of e-mail messages sent by one of their subscribers. Many are warning subscribers not to place such messages on the Internet.

Don't Copy That Download

Ownership of information on the Internet seems to be a fuzzy concept to many people. However, the principle of ownership of information on a WWW site is the same as ownership of information in a tangible form. All are governed by the same copyright and trademark laws. Copyright laws say that a copyright exists in any "original" artistic or literary work that is fixed in a "tangible medium of expression." "Original" does not mean novel, creative, or even good; it means that material was developed and formulated by an individual who did not copy it from anyone else. To qualify as a "literary work" requires that the work be more than a few words or a short phrase. A work is "fixed" in a tangible medium even if it exists only in a computer memory—even if it isn't printed on paper.

Information sent as e-mail messages is considered original work and entitled to copyright protection. The person who contributes a message to a discussion group is the owner of the message copyright. However, there is a growing body of legal opinion that says that when an individual subscribes to a discussion group, he or she gives an "implied" license to the discussion group to copy messages—subscribers' original works—to the other folks in the group.

It seems widely accepted that the owner of the discussion group owns the copyright to the compilation of messages. Thus, the compilations can be used by the group administrator as he or she sees fit, such as archiving them and making them available to group members or even to a broader public. A subscriber can't claim that his or her copyright on a single message prevents the administrator from forwarding that message to the group or adding it to the compilation of messages. In other words, a subscriber can't receive information from others, and expect that his/her messages won't be shared.

There is also a temptation among people using the Internet to assume that everything on the net, including messages from discussion groups and group archives, is "in the public domain." This isn't so. Just because an author offers material for the public to read on the Internet does not put the material in the "public domain." Materials in the public domain are strictly limited to government documents and material for which an author has clearly relinquished his/her copyright. Information (text, graphics, and photographs) found at Web sites (other than information from the government) is copyrighted material.

Often, attorneys will hear: "I downloaded this information because I was doing training on this subject for students or employees," or "My listserv was discussing this topic and I knew that this article would be relevant to the discussion." These excuses do not justify or provide a defense to unauthorized copying. Saying that you are using an article for "educational purposes" is not enough. Issues that must be considered are:

- Did you download the entire article, e-mail message, photograph, or graphic or just a small portion?

- Did you attribute the information to the author/owner?

- How widely did you disseminate the information? Did you share it with three people in the office or on a listserv of 1,000 people?

- What impact would this dissemination have on the market value of the information? U.S. trademark law permits individuals who sell goods and services to protect any words, phrases, symbols, or designs that identify or distinguish their goods and services from others'

products. For instance, NACE's "JobWeb" name and logo are trademarked. No one else can use the JobWeb logo to identify similar goods and services. Although the name and logo appear on the Internet and can be cut, pasted, or downloaded, this does not mean that the name or logo can be used by anyone else to identify goods or services. When you are surfing the net for interesting graphics to put on your home page, don't take someone's trademarked logo or design. (Pictures and photographs are protected by copyright law, so you can't take them either).

If you download information, pictures, works of art, or logos, you must obtain permission from the owner of the original work before you copy and distribute this material to others.

Use It Carefully

The Internet is a good tool for communication and research. However, it is important to remember that a few rules apply there. Simply speaking: If you wouldn't write the message on a billboard, don't write it in an e-mail message. If you wouldn't copy the information from a book and claim it as your own work, don't copy it from the Internet.

Reading Comprehension Questions

1. Briefly discuss the meaning of the title "You Are Not Alone: Beware of What You Say on the Internet." _____

2. How does the author define *defamation?* _____

3. How does the author define *harassment?* _____

4. What questions does the author suggest you answer before sending information in e-mail discussion groups? _____

5. Why can't you freely copy downloads and other information taken off the Internet? _____

6. Before you use an article for "educational purposes," what issues should you consider? _____

7. What two rules does the author say apply to sending or copying information on the Internet? _____

_____

WRITING: Technical Definitions

The reading passage included definitions of terms essential for understanding the topic. Without a precise definition of *defamation* and *harassment,* we might not understand the legal issues surrounding the use (and misuse) of e-mail on the Internet.

Definitions

Definitions of terms are the foundation of technical writing. A precise set of terms is used in technology, and only with a common understanding of those terms can information be communicated accurately.

Some terms used in technology have meanings entirely different from those with which you are familiar in everyday life. Examples of such words are *power, force,* and *communication.* For example, the term *communication* used in casual conversation can include speaking, listening, reading, writing, and body language. But to an electronics technician, if the message wasn't transferred electronically, it wasn't communicated at all. In fact, the study of communications systems begins with Samuel Morse's invention of the telegraph in 1837, even though we all know that throughout history, people have been sending verbal and nonverbal messages to anyone who would pay attention.

Some terms are used with more precision in technology than in everyday life. Words such as the following have precise meanings in technology and must be used carefully:

absolute	current	fundamental
critical	force	ground
intensity	power	specific
inversely	rate	static
potential	relative	uniform

Some terms are frequently confused. Can you state the difference between *force* and *power?* These are words that are used interchangeably in everyday language, but in technology the meanings are different. There are many such terms in technical writing.

Sometimes students are in such a hurry to do problems and assignments that they skip to the end of the chapter, referring to the chapter only as a last resort. These students are missing the "verbal" part of their field—how the concepts are explained in words. Don't be this kind of reader. Eventually, you will have to communicate what you know in words, either spoken or written. You won't be able to communicate entirely in numbers. Get used to how the experts, the authors, describe the principles of your technology. Learn the terms and how to use them correctly. Several examples of short and long, and formal and informal definitions are presented in this chapter to give you practice in this skill.

Informal Definitions

You can probably remember learning your first definition in your field. In electronics, it was probably

> *Resistance:* opposition to current flow.

This is an informal definition. A definition placed between commas or parentheses is usually an informal definition.

> A *potentiometer* (variable resistor) is used for volume controls.

If too many informal definitions are used, a report may become disjointed and distracting. Normally, a writer who plans on using more than two unfamiliar technical terms in a report will define the terms formally in the introduction or glossary.

Formal Definitions

A formal definition has two functions: it identifies the larger class (group or category) that the term belongs to, and it provides distinguishing characteristics:

term   >   class   >   characteristics

For example, consider the term *Porsche*. A Porsche is in the *class* called automobiles, or more specifically, German automobiles. The definition goes on to provide *distinguishing characteristics* or details about the term that make it different from other members of the group.

A Porsche, pronounced "pour-sha" (*term*), is a German-made automobile (*class*) with high-performance capabilities, a small, aerodynamic body design, and a price tag starting at $25,000 (*characteristics*).

A formal definition can be written for any technical term, and often the most difficult part is determining the class! For example, is resistance a device, a quality, a capability, or an action? Technicians must occasionally make such subtle distinctions.

Device	Quality	Capability	Action
resistor	resistance	resistivity	resist
module	modular	modularity	modulate

In which of the groups above do each of the following terms belong?

resonance   resonant
resonator   resonate

Resonator is an object, something you can touch, so it would be in the class of devices. *Resonate* is a verb, so it is a process or action. *Resonant* is how we describe an object, an adjective, so it is a quality. *Resonance* is the capability of performing the action.

Once the group has been determined, technicians usually don't have much trouble furnishing the distinguishing characteristics.

A resistor   is an electronic device
(term)   *(class)*
that is used in electronic circuits to oppose and control current flow. Its capacity to resist current is indicated by a color code or stamped values.
(distinguishing characteristics)

One final point to remember is to avoid using the term, or any variation of the term, in the remainder of the definition.

Wrong: A *resitor* is an electronic device that *resists* current flow.

In the example above, find a *synonym* (another word with the same meaning) for *resist,* such as *oppose* or *control.*

Technical writing can be efficient. Writers say things one time only—no repetition, no rewording. Technical terms and jargon have been defined and clarified by professionals in your field. Most jargon has been established because it describes an idea or concept in a few words. Imagine trying to describe waveforms without using jargon such as *sawtooth* or *square,* or to describe joint designs without *V-groove, J-groove,* or *scarf joint.* The result would be wordy and cumbersome.

The disadvantage of jargon is that it assumes that the reader also understands the technical meaning of the term. Writing for a nontechnical audience, those not expert in your technology, takes special attention. It requires explicit definitions of terms in clear, simple language.

Dictionaries

Dictionaries are written for certain audiences. Think for a minute about a car manual. It may be geared for general owners with only the basic operating needs, or for highly trained automotive specialists who need precise specifications. Likewise, dictionaries may be geared for everyday word usage or for highly specialized purposes.

Small, pocket-sized dictionaries provide only the most commonly used words and definitions. If you look up *resistance* in one of these dictionaries, you will probably only find the root word, *resist,* with several common endings, but no mention of current flow.

At the opposite end of the spectrum are technical or scientific dictionaries that offer *only* technical terms and definitions. One example is *Webster's New World Dictionary of Computer Terms* by Bryan Pfaffenberg (Que, 1999). If you look up *resistance* in this book, you will find only the electronics usage. This dictionary and others like it are useful guides for beginning technical students and people who need to read technical information.

"College editions" and large dictionaries include the commonly used definitions as well as an extensive number of technical definitions of terms that are used in different scientific disciplines. Keys such as *Elec.* or *Mech.* indicate the specific definition used in technology. If you look up *resistance,* you may find seven or more distinct definitions of how the word is used in different disciplines.

Exercise 4.1   *Write formal definitions for the following terms in your own words. The classes are provided for the first five terms. Find the technical definitions, noted in dictionaries by abbreviations in italics such as* Elec. *(electronics) or* Mech. *(mechanics).*

> ***Example***   Conductor—A conductor is a device or material that readily carries electricity, heat, or sound.

1. Electron (particle) _____

_____

2. Torque (twisting effect) _____

_____

3. Piston (sliding piece) _____

_____

4. Girder (structural beam) _____

_____

5. Fillet (concave junction or arc) _____

_____

6. Module _____

_____

7. Power _____

_____

8. Battery _____

_____

9. Chord _____

_____

10. Load _____

_____

Exercise 4.2   *Define 10 technical terms in your area of study. Write formal, one-sentence definitions of each term.*

1. _____

_____

2. _____

_____

3. _____

_____

4. _____

_____

5. _____

_____

6. _____

_____

7. _____

_____

8. _____

_____

9. _____

_____

10. _____

_____

Exercise 4.3   *Find a magazine article focused on your area of study. Locate and copy five technical definitions from various articles in the magazine. Then critique each definition, and revise it to include any elements needed to make it a formal definition.*

1. _____

_____

_____

_____

2. _____

_____

_____

_____

3. _____
_____
_____
_____
_____

4. _____
_____
_____
_____

5. _____
_____
_____
_____

Extended Definitions

Some objects or concepts require more than a one-sentence definition. An extended definition might require a paragraph or even several pages to fully define a complex concept or object. An extended definition includes the standard definition sentence, but also provides more details that describe the object. It can contain related definitions and examples that illustrate the term.

The following paragraph defines _harassment_ by providing not only a definition, but also two differing situations under which harassment can occur and the legal consequences of each one.

Harassment is words or actions that are designed to threaten, intimidate, and/or make a person's workplace or educational environment unbearable and intolerable. E-mail can be harassing. If harassment occurs in the workplace and is directed toward employees of a certain race, ethnic group, age, disability, religion, or gender, then it is a violation of state and local EEO laws. If harassment is in an educational setting and is directed at students, then civil rights laws prohibiting discrimination in educational and/or public institutions are violated.

More commonly, extended definitions include examples that illustrate and clarify the term or idea. For example, the author clarifies the definition of an e-mail discussion group by comparing it to an electronic roundtable:

An e-mail discussion group (newsgroup, listserv, chat room, etc.) is nothing more than an electronic roundtable. As with any roundtable, the participants share information about a particular topic. A subscriber to the group may ask other group members for a reference about a product or person, e.g., "Has anyone used (fill in name of product or service)?" or "Has anyone worked with (fill in name of person or organization)?" Sometimes a subscriber "warns" or "advises" other subscribers about a product or vendor.

Exercise 4.4 _Choose one term from Exercise 4.2, and write an extended definition in a paragraph. Write the definition sentence as your topic sentence. Include the details or examples that give meaning to the term._

SPELLING: Plurals

Plural Nouns

Nouns can be **singular** (one) or **plural** (more than one). Normally, to change a word from singular to plural, we simply add a final *s* to the end of the word. If the singular word ends with *s, x, z, ch,* or *sh,* we add *es* to make the word plural.

one noun	two nouns
one switch	three switches
a drillpress	many drillpresses

There are several exceptions, however, that show up frequently in writing. In this section, we review some of the **irregular plurals** that you may encounter. The one sure way of spelling a plural word correctly is to look the word up in a dictionary. Usually, an irregular plural ending will be noted early in the entry with *pl.* followed by the last syllable of the plural spelling. If two spellings are given, either is correct and the first is preferred. For example, the entry for *deer* will look similar to the following:

deer (dir) *n., pl.* deer, occas. deers

This tells you that *deer,* pronounced as "dir," is a noun, and the plural spelling is *deer* or, occasionally, *deers.*

If you look up a noun and no plural is given, you can assume that the plural is regular and just add a final *s.*

Y plurals

If a word ends with a *y,* and is preceded by a consonant, change the *y* to *i* and add *es.*

frequency	two frequencies
country	many countries

However, if the final *y* is preceded by a vowel (*a, e, i, o, u*), as in *delay,* do not change the *y.* Changing the *y* to an *i* would result in three vowels next to one another. We don't do this in English, because it makes words difficult to read; we simply add an *s* to the end.

Wrong: delaies

Correct: delays

Another exception to this rule relates to the plural form of proper nouns. Do not change the spelling of the noun, and do not add an apostrophe. Just add a final *s.*

two Jerrys   all the Rileys

In the case of abbreviations, numbers, letters, and **acronyms** (words made from one or more letters from each of several words in a term or phrase), to make a plural form, simply add an *s* or add an apostrophe and an *s* after the last letter.

several IBM PCs	two 1's
many M.D.s	all CRT

Note: We sometimes add an *s* to the end of verbs, also. If the verb ends in *y,* the same spelling rules as those noted above apply when changing the verb to its *s* form for a singular subject.

We carry	He carries
We say	He says

Exercise 4.5 *Make each word plural or the s form of a verb.*

1. array _____ **2.** memory _____

3. display _____ **4.** carry _____

5. specify _____ **6.** priority _____

7. Riley _____ **8.** alloy _____

9. theory _____ **10.** vary _____

F Plurals

Usually, when a word ends with *f* or *fe,* we change the *f* to a *v* and add *es.*

   a wife   all wives

Some common exceptions to this rule include the following:

one belief	many beliefs (*believes* is a verb)
one brief	many briefs
a handkerchief	many handkerchiefs
a roof	many roofs
a safe	many safes (*saves* is a verb)
one sheriff	many sheriffs

Use the dictionary to be sure.

Exercise 4.6 *Make each word plural.*

1. The _____ of writers differ.

               ability

2. He reshingled _____ in his spare time.

               roof

3. The speaker introduced himself to the _____.

               employee

4. The men escorted their _____ to the banquet.

               wife

5. The spreadsheet allowed several _____ of data.

               array

Hyphenated Words

For hyphenated words, *s* or *es* is added to the main word in the compound.

sister-in-law	sisters-in-law
son-in-law	sons-in-law
vice-president	vice-presidents
cross-examiner	cross-examiners

Irregular Plurals

Some words have structural changes in the plural form. This is especially true of foreign words and endings.

child	children
foot	feet

woman	women
medium (in communication)	media
datum/data (either is acceptable)	data
alumna (female)	alumnae
alumnus (male, mixed)	alumni
syllabus	syllabuses/syllabi

Sis to *Ses*

Adding an *es* to a word that ends in *is* would be a tongue-twister.

crisis	crises
thesis	theses
diagnosis	diagnoses
axis	axes

Exercise 4.7   *Write the correct plural.*

1. The two _____ brought their _____.
       daughter-in-law   *child*

2. The drafter could not decide between her two _____.
                                         axis

3. After graduation, a reception was held for all the _____.
                                   alumnus

4. We were alerted to all the _____ through the _____.
                 crisis   *medium*

5. The _____ were collected automatically.
       datum

VOCABULARY: Latin Number Roots

Latin number roots are found in many technical words. Most people can recognize these root words by themselves, but it takes special attention to find them in other words.

Latin Root	Number	Example	Meaning Using Latin Root
uni	1	unique	one of a kind
du	2	duplex	two-way communications
tri	3	tricycle	three-wheeled cycle
quatra	4	quarter	one-fourth
quint	5	quintet	five-piece musical group
sex	6	sextet	six-member group
sept	7	September	seventh month of old Roman calendar
oct	8	octopus	eight-armed sea creature
nov	9	novena	nine-day religious devotion
dec	10	decimal	number expressed in base 10
cent	100	centigrade	thermometric scale from 0 to 100 degrees
mill	1000	millimeter	1/1000 of a meter

Exercise 4.8   *Use the Latin number roots to complete the words.*

1. A number system in base 8 is called an _____al system.

2. A _____al-purpose machine can perform two functions.

3. An exam given every three weeks is called a _____-weekly exam.

4. The number system in base 10 is called the _____imal system.

5. One-thousandth of an ampere is called a _____iampere.

Exercise 4.9   *Write an informal definition for each of the following Latin-based words. Use complete sentences.*

Example

Century: A century is a 100-year period.

1. Duodecimal _____

2. Decathlon _____

3. Trilateral _____

4. Octogenarian _____

5. Quadrant _____

6. Unilateral _____

7. Sexennial _____

8. Million _____

9. Quintuplicate _____

10. Centimeter _____

WORD WATCH: *to, too,* and *two*

Two is the number (2) and is rarely confused with *too* or *to*.

> There were only *two* hours left.
> There could be *two* interviews.

Too is a modifier and can have *two* meanings. It can mean "more than enough" when it is in front of another modifier:

> The gray jacket was *too expensive*.
> The blue tie was *too long*.

Too can also mean "also." When used this way, *too* usually has commas around it:

> I'll interview with that company, *too*.
> They, *too*, have positions available.

To is used in all other cases, usually as a direction (preposition) or in front of a verb.

> I selected the suit *to* wear *to* the interview.
> I planned *to* arrive early.

Exercise 4.10   *Choose the right form of* to/too/two *for each sentence.*

1. I arrived only _____ minutes early for the interview.

2. I gave my résumé _____ the receptionist and introduced myself.

3. She said it wouldn't be _____ long before she would take me _____ the manager.

4. I noticed that she made _____ phone calls.

5. Suddenly, a door opened and a woman asked me _____ step into her office.

6. She introduced herself _____ me as the personnel manager.

7. I was _____ nervous _____ remember her name.

8. Fortunately, her nameplate was sitting just _____ feet in front of me.

9. Unfortunately, it gave just her first and last names, and I didn't know whether _____ call her "Miss" or "Mrs."

10. I used the title "Ms." and I was able _____ say her name _____ times during the interview.

CHAPTER 5

Technical Descriptions

- Write a technical description of an object.
- Write a technical description of a process.
- Spell numbers correctly.
- Use Greek number roots correctly.
- Use *wear, we're, were,* and *where* correctly.

READING: Writing Science Articles without a Ph.D.

by Malcolm Ritter

As a fledgling science writer, I interviewed a doctor about an experimental drug I'll call A. He wanted to see if it worked better than the standard drug B. So, he said he planned to round up patients for a comparison test and inject A in one arm and B in the other.

Gee, I said, if you inject A in one arm of your patients and B in the other arm, how will you tell which drug helps them get better?

That's when he started to laugh. And I learned something about writing science.

In fact, all my training in science writing has been from 20 years of experience. If you're a beginning science writer, or want to be, you probably have some questions about how to do it better. So take a moment for some answers from an old-timer. They'll make you a better writer than I was (though probably less of an object of amusement).

Where Can I Find Story Ideas?

By and large, you find science story ideas the same way you find other story ideas:

by paying attention. If you have a local or regional audience, here are a few ideas:

- Asks local doctors how some new medical study is going to change their practices. In fact, how much attention do they pay to these studies in general? What does it take to get them to change their ways?

- Local patient-support groups and advocates might tell you if anybody is trying a new experimental therapy. If researchers are advertising for volunteers for a study, ask what's going on.

- Hospitals, universities and other organizations should be happy to send you press releases on their research. If somebody local is about to present a paper at a scientific or medical meeting, that could be a story. Universities may also have an office that keeps track of research the faculty is doing.

But what if your audience is defined by its interest in some topic rather than geography?

- Find the big research meetings in that area (a handy practitioner or two should be able to tell you) and then ask meeting organizers for a press kit. Or

Writer's Digest, June 1999. Reprinted with permission of the author.

go over the program with somebody in the field (I'll discuss how to find them later) to find out what's hot.

- Chat with scientists or medical specialists about what the important research topics are in their fields. They might be especially productive if they've just come back from a professional meeting. Ask not only about what was on the meeting program, but also about the buzz in the hallways.

- Browse journals and popular science magazines like the weekly *Science News*. Journals like *Nature* and *Science* can be tough plowing, but there's often a section in front that summarizes the best stuff and provides background.

What makes a good story? The most obvious category is something that's important to the lives of your audience. But there are great stories that explain something in everyday life: one I wrote about why coffee stains form rings as they dry was hugely popular. And some stories just make a good, readable yarn, like the brief report I found in a journal about a woman so hungry for psychological support that she faked having breast cancer.

Keep these categories in mind as you sell your story ideas to an editor. If you collect supportive comments from experts in the course of your interviewing ("This research is a stunning development"), you can use them like book jacket blurbs to sell your story.

How Do I Prepare for Interviews?

Preparation is crucial. You'll ask better questions, understand the answers better and assure the researcher you're not a boob who's going to misquote him or her in print.

A librarian can help you find scientific literature and, with luck, popular articles on the topic. That can also reveal experts for background interviews to get an overview. If nothing else, a faculty member or postdoctoral student at your local college might help.

Do Scientists Speak Plain English?

Sure, if you ask your questions in plain English. Don't be afraid to say "Gosh" now and then. Sound like an interested and well-prepared layman, and your source should get the message to keep it simple.

Second, feel free to say things like, "You lost me there," or "Could you walk me through that again?" Don't feel embarrassed (I don't). You can also ask, "How can I explain this to my readers?" if you want to shift the burden of ignorance now and then.

Look for analogies. Saying, "So, it's kind of like a key fitting into a lock?" not only checks on whether you understand, but it may prompt your source to adopt that analogy or a better one. Similarly, while scientists are used to communicating concepts, you should press for examples. You'll certainly ask for definitions of jargon. But beware of words that mean one thing to you and quite another to your source. The most common example is *significant*.

If two groups of patients try different therapies, and the outcome is measured in some numerical scale, one group may do "significantly" better than the other. That doesn't necessarily mean they were noticeably better off. It only means that by the standards of statistics, the differences between the two groups was big enough to be real rather than a one-time fluke. So ask what the difference would mean in real life.

And keep your readers in mind. They haven't done all the preparing and interviewing you've done. So ask the naive questions they'll want answered, like whether this research in mice is going to cure cancer tomorrow. You can always start them with, "My editor will want to know . . ."

Is This Person's Conclusion Valid?

Let's find out. Even when the work appears reputable, your readers will want to know what other experts think.

Several services can refer you to experts. The Media Resource Service at 800/223-1730 or **www.mediaresource. org** is one. Or try ProfNet, at 800/PROF NET, via e-mail at profnet@profnet.com or at **www.profnet.com.**

Once you've found an expert or two, describe the research you're writing about and ask for an opinion. It's best if you can provide published research by your source. That might reveal details that make all the difference.

You want to know: Is this research onto something? Is this hypothesis at least credible? Is it new, or the best evidence yet for an idea that's been around a while? How strong is the evidence here? Are there other ways to explain these results? How credible are the conclusions? Does this work run counter to what scientists have thought before? In fact, is this new evidence clearly outweighed by a mound of previous work to the contrary?

If you get major criticism of the research, ask your original source for a reply. The details of the debate might be too technical for your readers, but you can decide how far to wade in.

One other thing: Find out who paid for the researcher's work. It may be a company or organization with a financial interest in the results. The scientist may also stand to make money from a patent, or be a consultant to an interested party, or hold shares of stock in a company that could benefit from the study results. Tell your readers about ties like this, and give the researcher a chance to deny that the ties made any difference.

How Do I Write This Stuff Clearly?

Remember all those analogies and examples you pressed your sources to supply? Now you can put them to work. They are your best friends in making scientific topics understandable to your readers.

A while ago, I wrote that an experimental vaccine "made multiple sclerosis patients build up a police squad of blood cells to stop vandalism in their nervous systems." As the story explained, it made them pump out more of one kind of T cell that prevented another kind of T cell from attacking the protective sheath around nerves in the brain and spinal cord. Diagnosis: complex idea. Prescription: metaphor.

Even a familiar concept can be introduced with an example. Researchers recently found evidence that a particular nerve helps the brain store emotionally charged events in long-term memory. My lead: "Why do you remember prom night so well when you don't have a clue what you did two nights later? In part, a study says, you can thank a nerve that runs to your brain from deep in your innards."

Use statistics sparingly, focusing on those that most clearly make the point. And if you can't explain something without lots of background, tell the reader it's time to fall back: "To understand why Smith's findings are so important, it helps to know a thing or two about objects called active galactic nuclei."

Finally, show your copy to a friend to see if it makes sense.

What Did You Learn From That Doc?

1. *Arm* is one of those words with a hidden technical meaning. In drug experiments, it means a group of participants. The patients getting drug A would be in one arm of the study, and those getting B would be in the other arm.

2. A scientist can think he communicated an idea, and you can think you got it, and you can both be wrong.

3. If something a source says doesn't quite add up, ask about it. Don't be embarrassed.

After all, it's better to look dumb in conversation than in print.

Reading Comprehension Questions

1. What is the fundamental way a science writer finds story ideas?_____

2. According to the author, what makes a good story? _____

3. Before the author describes a research study, what does he do to prepare? _____

4. During an interview, what does the author suggest as a way to confirm your understanding? _____

_____

5. Describe the techniques used by the author to make scientific topics understandable to nonscientific audiences. _____

_____

WRITING: Technical Descriptions

The reading article describes how a science writer reports scientific topics to a newspaper audience. To do this, he translates the medical and scientific terms and issues into everyday, common language for the general public. Writing to a nontechnical audience, as the author does, includes the daunting task of explaining difficult concepts and terms using simplified language. For this reason, many technical writers prefer to write to technical audiences, because they can freely use the abbreviated technical language in the industry.

A technical description can be part of a larger report or a report by itself. It is especially important when the report concerns a device, tool, process, or concept that is new or unfamiliar. Descriptions typically include a definition of the object or idea, an orientation to the overall characteristics, followed by detailed descriptions of the parts in a logical order. For example, to describe a device, a writer would first describe the function of the device (what it does and when it is needed). Next, the writer would describe the physical appearance of the object and its component parts, one by one, in the order in which they appear or play into the larger function of the device itself.

Comparisons

Technical descriptions sometimes compare unfamiliar objects or concepts to familiar objects or concepts. In technology, people need to express values, shapes, angles, and joints in concrete, meaningful terms. To do this, we use familiar or graphical terms to describe size, structure, and location.

We compare location and shape of parts to familiar anatomy: screws have heads, saws have teeth, and roads have shoulders. Circuits have elbows and legs. We use shapes of letters and symbols to define the shapes of parts and joints. Filters can be L-type, T-type, or π-type. The basic weld-joint designs include the square groove, bevel groove, U-groove, J-groove, and V-groove. Workers install the I-beam and construct the K-brace, J-strip, and P-trap.

Many fields use this abbreviated method of describing a complex design or shape to add exactness to their language. Examine terms in your own career field to find more examples of comparative terms.

The following example compares something technical (a word processor) to something familiar (an office).

A word processor is a piece of software that enables a computer to function like an automated office. With the software, the computer works as a typewriter so that you can create your own files: memos, letters, reports, and graphics. The computer memory, commonly a disk, acts as a file cabinet from which you can retrieve files, edit them when necessary, and save files indefinitely. Acting as a secretary and a copier, the word processor controls the format, type font, and number of copies to be printed on the printer. Some word processors also contain a dictionary and thesaurus to check spelling and word choice and a mail merge function for automatic addressing, which relieves secretaries of these duties.

Notice the comparisons:

OFFICE	WORD PROCESSOR
Typewriter	Computer
Memos, letters, reports	Files
File cabinet	Computer memory (disk)
Secretary and copier	Printing function
Secretary	Dictionary, thesaurus, mail merge

Comparisons are useful for explaining and understanding. Find some examples of comparisons in your technical reading.

Analogies

A senior professor once stated that teaching is simply a matter of making comparisons: relating new, complex ideas to familiar ones. Many writers and professors use a method called the analogy.

An analogy is a formal comparison based on the resemblance of two unrelated objects or ideas. For example, professors often compare current flow to water flow. This analogy continues throughout the study of electronics to make the principles and purposes of devices more concrete. An analogy is useful only if the two concepts have more than one similarity. For instance, in the water/current flow analogy, not only does flowing current behave like flowing water, but capacitors behave like water buckets that store and release water, and circuit gates are similar to floodgates.

The author of the reading article referred to the following article that uses an analogy to describe a new vaccine to a newspaper audience.

VACCINE STOPS MULTIPLE SCLEROSIS PROGRESSION IN SMALL EXPERIMENT

By Malcolm Ritter, AP Science Writer, 9/30/96. Reprinted with permission of Associated Press.

An experimental vaccine enabled multiple sclerosis patients to build up a police squad of blood cells to stop vandalism in their nervous systems, and that kept sufferers from getting sicker, a study found.

Scientists tested the vaccine against a kind of MS that gets progressively worse over months or years. None of the six patients who built up police-like cells in the blood got worse during the yearlong study, while 10 of 17 other patients did. The study had so few patients that it couldn't prove the vaccine would be useful. But experts said the vaccine's effect on the immune system was encouraging. "It's not a universal treatment at this point and should not be considered so until we have evidence in a lot more patients," said the study's author, Arthur Vandenbark, of the Veterans Affairs Medical Center in Portland, Ore., and the Oregon Health Sciences University.

About 300,000 Americans have MS. They have such symptoms as unusual tiredness, loss of balance and muscle coordination, slurred speech, tremors and difficulty walking. In severe cases, they are partly or completely paralyzed.

The vaccine was tested against chronic-progressive MS, which accounts for about 15 percent of cases. Nobody knows what causes MS. But scientists do know that the immune system mistakenly attacks the protective sheath around nerves in the brain and spinal cord. That causes the symptoms. This vandalism is caused by certain blood cells called T cells, which gang up at the sites of destruction.

People with MS naturally have some police-like T cells that can turn the vandalizing ones off, but not enough of them, Vandenbark said.

So his vaccine was aimed at getting the immune system to churn out more of these police T cells. Patients got the vaccine or a placebo injected weekly for four weeks, then monthly for 10 months. Researchers tracked them for a year. The vaccine mimicked a piece of a protein carried

---

VACCINE STOPS MULTIPLE SCLEROSIS PROGRESSION IN SMALL EXPERIMENT— continued

by some vandal cells. When it was injected, the police cells noticed the sharp rise in the number of these telltale protein pieces. In response, the police cells multiplied.

Five of nine patients who received one form of the vaccine showed a rise in their levels of policing T cells. A sixth patient showed the same result from another vaccine form.

In contrast to most of the other patients, these six retained their abilities over the yearlong study in tests of walking speed and use of hands and arms. In four of the six patients, the number of vandal T cells in the blood dropped. The two others had low levels to start with.

Abe Eastwood, director of the research and grants program at the National Multiple Sclerosis Society, called the vaccine "a very promising and interesting idea." But he said it's too soon to say whether it will be a useful treatment, since only six patients showed a response.

The author uses the analogy of police and vandals to describe the two types of T cells to make the findings of a research study understandable to a nonmedical audience. He uses terms such as "police-like T cells" multiplying to contain "vandalizing" T cells that "gang up at sites of destruction."

For descriptions written to technical audiences, writers include the specific details and terms used in the industry. For example, if the preceding article were written for people in the medical field, the analogy comparing T-cells to police and vandals would not be necessary. Instead the article would probably focus on the exact methodology of the study, vaccine ingredients, related research, statistical significance, data tables, interpretations of the results, and recommendations for further research—things most newspaper readers would neither understand nor care to read.

Exercise 5.1   *Identify an analogy used in your area of study, and construct a table showing at least three of the comparisons of the technical terms or ideas with common terms.*

Exercise 5.2   *Internet assignment: Use Malcolm Ritter's name as a keyword to find more science articles he wrote. Or use the name of another science writer, such as the science writer for a magazine or large-city newspaper. Print an example of a comparison or analogy the author uses to make scientific facts or principles easy to understand for a general-interest audience.*

Technical Slang

Another suggestion for clear technical reporting is to avoid technical slang, words used within a specialized area that are unfamiliar to the public. This is important when communicating to customers, superiors, or subordinates, who might be confused by the slang terms. As a communicator, it is your job to prevent misunderstandings, and you can do this by using the most common, yet accurate words possible.

Question:	"What size mill is that?"
Translation:	What size engine is that?
Answer:	"It's a pocket-rocket."
Translation:	It's a 1.6-liter turbo-charged engine.

Exercise 5.3   *Write five sentences that use slang from your area of study. Then exchange your sentences with another student to try to write an interpretation in clear language.*

Clichés

Some speakers and writers overuse comparisons to add color to their words. Overused comparisons are called **clichés** (pronounced *klee-shays*), and they should be avoided in technical writing. The following exercise will give you a chance to recognize common clichés.

Exercise 5.4   *Complete the phrases with common expressions.*

1. As quick as _____

2. As fast as _____

3. As smart as _____

4. As slow as _____

5. As sharp as _____

Physical Descriptions

The technical description of an object generally starts with the general information, and proceeds to specific information.

The following example describes the sound system on a music stage.

Once the stage is erected, the two main speaker towers are positioned on scaffolds 54 feet high. Ultimately, the Springsteen tour will present 3200 sq. ft. of loudspeakers to a single crowd. Each of the 160 speaker cabinets contains two 18-inch, low-frequency drivers; four 10-inch, lower/midrange assemblies; and two each of upper/midrange and high-frequency drivers. (Eskow, "The Heart of Rock 'n' Roll," *Popular Mechanics,* March 1986)

You were probably struck by the number of measurements included in the description, which left you with an impression of ear-shattering sound capability and technical sophistication. Notice how the details progress from the overall sound capability to the specific details of the individual speakers.

Regardless of the object being described, a physical description has the same purpose: to present the facts about the object. Technical writers use descriptive terms carefully and precisely, with exact terms. They use modifiers sparingly, but when they do, the modifiers are adjectives that add meaning, such as *parallel, perpendicular, cylinder,* or *grainy.* Avoid abstract and vague words such as *nice, very, really, a lot,* or *pretty.*

Certain types of information add meaning to physical descriptions:

Color   Size   Shape

Texture   Quantity   Part names

TIP   A typical outline for a physical description contains the following elements:

- **Orientation to the object or device,** including a technical definition, and when and why the device is used.

- **General description of the device,** including the overall dimensions, appearance, and components of the device.

- **Description of each component,** in sequential or logical order, including its physical appearance, purpose, and relationship to other components.

Exercise 5.5   *The following description is taken from the instructional manual for a Fluke digital multimeter, model 8010A / 8012A. Underline the physical details, nouns, and adjectives describing color, texture, size, quantity, shape, or part names.*

1-7. GETTING ACQUAINTED

1-8.   Let's take a brief look at your instrument before we discuss exactly how to operate it.

1-9.   The meter is light (2 pounds and 6 ounces for the standard model) with a low profile that hugs the work bench. The light gray case goes with any decor and is made of rugged, high-impact plastic. The handle can be rotated to eight positions to function as a handle for carrying the instrument or as a stand to tilt the front panel up for convenient operation. The handle can be rotated out of the way. To change the handle position, pull out on the round hubs where the handle joins the meter; then rotate the handle to the desired position.

1-10. On the rear of the meter are a Phillips screw and a power cord receptacle. The Phillips screw holds the outer cover in place.

1-11. The LCD (liquid crystal display) covers the left part of the front panel. The right-hand portion of the front panel contains two horizontal rows of controls and connectors. The top row consists of ten pushbuttons—the four switches on the left determine the measurement function of your multimeter and the other six switches determine the range of measurement. The bottom row consists of controls and the input terminals.

Exercise 5.6   *Write a paragraph describing an object, tool, or instrument. Include exact details in complete sentences. Include a definition of the device, and detailed descriptions of its appearance.*

Process Description

The technical description of a process describes how something works, beginning with general information about the overall function of the process, and proceeding to the specific materials or skills required. The description can include a flowchart or schematic to show the sequence of actions or decision points in the process.

The following example includes a description of what a robot arm does in a project.

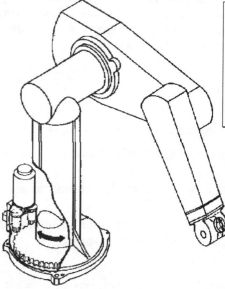

Puma Robot Drawing Project

How it works:

The project uses a Staubli Puma 562 robot arm donated by IBM and interfaces it using a number of software pieces written in C++, Perl, and VAL II. The objective is to manipulate the arm to draw on a canvas.

The remote user downloads the software and executes the programs. If successful, the user connects to the robot arm. The robot moves to a home position, activating the room lights. The user now selects a pen color and, through a series of programmatic commands, directs the robot to grab a pen and draw on a canvas.

FIGURE 5–1
Puma Robot Drawing Project

(Courtesy of Staubli Corporation.)

This process description includes a line drawing of the robot arm, and describes how the robot arm works in the project. The physical details of the robot arm are the subject of a descriptive report found in Chapter 10.

TIP   A typical outline for a process description contains the following elements:

- **Orientation to the process,** including a technical definition, and when and why the process is performed.

- **General operation of the process,** including the main divisions of the process; materials, skills, and time required; and pre-operation conditions.

- **Description of each step in the process,** including why and when it takes place, how long it lasts, and any human intervention required.

Exercise 5.7   *Write a paragraph describing how an object works. If practical, describe the process of the object you described physically in Exercise 5.6.*

SPELLING: Using Numbers

One of the effects of technology in our society is our frequent use of numbers in all types of communication. Technical writers are often unsure whether to use the word for a number or the figure for the number. The truth is that the rules are changing as fast as technology.

Style guides and textbooks differ about when and where to use figures instead of words, and some companies have their own policies about this question. Observe in you rown reading how different authors and publishers treat numbers.

There are a few general rules that you can follow, although even these rules, as always, have exceptions and even alternatives.

Rule 1:   Numbers that begin a sentence are ALWAYS written as words. No figures are used to begin a sentence because you cannot capitalize a figure.

> Fifty-seven dollars seemed too high for a multimeter.
> Eighty percent of the chips in that shipment were defective.
> Sixty-five is no longer the age of retirement.

Do not begin a sentence with a number and/or symbol that is more than four words long. Something this long is never written in words. Instead, rewrite the sentence to place the number somewhere else in the sentence.

Example

Wrong: 210° is the actual degree value.

Right: The actual degree value is 210°.

Wrong: Two hundred fifty thousand dollars was awarded as a grant to the research team

Right: A research grant of $250,000 was awarded to the team.

The following rules apply only if the numbers appear inside the sentence.

Rule 2:   Numbers are written as words if either of the following situations apply:

(a) they can be written in one or two words:

> He logged sixty hours of overtime.
> We received 325 new training manuals.

or   (b) They are below 100 (some style guides say below 10):

>   Only twenty of the manuals were needed.
>   We sent 300 manuals back.

Note: Put a hyphen (-) between two-word numbers, such as from *twenty-one* to *ninety-nine*. Also, if numbers over and under 100 are used in a series, all are written in figures.

Rule 3:   If one number follows another number, the first is written in words and the second in figures.

>   three 5-ohm resistors   three 2" × 4" boards
>   two 10% voltage drops   two 14-inch tires

Rule 4:   Precise measurements of time, distance, capacity, dimension, amount, and percent that need to be emphasized, noticed, or remembered are written as figures. In digital notation, use the figures 1 and 0.

>   The horizontal scale factor is 2 ms/cm.
>   The largest voltage drop, 70 V, occurs across the largest resistor.
>   The bill included $13.50 for parts and $50 for labor.

Exercise 5.8   *Rewrite the following sentences to correct any incorrect number expressions.*

1. Two hundred thirty-five feet of wire were used in the prototype.

_____

2. We completed 2 tests before we recognized the major problem.

_____

 3. Simply multiply a decimal by one hundred to change to a percentage.

_____

4. Connecting 2 400-V capacitors in series does not always provide eight hundred-volt capability. _____

5. The first compass, the lodestone, helped Chinese sailors over two thousand years ago. _____

Rule 5:   Specific dates and addresses are written in figures. If the street name is a number, use the rule that numbers below 100 are written in words.

>   August 15, 1972
>   2470 West Twenty-First Street
>   452 South 152 Avenue

Rule 6:   Pages, ages, and numbers of chapters, charts, and graphs are written as figures.

>   Chapter 7
>   Figures 4-3, 4-4
>   4-year-old machine
>   The company is 2 years old.

Rule 7:   Fractions that express general ideas or approximations are written as separate words, and they are usually hyphenated.

>   When the tests were three-fourths completed, we stopped.
>   Only two-thirds of the employees felt the stress.

Rule 8:   Fractions that express exact measurements are written as figures. Mixed numbers (a whole number and a fraction) are written either with a hyphen or a space between the whole number and the fraction. The best method is the one that best clarifies your information. Be consistent.

Hazard: In technical writing, decimals, degrees, and most percentages and fractions are written in figures.

90° angle	$1\frac{1}{8}$ inch wire
80% efficiency	$1\frac{1}{8}$ inches
10.3 hours	2.0 kW

Rule 9:   Spell out units on first usage, with abbreviations in parentheses. Leave one space between the numeral and the abbreviation, except when they are part of an adjective preceding a noun, and then use a hyphen.

First Usage	**Second Usage**
400 megahertz (MHz)	450 MHz or 450-MHz processor
4.3-gigbyte (GB) disk	5 GB or 5-GB disk drive

When using units as a noun in measurements, add *of* to form a prepositional phrase:

You must have 100 MB of hard-disk space to install the program

Exercise 5.9   *Rewrite the following sentences to correct any incorrect number expressions.*

1. The supervisor received 9/10ths of the credit. _____

_____

_____

2. He added five quarts of ten W forty oil. _____

_____

_____

3. The blueprint is shown in Figure 5.three. _____

_____

_____

4. Scott was hired on January fifth, 1999, and JoAnne was hired exactly one year later.

_____

_____

5. In the cse of a twenty-μ A movement, we would need five hundred kΩ between the terminals to make the ten-volt measurement. _____

_____

_____

VOCABULARY: Greek Number Roots

Many Greek number roots are used in everyday English. Since ancient Latin and Greek are sister languages, some numbers will be identical or similar to the Latin number roots.

Greek Number	English Number	Example	Meaning Using Greek Root
mono	1	monosyllable	one syllable
di	2	dioxide	two parts oxygen
tri	3	tricycle	three-wheeled toy
tetra	4	tetrachloride	four parts chlorine
penta	5	pentagon	five-sided figure
hexa	6	hexameter	six beats per line
hepta	7	heptarchy	government of seven rulers
oct	8	octopus	eight-armed sea creature
ennea	9	(this root is almost unused in English words)	
dec	10	December	tenth month of the old Roman calendar
hecto	100	hectogram	100 grams (metric)
kilo	1000	kilometer	1000 meters (metric)

Exercise 5.10   *Use the Greek number roots to complete the words.*

1. The number system in base 16 (six plus ten) is called _____imal. (Use two roots)

2. Laser light is _____chromatic since it produces only one color.

3. A _____watt is a 1000-watt unit of energy.

4. A situation with a choice between two unpleasant alternatives is a _____lemma.

5. The first five books of the Bible are called the _____teuch.

Exercise 5.11   *Write an informal definition using the Greek root. Write complete sentences.*

1. Monostable _____

2. Dialogue _____

3. Trilogy _____

4. Octant _____

5. Hexagon _____

6. Heptavalent _____

7. Pentagon _____

8. Tetragon _____

9. Kilovolt-ampere _____

10. Decade _____

WORD WATCH: *wear, we're, were,* and *where*

Use *wear* as a verb to mean "to cause to deteriorate by use," or "to have on the person," as clothing. As a noun, use it to mean "the act of wearing," or the "deterioration of something as a result of being used."

> The part will *wear* out from overuse.
> We had to *wear* protective clothing in the lab.

Use *were* as the past tense of the verb *to be.*

> The jurors *were* tired from deliberating all day.

Use *we're* as contraction for *we are.*

> *We're* not familiar with the study.

Use *where* as a reference to a place.

> The picture showed *where* to put the batteries.

Exercise 5.12   *Use* wear, we're, were, *or* where.

1. The scientists didn't _____ white lab coats to the meeting.

2. They _____ meeting to discuss the findings of the study.

3. The reporter sat _____ he could see the projector screen clearly.

4. There _____ no questions until the first speaker finished.

5. The subjects of the study _____ middle-aged men.

6. They lived in locations _____ the sun's ultra-violet rays _____ highest.

7. One sample group was allowed to _____ sunscreen.

8. Sunscreen can _____ off from sweat, humidity, and swimming.

9. If _____ careful, we can successfully block UV rays, no matter _____ we live.

10. Chances are decreased even more if we _____ hats and protective clothing.

CHAPTER 6

Summaries

- Write a summary of an article.
- Spell words with double letters correctly.
- Use *tele, phono, photo, graph,* and *gram* correctly.
- Use *loose, loosen, lose, loss,* and *lost* correctly.

READING: Hubble Expands the Universe

Imagine seeing stars and planets as if they were just outside your living room window. From this window seat to the universe, you could see the birth and death of stars and galaxies as they appeared billions of years ago. The Hubble Space Telescope is your window seat to the universe. Hubble has provided us with front row seats to fragments of a comet slamming into Jupiter and stars being born in huge craggy towers of cold dark gas.

A Front Row Seat to the Universe

Deployed April 25, 1990 from the space shuttle *Discovery,* Hubble is one of the largest and most complex satellites ever built. Hubble's deployment culminated more than 20 years of research by NASA and other scientists. The telescope is named for American astronomer Edwin P. Hubble, who first discovered that countless island cities of stars and galaxies dwell far beyond our Milky Way.

But NASA didn't launch the telescope into space to get closer to the stars. Hubble barely skims the Earth's atmosphere, orbiting just 380 miles above our planet. The nearest star, our sun, is 258,000 times farther away.

Hubble is in space because it can see the universe more clearly than we can from Earth. Looking at the heavens through a ground-based telescope is like trying to identify someone at poolside from the bottom of a swimming pool. Our vision is blurred. That's because we live at the bottom of the Earth's atmosphere, an ocean of air that smears and scatters starlight. That's why stars twinkle.

Scientists have known for several years that our atmosphere obscures and distorts light. The scientists who pioneered rocketry decades ago concluded that the best view of the universe is from above the Earth's atmosphere.

With Hubble, astronomers are getting a clearer picture of the universe. The tele-

FIGURE 6–1
Hubble Space Telescope

Reprinted courtesy of the Space Telescope Science Institute (STScI), Baltimore, Maryland. STScI is operated for the National Aeronautics and Space Administration (NASA) by the Association of Universities for Research in Astronomy, Inc. (AURA).

scope's stunning photos are showing the world about the wonders of space. Many of the world's foremost astronomers are using Hubble to probe the horizons of space and time.

Designed to last 15 years, Hubble is providing intriguing new clues to monster black holes, the birth of galaxies, and planetary systems around stars. To provide astronomers with the latest Hubble data, the Earth-circling observatory must be maintained by hundreds of scientists, engineers, and computer programmers at the Space Telescope Science Institute in Baltimore, MD, and the Goddard Space Flight Center in Greenbelt, MD.

All in a Day's Work

The Hubble Space Telescope is as large as a school bus and looks like a five-story tower of stacked silver canisters. Each canister houses important telescope equipment: the focusing mirrors, computers, imaging instruments, and pointing and control mechanisms. Extending from the telescope are solar panels for generating electricity and antennas for communicating with operators on the ground.

The 12-ton telescope collects faint starlight with an 8-foot-diameter mirror. The mirror—tucked inside a long, hollow tube that blocks the glare from the sun, Earth, and moon—is slightly curved to focus and magnify light.

Unlike ground-based telescopes, astronomers cannot look through Hubble's lens to see the universe. Instead, Hubble's scientific instruments are the astronomers' electronic eyes. The telescope's instruments include cameras and spectrographs. The cameras don't use photographic film, but rather electronic detectors similar to those used in home video cameras. The spectrographs collect data by separating starlight into its rainbow of colors, just as a prism does to sunlight. By closely studying the colors of light from a star, astronomers can decode the star's temperature, motion, composition, and age.

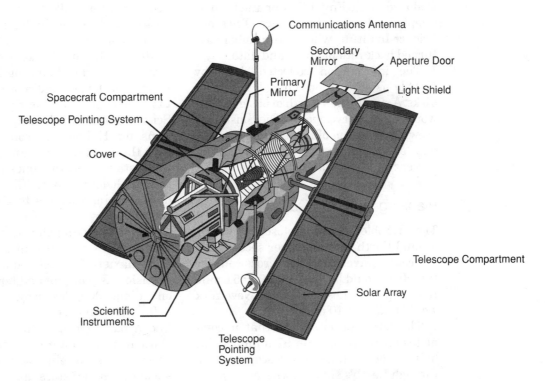

FIGURE 6–2

Hubble Space Telescope

Courtesy of NASA.

Hubble must maintain a steady position to take long exposures—sometimes hours—of the same subject to produce images of distant or faint objects. Otherwise the images will be blurred. To accomplish this mission, the telescope must battle such celestial elements as air drag, the sun's radiation, and the gravitational pull of objects.

For Hubble, maintaining proper direction is similar to a sailor fighting the wind and water to keep his sailboat on course. Hubble is successful because of its sophisticated pointing control system, which includes gyroscopes and Fine Guidance Sensors. Once the telescope locks onto an object, its sensors check for movement 40 times a second. If movement occurs, the wheels, which are constantly rotating, change speeds to smoothly move the telescope back into position.

Once Hubble gathers pictures and data on celestial objects, its computers turn the information into long strings of numbers that are beamed to Earth as radio signals. This information streams through a series of satellite relays to the Goddard Space Flight Center and then by telephone line to the Space Telescope Science Institute, where the numbers are turned back into pictures and data.

The information collected daily by Hubble is stored on optical computer disks. A single day's worth of observations would fill an encyclopedia. The constantly growing collection of Hubble pictures and data are a unique scientific resource for current and future astronomers.

Planning Hubble's Day

The Hubble Space Telescope whirls around Earth at 5 miles per second. If a car could travel that fast, cross-country traveling would be a breeze. A 2,451-mile road trip from Los Angeles to New York would take just 10 minutes.

The telescope is so fast that it completes an Earth orbit in 97 minutes. For half of each orbit, the telescope passes through Earth's shadow, where temperatures plunge to 250 degrees below zero. When Hubble swings back into sunlight, outside temperatures skyrocket to above boiling. The extreme temperature shift is almost like traveling from Antarctica to the Sahara Desert within minutes.

Hubble is protected from these temperature extremes by an exterior thermal blanket, which allows the telescope to maintain a constant temperature. But a steady temperature isn't the only requirement the telescope needs to function properly. Hubble's computers, pointing systems, and imaging instruments need electricity to operate. The telescope receives its power from a pair of rectangular 40-foot-long solar arrays. Each array is an 8-foot-wide blanket of solar cells. The flexible, lightweight arrays collect sunlight and convert it into 2,400 watts of electricity—enough to power two dozen household light bulbs.

Because Hubble is not always in direct communication with ground stations, operators cannot control the telescope every minute, as people do radio-controlled airplanes. Instead, every second of Hubble's activities in space must be planned in detail. Imagine if you had to plan every minute of every day. Get out of bed, put on a pair of slippers, walk to the bedroom door, walk down the hall to the bathroom, turn on the light, pick up the toothbrush, take the cap off the toothpaste, etc. It would take a whole day to plan a whole day. Planning Hubble's day is an even longer and more complicated task for experts at the Space Telescope Science Institute. Their detailed instructions for Hubble are converted into a code the spacecraft's main computer can understand. Several times a day, operators at Goddard Space Flight Center radio these instructions to Hubble's electronic "brain."

The telescope's pointing systems then find and lock onto distant planets, stars, and galaxies. This task requires the same precision as pointing and holding a laser on a dime 400 miles away.

To make an observation, Hubble must find a pair of bright stars called guide stars near each target planet, star, or galaxy. These bright stars are Hubble's anchors. Guide stars allow Hubble to maintain its steadiness on a target, just as anchors keep a ship from drifting. To find guide stars for nearly every object in the sky, mission planners use an immense catalog containing the sky addresses of 15 million stars.

Hubble's Top Science Findings

Since its launch in 1990, the Hubble Space Telescope has provided remarkable new views of the universe, which have revolutionized astronomers' thinking about many astronomical mysteries.

Postcards From the Edge of Space

Hubble's powerful capabilities have allowed astronomers to peer into the outer limits of the universe and uncover a variety of never-before-seen galaxies. The observations clearly show that different types of galaxies evolved at different rates. The giant elliptical galaxies formed shortly after the Big Bang and changed little; spiral galaxies like our Milky Way took longer to form and have undergone dramatic changes; and dim dwarf galaxies quickly appeared and then mysteriously vanished.

Black Holes: From Fiction to Fact

Hubble has uncovered convincing evidence for the existence of super-massive black holes in space. By using Hubble to measure the whirlpool-like motion of stars and gas in the cores of galaxies, astronomers have calculated how much matter is packed into a galaxy's hub. In the three galaxies probed so far by Hubble, the mass of hundreds of millions or billions of suns is compressed into a region of space no bigger than our solar system. The Hubble results fit the definition of a black hole: an extremely compact and massive object. A black hole is the simplest explanation for the observed phenomena.

Zeroing In On the Age of the Universe

Hubble is helping astronomers precisely calculate the age of the universe by providing accurate distances to galaxies, an important prerequisite for calculating age. Hubble measures the distances to neighboring galaxies by finding accurate "milepost markers," a special class of pulsating star called Cepheid variables. These, in turn, are being used to calibrate more remote milepost markers. Preliminary findings suggest that the universe may be only 9 billion years old, younger than previously thought; other researchers using Hubble argue that the age is more like 16 billion years. The research—and scientific debate—will continue for some time.

Hunting For Brown Dwarfs

Hubble has confirmed the existence of an elusive, long-sought class of object called a brown dwarf, an object too large to be a planet but too small to be a star. Astronomers using a ground-based telescope made the initial discovery. Hubble provided a sharper, follow-up image that clearly separated the brown dwarf from the star it was orbiting.

Planets Under Construction

While surveying the Orion nebula, a nearby star-forming region, Hubble returned images of pancake-shaped dust disks around dozens of embryonic stars. These disks may eventually condense and form planetary systems, and their abundance alone suggests that the conditions necessary to form planets are common elsewhere in the universe.

A Star is Born

Looking to neighboring stellar "maternity wards" to see a replay of the events that created our sun and planets, Hubble has uncovered remarkable new details of star birth.

FIGURE 6–3

In the direction of the constellation Canis Major, two spiral galaxies pass by each other like majestic ships in the night. The near-collision has been caught in images taken by NASA's Hubble Space Telescope and its Wide Field Planetary Camera 2 (11/4/99). (NASA and Hubble Heritage Team. Courtesy of STScI.)

Hubble revealed an eerie scene, illuminated by nearby hot stars, of huge stalagmite-like towers of cold, dark gas with finger-like protrusions containing embryonic stars just emerging from their incubation.

Planet Quest

Probing the inner edge of the dust disk around the star Beta Pictoris, Hubble found a curious warp in the disk, like the twist in an airplane propeller. The most likely explanation for the twist is that the disk is feeling the gravitational tug of an unseen planet, perhaps the size of Jupiter, orbiting the star at a slightly different angle from the disk. Astronomers have long suspected that the star Beta Pictoris has a planetary system.

Ring Around the Solar System

Hubble has provided definitive evidence for the existence of a vast belt of primordial icy debris around our solar system, a reservoir for comets flying through interplanetary space. Though astronomers using Earth-based telescopes had previously identified some of the largest objects in the belt, Hubble uncovered evidence for an underlying population of more than 100 million comets.

Smash Hits

Hubble offered a ringside seat to a once-in-a-millennia event when 21 fragments of comet Shoemaker-Levy 9 collided with Jupiter. As each comet fragment crashed into the giant planet, Hubble caught mushroom-shaped plumes along the limb of the planet, detailed views not possible using any other telescope. The largest fragment impact created an Earth-sized "bull's-eye" pattern on Jupiter.

Pluto Unveiled

Hubble provided the first direct look at the surface of the distant planet Pluto. The pictures show that Pluto has a remarkably varied surface, mottled with bright and dark regions. This will probably be our best look at the tiny planet until space probes venture to this "frontier outpost" of our solar system.

The Big Bang's Alchemy

Hubble detected what may be ancient helium gas that produced galaxies in the early universe. The space telescope found gas older than most stars. The discovery confirms the Big Bang theory's model: helium was produced with hydrogen in the first three minutes after the Big Bang. In addition, Hubble found that certain light elements created in the primeval universe, such as lithium, are in the exact quantity in space as expected if the Big Bang really happened.

Reading Comprehension Questions

1. What is the purpose of the Hubble Space Telescope? _____
_____

2. Underline the analogies and comparisons in the article that compare technical information, such as the size and speed of Hubble and the amount of information transmitted back to Earth, with ordinary objects. _____

3. Briefly (in one sentence), describe how Hubble collects and transmits data back to Earth. _____
_____

4. Briefly describe how scientists control what Hubble does each day. _____
_____

5. In the article, put a check mark next to the types of discoveries that Hubble has made. _____
_____

WRITING: Summaries

The article about the Hubble Space Telescope (HST) includes an overview of the purpose, design, and operation of the HST, as well as a summary of its key discoveries. For this summary, the author uses consistent headings, followed by a brief description of each discovery. Readers can quickly skim over the last section of the article and still have a fairly good idea of the wide variety of discoveries.

Headings, as well as the use of numbers, letters, dots, or dashes, make the material easier to read and remember because it usually means fewer words and concentrated information.

Summary
A condensed account of a report to recap the main points. Usually located at the end of the document.

A **summary** is a condensed account of the essential information included in a longer piece of writing. A summary usually appears at the end of an article or report. The function of a summary is similar to that of a schematic diagram, which gives a clear, brief presentation of a device without the clutter of the actual materials necessary to build the device.

For example, if you needed information on a report about the HST, you would find information in many sources, too many sources to actually read. You might find professional abstracts of journal (magazine) articles. By reading these brief summaries, you would be able to judge which articles would be most useful to you.

A summary answers the basic questions that readers want answered before they devote more time to reading the article or book. Many people who are interested in keeping up with technology do not have the time to read every article printed about their field. They often rely on professional abstracts to find the most useful articles.

A reader searching for information has predictable questions for each article:

What? Who? Where? When? Why? How?

Go back to the HST article, and underline the answer to each of these questions. They will normally be found early in a summary. Summaries include only the key facts, ideas, and conclusions.

TIPS Follow these helpful steps when writing a summary.

1. Read the article carefully—more than once—before starting to write. Use your pencil to mark key ideas, phrases, and conclusions.

2. Look for the author's own summaries at the beginning or end of the article. Often, boldface headings indicate a transition and a new key idea.

3. Note the author's organization—find the main idea of each paragraph or section.

4. The length of a summary is usually about 33 percent of the length of the article, although this is by no means a rule. Instructors seldom require more than one page, and professional abstracts are rarely longer than one paragraph, no matter how long the article.

5. Summarize each section (of longer articles) or paragraph (of shorter ones). Disregard figures of speech, examples, detailed descriptions, and discussions.

6. Do *not* include personal interpretations, agreements, or disagreements (no *I* statements). Write in the third person (*he, she, it, they*).

7. Read the article once more and compare it to your summary. Make any revisions that are necessary for clarity.

8. Format: the summary should contain the following information.

    a. Identification of the article being summarized (name of author, title of article, title of book or magazine, date of publication).

    b. Statement of the main idea of the article.

    c. Statements that explain all the important points used to support the main idea.

    d. Explanation or clarification of important points, if necessary.

Plagiarize
To copy exact sentences of someone else's work and pass it off as your own. Literary theft.

In a summary, writers reword and condense ideas. Copying exact sentences is considered plagiarism. Do not plagiarize other people's writing in a summary, or any other piece of documentation, for that matter.

If you reworded your responses to the reading comprehension questions on page 82 into a paragraph, you would have the basic information needed for a summary of the article. Compare your responses to the following summary of the Hubble article.

"Hubble Expands the Universe," obtained from the Web site for the Space Telescope Science Institute (STScI), Baltimore, MD, provides an overview of the purpose, operation, and discoveries of the Hubble Space Telescope. Hubble, orbiting Earth just above the atmosphere, can take clearer pictures of objects in the universe, without the distortion caused by the atmosphere. Hubble uses scientific instruments to take pictures with long exposures and beam them back to Earth as radio signals. Scientists provide detailed instructions to Hubble's "brain" using a computer code that directs and points Hubble at desired stars and planets. Important discoveries include information on new galaxies, black holes, the age of the universe, the formation of new planets and stars, astronomical collisions, and confirmation of the Big Bang model theory.

A summary of a research project would recap the purpose, results, conclusions, and recommendations, and would be written for a semitechnical or nontechnical audience. Two special types of summaries are the executive summary and abstract.

Executive Summary

Executive Summary
Modified summary located at the beginning of a report or document, used by upper management to preview the report.

An executive summary is a modified summary located at the beginning of a report or document. Its purpose is to highlight the bottom-line information needed by upper management to make a decision, including staffing, budget, and timeline considerations, sometimes in a bulleted list. It might also include a final recommendation or conclusion, depending on the purpose of the report. If the document describes a research project, the executive summary includes the purpose, background, results, conclusions, and recommendations, written for a semitechnical or nontechnical audience.

The following is an executive summary of the reading article. Note that a colon and a bulleted list are used to make the key ideas of the article stand out for readers who might not have the time to read the full article.

EXECUTIVE SUMMARY

"Hubble Expands the Universe," obtained from the Web site for the Space Telescope Science Institute (STScI), Baltimore, MD, provides an overview of the purpose, operation, and discoveries of the Hubble Space Telescope:

- Hubble, orbiting Earth just above the atmosphere, can take clearer pictures of objects in the universe, without the distortion caused by the atmosphere.

- Hubble uses scientific instruments to take pictures with long exposures and beam them back to Earth as radio signals.

- Scientists provide detailed instructions to Hubble's "brain" using a computer code that directs and points Hubble at desired stars and planets.

> **EXECUTIVE SUMMARY** *(Continued)*
>
> • Important discoveries include information on new galaxies, black holes, the age of the universe, the formation of new planets and stars, astronomical collisions, and confirmation of the Big Bang model theory.

Abstract

Abstract
A brief summary that previews an article, sometimes provided by a service.

An abstract is typically a one- or two-sentence summary that includes the author's name, publication and date, and keywords used by databases, librarians, abstracting services, and others to locate articles on specific topics. Abstracts sometimes contain related and alternative terms so that people can find the article using a word search, and they can use the related words to widen a computer search for a topic. People obtain abstracts to determine if they want to read the original article.

> **ABSTRACT**
>
> The Hubble Space Telescope, orbiting Earth, provides pictures of the universe using cameras and spectrographs that collect data. Scientists and astronomers have made many new discoveries based on data radioed back from the HST.
>
> Copyright STScI, http://oposite.stsci.edu/pubinfo/spacecraft/Primer/, Oct. 1998

Abstracts can include the author(s), title and subtitle, source (such as the magazine and date), description of the article, and identifier keywords related to the topic. Services and databases include a record number for easier retrieval. Academic abstracts also include the affiliation (university or institution) of the first (lead) author.

To conduct extensive research on a topic, ask a librarian or search the Internet for an abstracting service. Most services charge a fee and specialize in categories of information, such as astronomy or current events. When you enroll, you can provide authors' names, publication dates, or keywords and combinations of keywords to the service. Then the service provides you a list of abstracts that match your entries. From the list of abstracts, you select the articles you want to read.

Exercise 6.1 *Write a one-paragraph summary of the reading article in Chapter 5: "Writing Science Articles Without a Ph.D." In addition, add a final sentence that describes your reaction to the article.*

Exercise 6.2 *Internet assignment: Using the keywords* Hubble Space Telescope, Space Telescope Science Institute, *or STScI,* find a current discovery by Hubble or information about the instruments or astronaut visits to Hubble to install or repair its equipment. Write a summary of the information.

SPELLING: Double Trouble

In this chapter, rather than review a spelling pattern, we are going to concentrate on certain technical words that present writers with spelling problems because of a troublesome double consonant. Doubled letters are unvoiced, so we have no audible clues to remind us of the letters.

The only way to remember the correct spellings of these words is to practice spelling them correctly and observe them carefully to form a mental picture of the words. Soon you will recognize when a misspelled version "doesn't look right." Whenever you are unsure of a spelling, use the dictionary.

Exercise 6.3   *Draw a box around the double letters in each word. Then write the word twice.*

1. accessible
2. accomplish
3. antenna
4. approximate
5. assemble
6. battery
7. collapse
8. collector
9. communicate
10. connect
11. current
12. dissipated
13. efficient
14. installation
15. metallic
16. parallel
17. personnel (people)
18. profession
19. symmetrical
20. transmission

Exercise 6.4   *Pick out five words from the list above that you had trouble spelling, and use each in a sentence.*

1. _____
2. _____
3. _____
4. _____
5. _____

Exercise 6.5   *From the list of words above, fill in a word to complete each sentence. Each word may be used only once or not at all.*

1. Wire is used to _____ two points in a circuit.
2. The crystal had perfectly _____ faces.
3. Lynn read the manual to _____ the bicycle.
4. Howard received an application in the _____ office.
5. Metal fatigue may cause the bridge to _____.
6. The lawyer claims that the building is not _____ to the handicapped.
7. We guessed at the _____ distance.
8. The _____ paint on the car reflected the light.
9. The sales and service staff must listen carefully and _____ clearly.
10. The fan _____ the heat and cooled the motor.

VOCABULARY: *tele, phono, photo, graph,* and *gram*

The study of electronic or mechanical communications frequently uses five Greek root words. *Tele* means "far off" or "at, over, to, or from a distance."

      telecommunication
      telescope

Phono or *phone* means "sound, tone, or speech."

> telephone
> phonograph

Photo means "a light" or "produced by a light."

> photograph
> telephoto lens

Graph means "something that writes or records" or "something written."

> graphics
> telegraph

Gram or *Gramma* means "something written down or recorded."

> grammar
> electrocardiogram

Note: Do not confuse *gram,* the root word, with *gram,* the metric unit of weight.

Exercise 6.6   *Write a brief meaning of each word as it relates to the Greek root. Use the dictionary only as a last resort.*

Example: telecommunication—communicating from a distance

1. telescope _____
2. telephonic _____
3. phonograph _____
4. photograph _____
5. telephoto lens _____
6. graphics _____
7. telegraph _____
8. grammar _____
9. photosensitive _____
10. telecast _____

WORD WATCH: *lose, lost, loss, loose,* and *loosen*

The words *lose, lost, loss,* and *loose* can be confusing. With attention and a crutch, you can be sure of the correct use of each of these words.

Lose (pronounced *looz*) is a present-tense verb meaning "misplace," "be deprived of," "give up," "waste," or "bring to ruin." The other tenses are *lost* (past, past participle) and *losing* (present participle). There is no such word as *losed.*

> I tried not to *lose* the phone number.
> I *lost* it anyway.
> I *have lost* several valuable phone numbers.
> I think I am *losing* my mind.

Lost can also be an adjective describing something that is hopeless or missing.

> It seems to be a *lost* cause.
> I had to make up for *lost* time.

Loss is a noun meaning something that is lost. If *a, an,* or *the* can be placed in front of the word, use the noun form.

> The missing phone number was a terrible *loss.*
> The company took a *loss* on the sale.

Loose (pronounced *loos*) is completely unrelated to the other similarly spelled words. It is most commonly used as an adjective meaning "free" or "unrestrained." As a crutch, think of *too loose* (both have *oo*).

> I searched through my *loose* change.
> We finished up all the *loose* ends.

Loosen is the verb form of "loose," meaning "the act of making something free or unrestrained." The tenses are *loosened* (past, past participle) and *loosening* (present participle).

> I always *loosen* my tie when I arrive at work.
> I *loosened* my seat belt after the plane took off.
> I *have loosened* all the screws on the casing.
> While I *was loosening* the screws, the case fell off.

Exercise 6.7   *Use* lose, lost, loss, loose, loosen, *or* loosened *to fill in the missing words in the paragraph.*

The factory experienced a power _____ every day at 10 A.M. About one hour was _____ every day, and the management was at a _____ to explain the mysterious "downtime." The maintenance staff _____ all the power outlets to look for _____ connections. This resulted in even more _____ time, but the foreman decided that it was more cost-efficient to _____ a few hours while methodically troubleshooting and correcting the problem than to suffer an ongoing _____.

Exercise 6.8   *Use each of the following words in a sentence.*

  1. lose _____

  2. lost (verb) _____

  3. have lost _____

  4. lost (adjective) _____

  5. losing _____

  6. loss _____

  7. loose (adjective) _____

  8. loosen _____

  9. loosened _____

  10. loosening _____

CHAPTER 7

Graphics

- Write a functional description using a graphic.
- Prepare a graphic to include in a technical description of the device.
- Prepare a block diagram.
- Prepare a bar graph or line graph.
- Insert an electronic (computer) graphic into a document.
- Spell words with *ie* and *ei* correctly.
- Use roots *spec* and *son* correctly.
- Use *they're, their* and *there* correctly.

READING: Is There Safety in Numbers? It Depends on the Statistics

by Elaine Dickinson

A 59% decline in the number of boating fatalities over the course of the past 25 years and an absolute decline in 15 of the last 25 years are plenty of reasons for safety experts to cheer. But precisely why this has occurred and what should be done to reduce fatalities below its current plateau of about 800 per year, as well as get a handle on non-fatal accidents and injuries, remain somewhat of a mystery.

The reason is fairly simple: The information on boating accidents that was collected in 1973 is still pretty much the same, limited information reported today. Even in an "Information Age" with a whole new world of computer technology at our fingertips, boating statistics have not much improved or expanded over the years. This greatly limits the amount of analysis and comparison that can be done. Experts struggle to spot trends in the figures but even these can be misleading.

"Essentially all public policy decisions on boating safety are based on the same annual accident statistics. We won't make any more headway until we get

Boat/U.S. Magazine, November, 1999.

some better data," said BOAT/U.S. President Richard Schwartz. "While there was a modest boating survey done in 1989, the last major comprehensive survey of boaters was done by the federal government in 1975 and boating has changed dramatically since then."

Everyone in the safety field—from the U.S. Coast Guard, National Transportation Safety Board and state agencies to organizations such as BOAT/U.S., the Red Cross, the Coast Guard Auxiliary and U.S. Power Squadrons—relies upon the same "snapshot" we got in the 1970s and '80s. What has improved is the actual reporting because better training of marine police has, in turn, led to more professional and complete accident investigations and reports.

No Level Playing Field

What's lacking is depth. Some might argue that the picture we get each year is all we need to know to address the major issues. We know that 821 people died in boating accidents in 1997. When any person dies in a boating accident, it is one too many and no one in boating safety would disagree. But millions of dollars of boaters' tax money

FIGURE 7–1   **1997 Fatalities - Victim Activity**

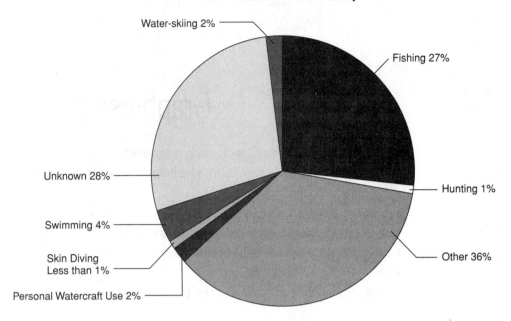

are spent every year on federal grants for studies, meetings, media campaigns, outreach efforts and law enforcement, with limited ability to ascertain if any or all of these are producing positive results. Are we spending our limited boating safety funds on the right things? What are we missing?

• Relative Risk—Because the level of participation in boating and the amount of time spent boating have not been adequately measured in a recent or reliable national survey, we have no idea if boating is safer or more dangerous than it was 10 or 20 years ago, or safer or more risky than other outdoors activities or other types of transportation.

"Raw" numbers such as we now get only tell you so much. They have to be put into context, called "normalizing" by statisticians. The raw data need to be seen against a backdrop of exposure to risk. In accident statistics for other activities, such as those for private aircraft accidents, the normalizing factor is hours spent in flight, i.e. accidents per 100,000 hours of flight time. This levels the playing field for the accident figures.

Without any data on time spent boating, our accident statistics are not only superficial but could be misleading. For example, a year with a spike in boating accidents could be the result of more boating due to good weather or simply more people filing reports. Likewise when fatalities go down, we're heartened, but could it be because we had a summer of bad weather and fewer people went boating?

Some might also ask, as long as fatalities appear to be on the decline, isn't that good enough? Yes and no. If a comprehensive national survey of boating found out, for instance, that even though there are more registered boats, people were spending fewer hours on the water, that would change the overall picture of a fatality "rate."

• Lack of Reporting—Many non-fatal boating accidents are not being reported by boaters, as required by law. An accident report is required to be filed when damages are over $500, there is a total loss of the boat, or when injuries require more than first aid. Only an estimated 2–3% of boating accidents are reported, according to the Coast Guard, so we basically know nothing about 98% of the non-fatal accidents that occur. Injuries are a good example of where more and better

FIGURE 7–2

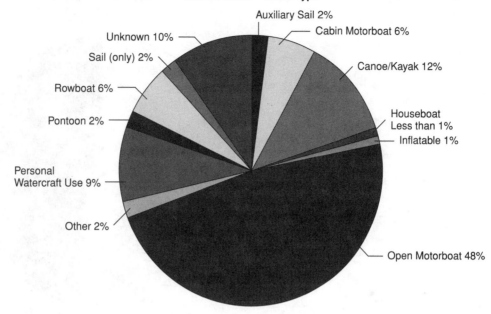

1997 Fatalities - Vessel Type

- Auxiliary Sail 2%
- Cabin Motorboat 6%
- Canoe/Kayak 12%
- Houseboat Less than 1%
- Inflatable 1%
- Open Motorboat 48%
- Other 2%
- Personal Watercraft Use 9%
- Pontoon 2%
- Rowboat 6%
- Sail (only) 2%
- Unknown 10%

information could help direct safety efforts. Until recently, a reported boating injury could be either three stitches or an amputated limb. Only in the past year or so have data about the specific type of injury been reported.

Since the non-fatal accidents and injuries that are reported are what's called "self-selected"—the boaters involved chose to file a report—no real conclusions can be drawn from these figures because they do not represent a valid random sample of boaters. These days, when professional pollsters can accurately predict the outcome of a national election based on a random sample telephone survey of a handful of voters, you'd think basic boating statistics could be brought up to speed.

It's unlikely the public is going to change it reluctance to self-report, so it's time to get at this information through a different method, using basic survey methods. Random sampling could be it.

On the plus side, the lack of a national survey should be addressed soon. BOAT/U.S. lobbied hard in Congress for some $2 million in additional funding for a comprehensive national boating survey, plus an update every five years.

According to Capt. Mike Holmes, chief of boating safety at the Coast Guard, a previous national survey conducted under a grant is being finished up this year, and its strengths and weaknesses will be used to refine the national survey in 2000. He expects it will take at least a year to complete the next survey and it will be done by an outside contractor. Holmes said he expects to finally get valid data on hours and days spent boating as well as better accident data.

What Do We Know?

Since it is generally accepted that virtually all boating fatalities are reported to police, we do know what causes the majority of boating deaths. Spectacular boat crashes and high-seas sinkings are simply not the norm. Without a doubt, the vast majority of people who die in boating accidents drown, most often in calm water and fair weather and on lakes and rivers. The victims most often either fell overboard or the boat capsized and they weren't wearing a life jacket.

Nearly half of all boating deaths occur in boats of 16 feet or less, and just over one-fourth of fatalities involve alcohol.

FIGURE 7–3

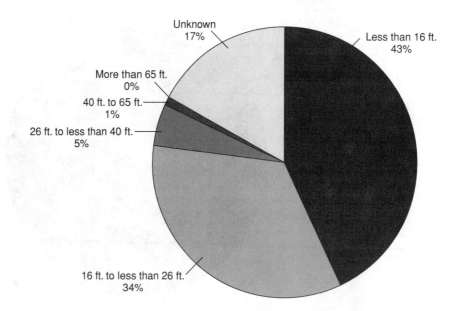

1997 Fatalities - Vessel Length

Of the boaters who died in accidents in 1997, 86% never had any boating instruction. They don't make for exciting news stories, but the raw statistics that we read year after year are numbingly consistent.

The best news is that fatalities since the 1970s have been reduced by more than half, dropping from an all-time high of 1,750 to an all-time low of 709 in 1996.

But since 1992, the number of fatalities has hovered between 709 and 829, forming a plateau that seems hard to get below without some drastic action, such as requiring those involved in the highest risk types of boating to wear life jackets. This option appears unlikely, however, given the anti-regulatory political climate and the low priority marine patrols get in state and local funding.

Where Are We Going?

Knowing how the most common types of fatal boating accidents occur has enabled organizations to target their education and outreach messages at the most critical problems. State agencies have also taken the lead role in enacting new laws to address their major problems. The three that are likely to continue to be the focus of safety efforts for years to come are: drink-

ing and boating, life jackets, especially for children, and boater education.

Alcohol: Every single state except Iowa has a specific boating-while-intoxicated law in effect, a huge improvement from 15 years ago when only a dozen or so states had even set a legal blood alcohol level for intoxicated boating.

While the reporting still may be less than complete, where it was routinely estimated that alcohol involvement in fatal accidents was as high as 50%, the 1997 figures indicate a lower rate of 27%. The national focus on drunk driving has shown that this is one problem where media campaigns really can change behavior over the long run.

Life Jackets are the next frontier for the new millennium. Already the idea of requiring people in boats less than 16 feet to wear life jackets is being seriously explored by the Coast Guard. Would such a law cut the fatalities by half? Would it get the attention of the scores of duck hunters and fishermen who drown each year because they don't think of themselves as "boaters" and often don't take the most basic precautions?

Some 31 states have laws requiring life jackets on children, and there are early signs that the number of child fatalities is declining.

The 15-year effort to get inflatable life jackets approved by the Coast Guard and into the hands of consumers at a reasonable price may also improve the drowning statistics. Manually inflated devices are now approved with automatics soon to follow. Groups such as BOAT/U.S. will continue to keep the pressure on until these more comfortable life jackets meet all carriage requirements.

Mandatory Education is also certain to be in the forefront as each year another state joins the ranks of those that have already passed laws requiring some form of education for all boaters. Where once marine industry opposition to any such requirement was fierce, as we head into 2000, the notion of requiring a boat operator to take a boating course draws fewer and fewer critics. BOAT/U.S. members overwhelmingly favor mandatory education. Our work in the coming years will be to see that states adopt education laws that are workable and make sense.

Clearly great strides have been made in reducing boating fatalities. How much better can we do? How many traumatic injuries can be eliminated? What new technology can be applied to boating? With exposure to risk data finally in hand, we'd love to publish a story in 2000 or beyond reporting that boating fatalities have dropped below 700 for the first time ever.

Reading Comprehension Questions

1. When were the original surveys for data concerning boating fatalities completed?
2. What are the two concerns of the author about the validity of data currently being reported?
3. What technique do statisticians use to provide a better context for raw numbers?
4. According to the text of the article, describe the conditions in which most boating fatalities occur.
5. According to the graphic of 1997 fatalities by victim activity, describe the conditions in which most boating fatalities occur. Then describe those in which the fewest occur.
6. According to the graphic of 1997 fatalities by vessel type, describe the type of vessel in which most boating fatalities occur. Then describe in which type the fewest occur.
7. According to the graphic of 1997 fatalities by vessel length, describe the type of vessel in which most boating fatalities occur. Then describe in which type the fewest occur.
8. What are the three current efforts to increase boating safety?

WRITING: Preparing Graphics

Graphics are visual representations of objects, numbers, and other data in the form of pictures, diagrams, graphs, and charts. As demonstrated in the article, the purpose of using graphics is to present information in a visual way and to clarify concepts such as location, size, relationship, and comparisons. For example, a photograph can display the overall appearance of an object, or a bar graph can show the relationship of numbers in two or three dimensions. In addition, graphics can increase interest and readability of documents for readers who might shy away from blocks of solid text.

At one time, writers had to draw charts and graphs by hand or rely on graphic artists or even programmers to provide realistic drawings and images. Now, with the aid of a computer and graphics software programs, you can create many types of electronic graphics and insert them into your documents for a professional and polished look.

This chapter provides guidelines for using several types of graphics, as well as special techniques for creating computer images.

General Information on Graphics

Besides digitized images, writers create and insert a wide variety of graphics that supplement written material. Graphs are representations of numbers and data in one, two, or three dimensions. Graphs must be complete as well as accurate. The article points out that missing (or unreported) numerical information can lead to false conclusions, sometimes resulting in wasted time, money, and effort.

As with images, at one time, creating graphics was left in the capable hands of artists. Because of the time and expense involved, figures and drawings were infrequent and functional. With the increased capabilities of computer hardware, software, and printers, technology has given us the ability to create, touch-up, and maintain many types of graphs.

The creative and colorful nature of computer drawings makes them appealing and eye-catching, not only adding to our understanding and interpretation of information, but entertaining us as well. Unfortunately, graphics can be used to mislead readers, as described later in this chapter. Interpreting graphics correctly is as important as preparing them. In this chapter we do both.

The main types of graphics used in technical writing are photographs, line drawings, graphs, and tables. The purpose of adding graphics to technical reports is to supplement the written material. Graphics are not used to repeat information that is already clear or to impress the reader, nor are graphics used to lengthen a report. Effective graphics can clarify information, organize data, and emphasize important points. The measures of effective graphics are simplicity and usefulness.

It is essential to plan your graphics as you outline a report. Add graphics where they are logical and useful for the reader. Explain every graphic in the text. And by all means, label each graphic with a number and a title even if your explanation is directly above or below the graphic. Add enough information in the title so that a person skimming through the report would have a clear idea of the nature of the illustration—remember that some readers look at the pictures first. Normally, graphics are placed in the document just below their written explanation; however, some instructors or companies require that all graphics larger than half a page be placed in an appendix.

Each graphic, except a table, is referred to in the text as a "figure," and each is numbered starting from Figure 1 and continuing on to the last figure of the report. A table is referred to as a "table" and is also numbered starting from Table 1. Capitalize the first letter of the reference if a number follows (Table 1 or Figure 1) since it is similar to a proper noun. The examples below show a few ways to refer to figures or tables.

A satellite dish collects signals (see Figure 1).
The satellite dish in Fig. 1 collects signals.
The figure shows a satellite dish that collects signals.
Table 1 shows the values.

If drawing by hand, use a template or ruler to draw graphics. All labels should be typed or printed neatly. Be precise but brief. If you copy your graphics from another drawing, you must footnote your drawing (use the standard footnote format shown in Appendix 3). Some original drawings credit the source of the data used in the drawing. You must also credit that source below the figure or table.

Figure 1   Satellite dish antenna.
Source: Scientific Atlanta

FIGURE 7–4
Scientific-Atlanta's IBT-1200.
Ku-band transmit/receive digital
earth station.

Courtesy of Scientific-Atlanta.

Because the style for each type of graphic varies, we will review some general guidelines and examples of each.

Photographs

Camera-produced graphics are easy to insert into reports using a scanner or digital reprints from negatives. They may be useful to show the overall appearance of an object, but line drawings also serve this purpose. If used, photos should always be clearly focused and include only the intended object—easier said than done. Because photographs do not necessarily indicate size, some photographs show the object next to something common, as in Figure 7–4. Here the reader can see the size of the satellite dish antenna in relation to a person.

Line Drawings

Line drawings include the vast majority of the graphics in your textbooks. Schematics, drawings of components, and block diagrams are all examples of line drawings. Follow these general guidelines:

1. Label all the significant parts of a drawing. If you use arrows or lines, they should touch the specific parts to which they point.

2. Use standard abbreviations, symbols, and terms in labels and explanations of figures. Be sure that the terms are consistent with those used in the text. Add a legend (a key for unfamiliar terms or symbols) if you are writing for a general audience.

    V = volts   A = amperes
    Hz = hertz   ac = alternating current

3. Add enough white space so that neither the drawing nor the labels will be too crowded.

4. Use the type of line drawing that best fits your subject. These include front, side, exploded, and cutaway views; cross sections; and block diagrams.

FIGURE 7–5

A simple electric circuit may be represented by a pictorial diagram (A), which involves drawings of the electrical components, or by a schematic diagram (B), which consists of interconnected standard symbols used by electricians to depict specific components.

Reprinted with permission of *Academic American Encyclopedia,* © 1986 by Grolier, Inc.

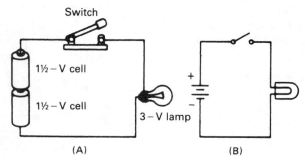

FIGURE 7–6

Cutaway drawing of waterproof cable design.

"Communication Cables," Vol. 3, p. 442. Reprinted with permission of *McGraw-Hill Encyclopedia of Science and Technology,* © 1982 by McGraw-Hill Book Company.

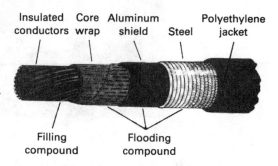

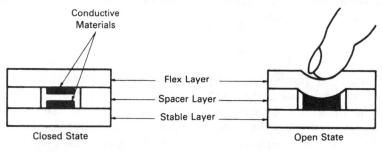

FIGURE 7–7

Operation of a conventional flat-panel membrane switch.

Courtesy of Dupont.

In Figure 7–5 you see two diagrams of an electric circuit. The pictorial diagram (A) is labeled to make it understandable to nontechnical people. The schematic drawing (B) is unlabeled because it would be read by technical people who understand the symbols used to represent specific components.

In Figure 7–6 you see a cutaway drawing of a waterproof communications cable. The labels are close to the parts. Outer layers are cut away in the diagram to expose the inner structure.

In Figure 7–7 you see a process drawing of a membrane touchpad. The second stage, as the pad is touched, displays the internal change of the pad. All parts should be labeled in the first drawing; similar drawings do not necessarily have to be labeled. Consistent labels and terms are important to show how parts change or move in a series of steps. The explanatory notes describe the process while referring to the drawing.

In Figure 7–8 you see an exploded view of the Hubble Space Telescope, which you read about in Chapter 6. Exploded views take devices apart to show internal structure and how parts fit together. The view shows 14 separate parts of the satellite. The parts are labeled.

Figure 7–9 is a block diagram, or signal flow chart, of a color TV transmitter. Many electronic and mechanical processes are more simply explained by using block dia-

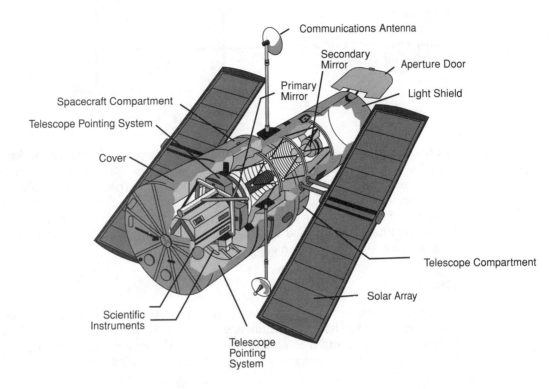

FIGURE 7–8

Hubble Space Telescope.

Courtesy of NASA.

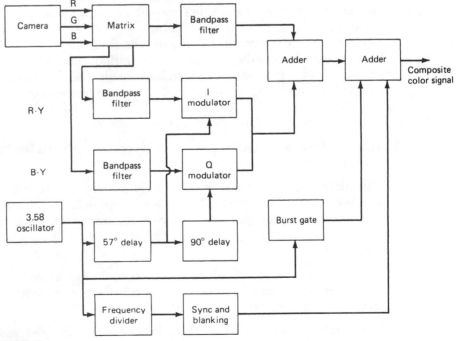

FIGURE 7–9

Block diagram of color TV transmitter.

Courtesy of Fred Kerr.

FIGURE 7–10
Lines that clearly denote
whether they make a connection
(dot) or not (arch).

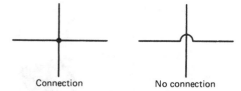

grams. Each block represents a functional unit or step, but the details of the unit are not included. The signals (lines, links, or arrows) show communications or relationships between units. Follow standard conventions when preparing these displays. Lines that cross should clearly indicate whether the lines make a connection (denoted with a darkened dot) or do not make a connection (denoted with a semicircle or arch in one of the intersecting lines). Use dashed lines inside blocks to show subunits. Provide just enough notation to make the diagram easy to understand.

Graphs

Graphs include displays of numerical data using bars, lines, curves, and circles. The purpose of graphs is to help your audience visualize the effects of a changed variable on a subject. The display often emphasizes a trend or illustrates the results of an experiment. The graphs you know best are waveforms and exponential curves.

Creating Graphs

Writers can quickly create and revise many types of graphs and charts using popular database or spreadsheet programs, such as Microsoft Excel, Lotus 1-2-3, or Quatro Pro. Read the user guide or online help for detailed instructions. Generally, you begin by entering numeric values in cells on a worksheet.

As you plan how to display the values in a graph, you must also plan the axis or value ranges, labels, and titles. In addition, you can add text to the final graphic. The chief advantage of using a database is that you can revise the numeric values in each cell at any time and reproduce the graph showing the changed values.

Finally, select how to display the values. Options for displaying values vary by program but usually include many standard single and multidimensional graphs, pie charts, and legends for figures or maps. You can also create a customized graph using features in the program.

For your final product, use the following guidelines:

1. The horizontal and vertical axes should be clearly labeled (such as frequencies, voltages, and time).

2. The increments should also be marked and labeled clearly. Increments should be regular (every 10 ms, every 20 mA). The lowest value (origin) is usually zero. The maximum value is usually just one increment higher than the highest value to be represented on the graph. An arrow at the outer point of an axis represents infinite increments.

3. Graphs sometimes contain shaded, colored, or figured areas to provide emphasis or a visual contrast to sets of numbers: one shaded side represents one set, and the other shade represents another.

Bar graphs show evenly spaced bars extending vertically or horizontally. Some writers print the exact value inside each bar, which is especially helpful when precision counts. The bar graph in Figure 7–11 shows the billions of dollars of projected sales for

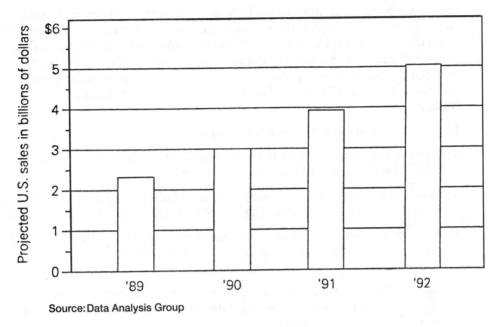

A Growing Market for Laptop and Portable Personal Computers

Source: Data Analysis Group

FIGURE 7–11
Bar graph of laptop and portable personal computer sales.

Reprinted by permission of the *Atlanta Constitution,* 9/7/89.

laptop and portable personal computers for 1989 through 1992. The Data Analysis Group projected sales to be about $2.2 billion for 1989, with a steady increase to about $5 billion for 1992.

Line graphs and waveforms are made up of dots placed at coordinates according to fixed increments on the vertical and horizontal axes. The dots are then connected by straight lines or smooth curves to show the subject's response to changing conditions. The horizontal axis usually represents the changed condition, and the vertical axis usually represents the subject's response or activity. Line graphs are particularly useful for displaying patterns and for predicting future activity. Normally, both axes begin with zero in the lower left corner, but in electronics, this is not always the case.

In Figure 7–12, a line graph represents the current (I_c) and voltage (V_{ce}) relationship in an electric circuit. It is clear that as current increases, voltage decreases. The increments are even, and the origin is zero on both axes. Dashed lines are used to plot current and voltage combinations (see Table 7–1 for another method of representing these values).

Pie charts are partitioned circles in which each partition represents a percentage or proportion of the category. The first segment usually begins at a line from the center to the top of the circle. The segments are automatically arranged in alphabetical order starting at the top (or midnight position), and proceed in a clockwise direction. If needed, label the last segment "others" to include all the remaining segments, and itemize the "others" below the circle. Print explanatory information horizontally inside the segment, if possible. If the segment is small, draw a line from it to a space outside the circle and explain it there. The pie chart in Figure 7–13 shows the activities in which victims of 1997 boating fatalities were participating at the time of their deaths.

Tables

Tables are displays of information in columns and rows. There is no limit to the number of columns or rows, and the values within the tables are precise. Tables are used to compare and contrast the features of two or more objects. Information is classified and organized into column headings. Columns are arranged horizontally with vertical lines between the columns. A subtotal or grand total can be placed at the bottom of a column and is emphasized with one or two horizontal lines separating it from the rest of the column.

1. Separate tables from the rest of the report with horizontal lines or a box.

2. Label the table at the top of the box. Include a title that explains the contents of the table. Cite the source at the bottom of the box.

3. Label each column. Provide adequate space between columns to accommodate the longest value in each column. Capitalize all major words in column labels.

4. Keep values listed in a column in a consistent form. Line up decimal points, and add zeros to fill out numbers to the uniform length. (It is conventional to add a zero before decimals less than one, as in 0.85.)

5. Use division lines (horizontal or vertical) to group columns or rows if they make the organization more understandable. Remember, however, that too many divisions make the organization unclear.

Table 7–1 shows the possible combinations of current (I_c) and voltage (V_{ce}), as displayed in Figure 7–12. The headings of the columns match the axes labels. Note that the units of measurement (mA and V) are also included in the headings for clarity.

Let the Reader Beware: Figures Can Lie

It is important to be aware that graphical representations of data can be misleading. The graphics can be constructed in ways that visually distort the data without being exactly dishonest. Books, newspapers, and technical journals will occasionally present line and bar graphs, for instance, that appear to offer unmistakable evidence of something, when, upon closer inspection, the evidence is unconvincing or nonexistent. The following examples are only a few of the methods used to misrepresent data.

In Figure 7–14 the origin of Figure 7–11 has been changed from zero to $2 million. By not starting at zero, the differences are overemphasized.

In Figure 7–15 the line graph uses a large increment (amperes rather than milliamperes) to deemphasize the difference.

In Figure 7–16 the line graph suggests a relationship in which the cause and effect are not clear, or which may be linked by other variables, such as population.

Examine all graphics carefully, particularly when you are using them to draw a conclusion. Prepare your own graphics with integrity.

TABLE 7–1 I_c vs. V_{ce} Combinations

I_c (mA)	V_{ce} (V)
1	9
2	8
5	5

Source: Introductory Electronic Devices and Circuits by Robert Paynter, © 1989, p. 135.

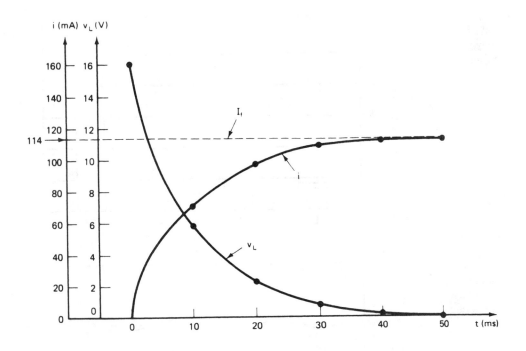

Vertical scale of rising curve:
20 mA/cm

Vertical scale of falling curve:
2 V/cm

Horizontal scale (sweep speed):
5 ms/cm

FIGURE 7–12
The direct-current loadline.

Robert Paynter, *Introductory Electronic Devices and Circuits,* © 1989, p. 135. Reprinted by permission of Prentice-Hall, Inc., Upper Saddle River, NJ.

FIGURE 7–13
Pie chart of 1997
Fatalities-Victim
Activity

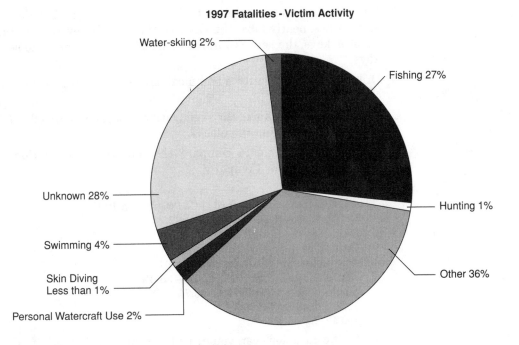

FIGURE 7–14
Origin of $2 million (rather than zero) overemphasizes differences. Compare with Figure 7–11.

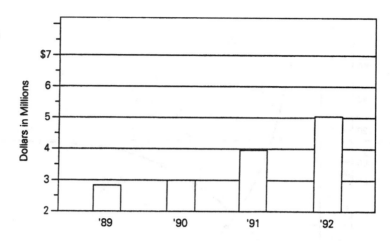

FIGURE 7–15
Ampere units (rather than conventional milliamperes) deemphasize differences.

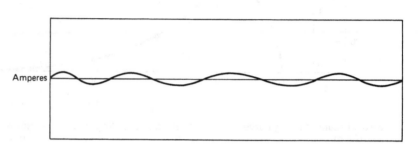

Exercise 7.1   *Using what you have just learned about graphics, complete each assignment.*

A. Write a functional description of the calculator pictured in Figure 7–17. Using lines and arrows, neatly label the ON key, the SHIFT key, the digital display, the set of number keys, the set of common arithmetical keys, and the set of special function keys.

B. Draw, label, and describe a technical device or instrument used in one of your other classes.

C. Draw a block diagram of the registration process at your school. Write a descriptive paragraph explaining the blocks.

D. 1. Complete a table to compare the following information from the Business Communications Company.

  The U.S. printed circuit board (PCB) market in 1999:
    Double-sided—54%, multilayer—32.7%
    Single-sided—8.3%, flexible—5%
    Total sales—$5.41 billion
  The projected U.S. PCB market for 2001:
    Double-sided—52.7%, multilayer—38.3%
    Single-sided—5.3%, flexible—3.7%
    Total sales—$10.4 billion

  What conclusions can you draw from these statistics?

  2. Draw pie charts to represent the statistics in part 1.

FIGURE 7–16
Unrelated cause-and-effect relationship.

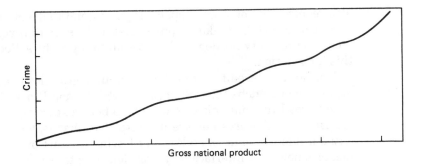

FIGURE 7–17
Scientific calculator.

Reprinted with permission of Radio Shack, Tandy Corporation.

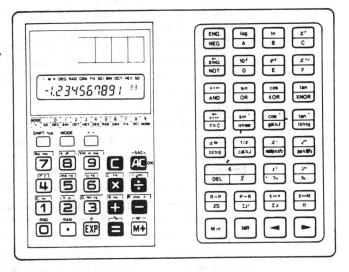

E. Draw a bar graph and a line graph to display the following information.

> According to the Acme Business Bureau, the Hot Circuit Company has experienced the following annual sales:
> In 1992, its sales totaled $2 million.
> In 1993, its sales totaled $8 million.
> In 1994, its sales totaled $21 million.
> In 1995, its sales totaled $50 million.
> In 1996, its sales totaled $74 million.
> In 1997, its sales totaled $103 million.
> In 1998, its sales totaled $97 million.
> In 1999, its sales totaled $95 million.

Write a one-paragraph description of the eight-year sales history of the company.

F. Bring in three examples of graphics from a newspaper or magazine article or advertisement. Write a one-paragraph interpretation of each graphic and include an evaluation of its purpose, effectiveness, and honesty.

Computer Images

Writers eventually learn how to create and process their own computer graphics images, sometimes through classroom training, mentoring, or just plain trial and error. Computer graphics can be very demanding of your computer's capabilities. Graphics also require a lot of storage space. If you intend to devote your time and creativity to producing colorful, lifelike graphics, first make sure that your computer has the equipment and memory needed for high-resolution graphics. Seek the advice of experts for this information.

You can create, edit, or reformat computer graphics using a graphics software program, such as Adobe PhotoShop, JASC Paint Shop Pro, or Microsoft Paint. Take some time to read the manuals or online help before starting a big project. Most programs are rich with features but are not intuitive; that is, you might not figure things out without making some mistakes and having to start over. And every upgrade usually means a new user interface and more features to try.

Graphics software programs encode, store, manipulate, and transmit graphical data to produce images. You can change the attributes or size of an image and even "doctor" or edit the pixels—the individual bits of data—in the graphic, by deleting lines, changing colors, or enhancing a shape.

The process becomes more complicated, however, when you see the number of graphics file formats available, or you find that your word processor or Web page cannot reproduce a file format that you painstakingly created and inserted. Why are there so many graphics file formats? How are they different from each other? And which one is best for your needs?

The problem began when individual graphics-software developers each created file formats to meet their own graphics applications without making them compatible with other graphics applications. Despite a clear need for one universal file format that can be used in all applications, none has been developed.

Writers must balance considerations for detail, color, type of source image, storage size, and compatibility between applications, such as a word processor and printer. Fortunately, graphics software programs allow you to convert graphics from one format to another, although they might not include all the formats you expect. So if you don't like the quality of a graphic, or an application doesn't recognize it, open the graphic in a graphics program and save it in a different file format.

TIP   As a general guideline on file formats:

- Use a **.bmp** (bitmap) for a Windows platform only, and when storage is not a problem. Bitmaps are not compressed and can become quite large with increased detail, color, and size.

- Use a **.gif** (Graphics Interchange Format) for platform-independent uses in desktop publishing (such as text and lines). The compression ratio keeps the size small.

- Use a **.jpg** (Joint Photographic Experts Group), pronounced "jay-peg," for photographs and images with natural (nonstraight) edges and colors that softly blend into each other. Although some details might be lost in the compression, most of us will not notice it.

- Use a **.tif** (Tagged Interchange File Format) for images with straight edges, such as line drawings. It also uses a compression ratio, which keeps the size small. However, check to make sure your application will recognize it.

You can also adjust the number of colors and dots per inch (dpi) on your printer to increase or decrease the resolution of the printed graphics.

For best results, save your graphics in a few different file formats until you find the right combination of quality, storage size, resolution, and compatibility. For more information on graphics file formats and how computers encode, collect, store, and compress data, refer to current books or manuals on the subject.

There are many alternatives to creating your own graphics, including clip art, graphics editors, screen captures, and scanners.

Clip Art

Many word processing programs and Internet sources provide batches of images that you can copy and paste, drag, or insert into your documents. Some include graphics with sound and motion (animation). Read the online help with your word processing program for instructions on how to insert clip art into your documents.

Graphics Editors

Graphics programs, such as Adobe Photoshop, JASC Paint Shop Pro, and Microsoft Paint, usually include an editor for creating new images and working with existing images. Depending on the program, you can open or draw an image, edit the features or colors, mask areas of the image, crop the visible area, adjust the size and orientation, convert the image to another format, and even animate the image (such as making a hurricane spin). Read the online help or manual to learn how to use the program efficiently. Terms and features vary among the different programs, but mastery of one program usually makes learning a second one easier.

Scanners

A scanner is an electronic device that you connect to your computer to reproduce a hard-copy image (such as a photograph) as an electronic image (such as a bitmap). When you place a photograph on the bed of the scanner, the device converts the image into an electronic format, and you can save it on your computer in the file format you prefer. Then you can open the scanned image in a graphics editor, if desired, or insert it directly into a document. You can also scan text as an object (which you insert just like any other graphic) or as a file that can be opened in a word processor.

Screen Captures

If you want to reproduce an image visible on the computer screen, you can take a "screen capture" of the image on the screen, similar to taking a photograph of the screen. Then you can paste the capture in your graphics editor, crop it to show just the section you want, and save it. To take a screen capture, follow these steps:

1. On a personal computer, press the Print Screen button to copy (capture) a full computer screen (to capture the "active" window only, the window with a dark title bar, hold down the Alt key when you press Print Screen). On a Macintosh, press and hold the Shift + Apple button and press 3.

2. Open a graphics editor, such as Adobe Photoshop or Microsoft Paint. Using the mouse, click Edit > Paste to open the screen capture in the editor's window.

3. Use the features of the editor to crop the image (drag the mouse from the top left corner of the area to the lower right corner) to view just the portion of the screen you want as your image.

4. Click Edit > Cut.

5. Open a new, empty screen in the graphics editor (this procedure varies among programs), and click Edit > Paste to show the cropped graphic, and save it in your preferred file format.

A NOTE ABOUT TRADEMARKED LOGOS

Do not capture company logos from Internet sites and insert them into your documents without the explicit permission of the company that holds the trademark. Violation of trademark law is a serious offense.

Inserting Graphics into a Document

Most word processing programs let you insert graphics directly into the document and drag them into position. For example, on a personal computer using Microsoft Word 2000, place the cursor in the document where you want to insert the graphic. Click Insert > Picture > From File. Browse to find the graphic, select it, and click Insert. Use the cursor to drag the image to the desired location. Right-click the image, and click Format Picture to modify the wrapping, color, size, and other available options. Read the online help of your word processing program for detailed instructions.

Exercise 7.2   *Using your word processor:*

- Find a Clip Art image of a computer, insert it into a page, and print it.

- Using the Internet, take a screen capture of a Web page with a computer pictured on it, open the screen capture in a graphics editor, crop just the computer portion of the image, save it as a bitmap, insert it into a document page, and print it.

- Scan a photograph of a computer, save it as a bitmap, insert it into a document, and print it.

SPELLING: *ie* and *ei*

You probably remember learning the verse that handles most of the *ie*/*ei* problems:

> i *before* e,
> *Except after* c.

believe	conceive
achieve	receiver
relief	deceit

> *Or when sounded like* a,
> *As in* neighbor *and* weigh.

sleigh	eight
their	freight
neighborhood	weight

As you would expect, there are some exceptions to these rules. The most common exceptions are listed below, and you will have to remember them. Notice that the exceptions are all *ei* spellings.

either	neither
seize	seizure
weird	leisure
counterfeit	forfeit
foreign	height

Do not confuse *ei*/*ie* vowel pairs with "unrelated" combinations, such as in *science* or *reinforce,* in which both vowels have a distinct sound, and thus a logical order.

Exercise 7.3   *Add* ei *or* ie *to spell each word correctly.*

1. Credit-card companies have been plagued by counterf_____ters.
2. It was a rel_____f when the Nevada test results y_____lded positive results.
3. N_____ther sc_____ntists nor engineers predicted the applications of lasers in l_____sure and recreation industries.
4. Dr. Maiman has rec_____ved acknowledgment for his ach_____vement.
5. Lasers have worked th_____r way into many facets of soc_____ty.

Exercise 7.4   *Proofread the following paragraph for* ie/ei *errors. There are six errors. Write the correct spellings above the incorrect words.*

  Passing through an airport customs line in a foreign country can be a wierd experience. Travelers can expect that officials may sieze suspicious-looking items for examination. If the traveler has purchased anything illegal, even unintentionally, he or she will have to forfeit the item. Freinds patiently prepare each other for thier inspections. It is always a releif to see each peice of luggage pass through an inspection.

Exercise 7.5   *Write the* ei/ie *verse from memory. Make sure that you include the second part.*

VOCABULARY: Roots *spec* and *son*

The Latin root *spec* (*spect*) means "look at." For example, *spectrum* means "visible light waves."

  The Latin root *son* means "sound." Combined with the suffix *ic*, meaning "having to do with," we form the word *sonic,* meaning "having to do with sound."

Exercise 7.6   *Using* spec *and* son, *complete the words.*

1. A person who watches or observes is called a _____tator.
2. An instrument to aid vision is called a pair of _____acles.
3. A musical composition written in three or four movements is called a _____ata.
4. A remarkable sight that attracts onlookers is called a _____acle.
5. Several people singing one melody are said to be singing in uni_____.
6. A close examination of an item is called an in_____tion.
7. A descriptive statement issued by a new company is called a pro_____tus.
8. A noise out of harmony is described as being dis_____ant.
9. A device used to increase vibrations is called a re_____ator.
10. A thought or conjecture formed from thinking about various aspects of a subject is called a _____ulation.

Exercise 7.7   *Using the Latin meanings of* spec *and* son, *write definitions of the following words as they apply to technology. Use the dictionary if necessary. Remember to define the words by using the term, class, and characteristics.*

1. specifications (specs) _____

  _____

2. spectrometer _____

_____

3. supersonic _____

_____

4. sonic boom _____

_____

5. resonance _____

_____

WORD WATCH: *they're, their,* and *there*

They're is the contraction of *they are.* If a substitution of *they are* is logical and appropriate, use the contraction form.

> *They're* the people I told you about.
>   (*They are* the people I told you about.)

Their is a possessive. It is followed by a noun.

> It is *their* turn.
> We drove to *their* house.

There is an adverb often used at the beginning of sentences (but it is not a subject) or a noun that indicates location, direction, or time, the opposite of *here.* You can see the shorter word *here,* which provides a clue for when to use the word.

> *There* are two answers. (*adverb use*)
> Put it down over *there.* (*noun use*)
> ("Here" could be substituted in both sentences.)

Note: All three forms are spelled beginning with *t-h-e.*

Exercise 7.8   *Complete the following sentences with the correct form of* they're/there/their.

The first time I attended a meeting of the Robotics Club, I watched the other members closely. I was interested in _____ projects and goals. I sat _____, not intending to speak, when I was called on to introduce myself. After standing _____ speechless for a few seconds, I finally mumbled my name and quickly sat down. Now, after attending several meetings, I realize that _____ more knowledgeable about robotics than I am, but that _____ also interested in my ideas. It's not just _____ club, but it's my club, too.

Exercise 7.9   *Write two sentences using each spelling.*

They're (Use as the subject and verb of the sentence.)

1. _____

2. _____

Their (Follow with a noun.)

3. _____

4. _____

There (Use as a direction or the beginning of a sentence.)

5. _____

6. _____

CHAPTER 8

Instructions

- Write travel directions.
- Write simple instructions for a process.
- Write formal instructions for a process.
- Spell words ending in *ly* and *ally* correctly.
- Use *micro* and *macro* correctly.
- Use *advice* and *advise* correctly.

READING: Easy as 1-2-3

by Mary Anne Donovan-Wright

Ever try to assemble a toy, bookcase or computer and think, "I could write better instructions than these!" If so, read on.

You've been grappling with a new bottle of aspirin for 20 minutes and you can't get it open. Then it dawns on you—this is a childproof cap! "Aha," you say, "adultproof, too." So, you sit down and read the instructions on the top of the cap, which tell you to "Push down firmly and turn while pressing on tab." A little arrow shows you which direction to turn the cap. Simple enough. You follow the directions and voilà, your headache's well on its way to being cured. You can thank a good technical writer for this cure to your aggravation—and the cure to your headache!

Later, you're putting your child's new swing set together. It's small, two swings and a tiny teeter-totter. No problem. But when you read the instructions that come with it, you wonder if you need to go to MIT first—"Insert slot D in receptacle A to midpoint of gradient W until post R meets ring C halfway to the end cap of bearing 2V. . . ." Where's that bottle of aspirin? This time you can thank a bad technical writer for your aggravation—and your new headache.

Technical writing is an art and a science. It's an art because it takes information and conveys it to the reader in a way that enables understanding and action. It's a science because it deals with methods, systems, design, theory and repeatable outcomes.

But Instructions Are Boring . . .

What is technical writing? Would you believe textbooks, phone books and cookbooks? How about office memos, trip reports, meeting minutes and project status reports? And, of course, there are the classics—office procedure manuals, journal articles and owner's and user's manuals. All of these documents aim to convey information that the reader can understand and act upon.

In the last 20 years, as technology has exploded and computers have become a staple in most homes, the demand for clear, understandable and effective documentation is greater than ever. That demand has changed the field of technical writing, says JoAnn Hackos, president of Comtech Services, a technical communications company based in Denver, and manager of the

Writer's Digest, August 1999

Society for Technical Communication's Communication Trends Committee. First, women are entering what was a male-dominated field and second, documentation now has to be written with a broader audience in mind than technicians. There are also more opportunities for freelance technical writers; with companies downsizing, many inhouse communication departments are being disbanded and documentation writing is being outsourced.

Most important to being a successful technical writer, Hackos says, is understanding how people interact with text and knowing the end users' goals. In addition to honing their writing skills, Hackos advises developing technical expertise.

"Know something about computer programming—it's going to make a difference," she says. In addition, she encourages picking up a course or two in cognitive psychology or human factors to "become more knowledgeable about users."

"There's lots of opportunity for people who want to do something interesting and have a variety of experiences," says Hackos.

How Do I Get Started?

The first step in a technical writing project is identifying what you need to communicate and why you need to communicate it. That's called an "objective."

Always write down your "objective" in a way that keeps you focused on what information must be conveyed.

Now that you know what needs to be conveyed, you hone in on to whom you're going to convey it. This is called "audience analysis," but it's really a combination of common sense and proven research tools.

To write effective technical documentation, you need to know as much about your audience as possible, including education, attitude toward you as the writer, attitude toward the documentation, prior training, experience in the subject of the documentation, professional or job responsibilities, any cultural characteristics, and their intended use of the document. Don't make the mistake many technical writers make—you are not the typical target reader. Undoubtedly, the person who wrote those teeter-totter instructions understood them perfectly!

How do you get information about your readers, especially in cases where you don't have direct access to them? Technical communicators employ a variety of tools to gather information about potential readers, including interviews, surveys, questionnaires, observation and letters. In addition to your own research, clients or bosses may provide useful information.

The next step is to measure the difference between your objective and your audience's skills, attitudes, knowledge and so forth.

Example: You'll be writing documentation for a copier targeted to home-based business owners. A telephone survey of home-based business owners in several states found that these enterprises range from highly technical (computer programming consultants, for instance) to highly untechnical (perhaps maid services). The research also found that the copier is going to be used frequently by other members of the household, including schoolchildren, and that it will be heavily relied upon an average of three hours a day, seven days a week. You know that the educational backgrounds of these potential users range from high school to advanced degrees. But the most important fact about this audience isn't something you need research to know, it's a matter of common sense: Time is money to these hardworking business owners. If they run into a knotty problem setting up their copier, repairing it or dealing with technical support, they're going to think twice about doing business with the copier supplier again.

All of these audience characteristics translate into the need for you to create efficient, easy-to-read, easy-to-reference documentation that uses lots of illustrations instead of lots of words to tell the story about how to use and maintain the copier.

What About the Writing?

Phew. You've done a lot of work already and you haven't written a word! By this time you know the difference between what your reader knows/feels/thinks/does now and what she needs to know/feel/think/do as a result of reading what you write. Often, you'll need to work hand-in-hand with experts—perhaps product engineers or designers—to gather your content, so your ability to work collaboratively is important. In addition to gathering your information, you must also structure it so that it's clear, concise, accurate, understandable, complete, well-organized, consistent and interesting! You may even be in charge of layout and graphics. Quite a tall

order, isn't it? Beyond that, as the writer of instructions, you may be just as, if not more, liable than the manufacturer should a user be injured because your instructions are deemed faulty or incorrect.

Many of the techniques used to write effective fiction or nonfiction have a place in technical writing. Keep these guidelines in mind as you write the documentation:

Your instructions and procedures should have clear, concise and explicit titles. The title should answer the question, "What will I be able to do using these instructions?" It should not be too broad: "Operating the Sewing Machine," or too vague: "Washing Clothes." It should describe precisely the contents and user outcome: "How to insert the bobbin into the bobbin case."

Perform the procedure or instructions yourself. Know what your reader will be up against. Make sure you can perform the procedure correctly before you sit down to write.

Use an effective design. Whether the delivery medium for your instructions is electronic or print, it's your job to ensure an effective design. Think about all design elements including use of headings, typeface, layout, graphics and illustrations. When using illustrations, label parts and components.

Begin with a statement of context or objective for the instructions. Before launching directly into the procedure steps, tell the reader the purpose of and objective for the instructions. Under a title of "Loading Film," you might write this purpose sentence: "This procedure tells you how to load film into the camera when you are under water."
As you write the instructions:

- Use the imperative mood: "Insert the tab into the slot." "Click on the icon."

  The imperative voice is direct, it puts the focus on the action the reader must perform and it is authoritative.

- Start each step with an action verb. Like the imperative mood, beginning each step with a verb focuses on the tasks the reader must do.

- Use examples whenever you can. Examples clarify your instructions.

- Use parallel construction.
- Number the steps to eliminate any ambiguity and to ensure the procedure is performed correctly.
- Design each step so it contains a single task. Otherwise, users may be confused.
- Err on the side of including too much information.

How Do I Check My Instructions?

You've prepared your first draft. Now what? Let's look at those operating instructions. You think you've done a good job, but no one has actually sat down with the copier and your instructions to see if they work. So, your document needs to be validated. The most effective way to do this is to ask someone within your target audience to try to operate the equipment with only your instructions in hand. You find just such a person in your department—the husband of a co-worker, for example. When the agreed-upon day and time arrives, you arrive with the copier and instructions and hand everything over to your tester. Your job now is merely to observe. Write down your observations. Record any questions. Then, incorporate that feedback into your second draft.

Should I Get the Experts' Opinions?

Once you've validated your technical document, you need to have it reviewed. People who should review your document include the engineers who gave you input, anyone the engineers ask you to send the document to, your manager(s), perhaps a lawyer to prevent any potential liability, the project manager and anyone else you've worked closely with on the project. It is also helpful to ask a peer or colleague for his or her thoughts. The more the merrier—and the more who've responded, in any way, to your document, the more assured you will be of its effectiveness and accuracy.

When you send the document out for review, include a cover sheet listing the names of all reviewers with a signature line next to each name. Insist on getting the signature of each reviewer.

Every document should be thoroughly tested to make sure it meets the users' needs. How do you test your finished documentation? Formally, use focus groups, surveys and questionnaires. Informally, call or visit the people who use your documents and talk to them, or ideally, watch them in action.

Hey, My Headache's Gone!

Writers who create effective documentation don't win the Pulitzer Prize or the National Book Award. For the most part, they toil in anonymity. Their reward, beyond payment for the project and additional work, comes when people are able to put together shelving units and beds, prepare meals, and figure out how to turn the lights on in their new car with a minimum of fuss. Their best work is when they're least noticed.

Reading Comprehension Questions

1. How is technical writing an art? How is it a science? _____

_____

2. Define what the author means by an "objective." _____

_____

3. How can a writer conduct an "audience analysis"? _____

_____

4. Why must writers ensure their instructions are accurate? _____

_____

5. What is the "imperative mood"? _____

_____

6. How does a writer validate a document? _____

_____

7. Who should be asked to review a final document? _____

_____

WRITING: Instructions

We are all familiar with the statement, "If all else fails, read the directions." Why is it so common to avoid reading the directions? Why do people risk the trial-and-error approach rather than reading the instructions before beginning? Possibly it's because they are impatient to get started. Or possibly the instructions seem incomprehensible, as though the writer assumes the reader has technical training and fails to define terms, locate parts with graphics, or provide basic information.

To make matters worse, some instructions sound ridiculous, possibly due to poor translation from other languages:

> On a kitchen knife: *Warning keep out of children.*
> On a string of Christmas lights: *For indoor or outdoor use only.*
> On a food processor: *Not to be used for any other use.*

Today, as the article states, people expect readable, accurate instructions, ranging from easy-to-follow steps to multivolume user manuals and service guides. Companies that fail to provide good instructions for customers lose revenue due to technical support calls and product returns. For this reason, many companies hire technical writers—those who write documentation as their primary job—to complete the finished user information. These writers may or may not have any technical training, but they know the art of writing understandable instructions.

However, as indicated in the article, technical writers cannot do the job alone. Your employer might one day ask you to provide input into the documentation provided to your customers, either as a technical expert or as a reviewer.

Some technicians and engineers turn to technical writing as a career, drawing on their training and experience to make instructions and manuals more accurate. These people must take special care to understand the audience, listen to users' questions, and add the basic information sometimes unintentionally overlooked by technical experts. More commonly, you will be asked to write informal instructions to coworkers and customers. You could be asked to write instructions to install a software program, assemble a piece of equipment, troubleshoot a problem, or drive from the nearest major airport to your office.

Following a few, simple guidelines can improve the readability and usefulness of your directions. You can adapt these guidelines for different types of instructions, as needed.

Elements of Formal Instructions

TIP Formal instructions should include the following elements:
- **Orientation.** Provide an overview of the device, such as the purpose of the instructions or function and starting state of the device.
- **List of materials.** Specify the materials, including sizes, part numbers, and quantities, needed to complete the procedure.

- **Step-by-step instructions.** In each step (usually numbered), use the active, imperative voice with the implied "you."
- **Graphics.** When needed for clarity or understanding, add numbered or labeled graphics. Reference each numbered graphic in the text, or add a description under the graphic to orient the reader to the figure. Use consistent terms in the text and the description.
- **Conclusion or summary.** Describe the final state after following the instructions so readers can complete the project, and add information for any additional or optional procedures.

Some instructions for complex devices or procedures also include troubleshooting tips or frequently asked questions. This information should address typical problems with the procedures, based on your own testing and customer feedback.

Guidelines for Writing Instructions

The following guidelines describe the general process of writing instructions. As an example, consider the task of writing instructions for using a new software program. Whether the end product will be a one-page "Fast Track" for colleagues or a full manual for users, the general guidelines are the same.

Step 1: Perform the procedure yourself. Learn all you can about the process or product before you start writing anything. This might include observing others performing the procedure or talking to the experts (such as a developer or engineer) or other involved people. Read the product specification or user manual. Become familiar with all the features and terminology.

Perform the procedures, logging all the steps you complete. It is easy to miss the small steps unless you perform them yourself. Don't rely on others to tell you how something is supposed to work.

If you experience problems, log the scenario in which the problem occurred. Then log what you did to correct the problem. Use your log to add information in your steps at the appropriate spot to prevent those same problems. Or, for longer documents, add a troubleshooting section with the problems you encountered and a description of what you did to correct (or prevent) them.

Step 2: Prepare a working draft. Write a draft of the numbered steps. Focus on the behaviors. Don't worry about spelling, grammar, or even complete sentences at this point. Let others read your draft, following each step. Their feedback will uncover missing or confusing information. Often, test subjects uncover confusing wording, such as: *"When you said to 'Close all applications before starting the installation,' did you mean I have to close Windows, too, or just the programs running on my Windows desktop?"*

Thinking like a new user can be the hardest part of writing instructions. If you write for the broadest audience (nontechnical), your instructions should pass the "6th-grader test," meaning that the instructions should be clear enough for the average 6th-grade student to follow. Revise your initial draft to clarify the instructions.

Step 3: Write the steps using simple, direct language. Now start to refine the language. If you struggle with describing a step, do more research and experimentation. Ask others to suggest alternatives for vague words or confusing sentences. Rewrite the instructions using an action verb and the implied "you." Don't worry about

FIGURE 8–1
Sample graphic

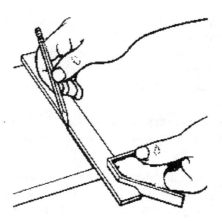

1. Use the T-guide to mark crosscuts.

sounding demanding or impolite. The imperative voice is the clearest to read and understand. Focus on the action. Compare the following sentences:

Precise crosscuts can be made by making a T-guide for your circular saw.
(Least clear: passive voice, indirect, and wordy)
Produce precise crosscuts by making a T-guide for your circular saw.
(Still unclear: active voice, but not focusing on the action)
Make a T-guide for your circular saw to produce precise cuts.
(Most clear: active voice, direct, focuses on the action)
For precise crosscuts, make a T-guide for your circular saw.
(Also clear: orients the user first, active voice, and direct)

Use standard terminology that is appropriate for the audience. If you are writing for coworkers, use the terms or acronyms common within your company or industry. If you are writing for a general audience, use simple language as much as possible. If a technical term is unavoidable, define it the first time you use the term. When possible, use terms that are common in the industry rather than terms used solely by your company.

Step 4: Include graphics if needed. Not all instructions require graphics. However, if the steps include parts' names, users will appreciate a graphic that illustrates or labels the trickier parts as shown in figure 8.1. If you include callouts to labeled parts, be sure to use consistent terms in the callout and in the text.

If a step includes details difficult to describe clearly, an illustration can show the details in pictures to supplement the discussion. A flow chart with decision points can illustrate different paths or options, and better orient the reader.

Step 5: Format the instructions to identify the organization. Select a layout and format that make the instructions clear and easy to follow. For example, if your steps are brief, simple numbering is appropriate. If the process has several distinct stages with steps within each stage, use headings to identify each stage. Typically, writers increase the font size of headings and make them bold.

Use a consistent format for headings and text to clarify the organization. For example, use the same font, font size, and paragraph spacing for all main headings, use a slightly different appearance for subheadings, and so on.

The following (abridged) example illustrates consistent headings, numbering, and spacing.

Cutting the Parts

1. Make a simple, self-aligning T-guide for your circular saw . . .

2. Butt the crossbar of the T-guide . . .

Assembly

1. Hook your tape measure to the top edge . . .

2. Extend the tape and place marks at the following locations . . .

Adding the Fascia

1. Mark the stock for crosscutting to fit the 1 × 2 fascia over the case front edges.

2. Make the vertical pieces 48″ long to match the sides . . .

Avoid overuse of bold, italics, underlining, font changes, and unimportant graphics. When overused, these elements add "noise" that distracts readers from the information and overall organization.

Use numbered steps when they must be followed in sequence (and check that the numbers are sequential—numbers can easily get mixed up during revisions). Become familiar with numbering formats available with your word processor, because they can provide a consistent style (font and margins) for numbered steps. Typically, writers indent the entire step so the numbers are easy to spot. Add white space between steps. If sequential steps require more than one paragraph with long explanations of each step, a numbered format is sometimes not practical. Instead, use a standard paragraph format with other methods to identify separate steps, such as clear transition words that signal sequence. For example, use words such as *first, next, then,* and *finally* at the beginning of a paragraph starting a new step. Start a new paragraph for each new step. For exceptionally long steps, consider using headings and subheading for steps instead of numbers.

Use bulleted lists for alternatives within a step or for steps that do not have to be followed in order. Lists are easier to read than paragraphs, but they are most effective with only one or two sentences per bullet. Although most writers stick to the standard round or square bullet symbol, you can use other symbols. For other options, check your clip art set or the symbols set available with your word processor. Use consistent tabs and spacing between bullets to improve the appearance of the list.

Experiment with other layout techniques, such as centering headings, changing the font type or size, and manipulating the line spacing to increase the visual clarity.

Step 6: Write an introduction to orient the reader.   Discuss who should follow this procedure and why, when or where to follow it, what it does, and where to go for further information or questions. This is the information that lays the groundwork for the instructions and identifies the intended audience and outcome. Writers sometimes write the introduction last because they don't have all the information themselves until the end.

Building a Bookcase

Our pine bookcase features a simple design to be built with basic woodworking tools. We made the case out of materials available at most lumberyards. These instructions will produce a 4-shelf bookcase with overall dimensions of 10″ deep × 34″ wide × 48″ tall. While the depth of the case is directly tied to the 1 × 10 stock, you can vary the height. . . .

Some writers also include a summary or conclusion at the end of the document, which further clarifies the expected outcome, or briefly describes any remaining steps not included with the instructions.

Adding the Finishing Touch

If you plan to paint your bookcase, first apply two coats of shellac over each knot to prevent the knots from bleeding through the final paint job. Then prime and paint the bookcase according to the manufacturer's instructions. . . .

Step 7: Add a materials or requirements list.   Include all equipment, tools, or minimum requirements needed for the complete project. Many writers include this information in a listed or bulleted format to make the items easy to spot. Include precise quantities, sizes, and part numbers. Categorize the materials into logical groups, especially if different types of materials are needed:

Materials List

Basic tools:	**Lumber:**
Portable circular saw	(1) 1/2 × 1/4″ parting strip
Block plane	(2) pieces of 1 × 4 pine
Combination square. . .	(5) pieces of 1 × 10 pine . . .

Step 8: Identify notes and warnings.   Use clear wording and formats to highlight information that has special importance to the user. Use Notes, Tips, Cautions, and Warnings, according to the conventions of your style guide. Position the information prior to or within the relevant step, not hidden at the end of the document. General warnings belong at the beginning of the document. Use a box, bold heading, or special symbol to catch the reader's eye. Many companies have a guideline for the correct usage of each type. For example, the conventions at one company are the following:

- Use "**Note**" to emphasize information or supplement information that was already provided in the instructions but might apply only to certain situations.

Note: Becoming a registered user makes you eligible for discounts, updates, and free technical support.

- Use "**Tip**" to provide shortcuts, alternative methods, or techniques for performing an action, but not for essential information.

Tip: If you choose not to include this program in your Startup folder, you can start it using Start > Programs.

- Use "**Caution**" when an action or failure to take an action could result in loss of data.

Caution! Back up each file weekly for archival purposes.

- Use "**Warning**" when an action or failure to take an action could result in harm to the user or damage to the hardware.

Warning! Unplug the power cable before opening the console casing.

Step 9: Edit, revise, and refine your language.   Review your document, and, if possible, ask other people to review it. Aim for two types of reviews: one for language and one for technical accuracy. Each reviewer might suggest different revisions. Most professionals consider the review process an integral part of document development—a time to fine-tune the instructions before they "go out the door." From time to time, reviewers disagree with each other or write conflicting edits. Sometimes you might disagree with their edits. When this happens, focus on the audience, your intended readers. Discuss disagreements with your reviewers, letting them understand the other viewpoint.

This is also the time to examine grammar and spelling, reduce wordiness, eliminate repetition, and sharpen your language. Revise the steps as needed, based on the review process.

Step 10: Observe someone follow your instructions. The best way to test instructions is to ask someone (a test subject) to read and follow your instructions. Avoid helping the subject; let the person rely only on the written information. Keep a log of difficulties experienced by the tester—they are bound to occur at unanticipated spots. Note where, when, and why the test subject had problems.

Then revise your document accordingly, clarifying misunderstood steps and adding missing information in places where the test subject had problems. If possible, ask another person to test your revised document.

Step 11: Put the final touches on the instructions. Complete the final revisions based on the testing. Take a last look at the layout, send the document to the printer, and give yourself a standing ovation!

Writing instructions might not be as easy as one, two, three, but you can feel a great deal of satisfaction from producing instructions that work.

The following (abridged) example illustrates a set of organized instructions. For an example of complete instructions, see Appendix 2.

Building a Bookcase

Our pine bookcase features a simple design to be built with basic woodworking tools. We made the case out of materials available at most lumberyards. These instructions will produce a 4-shelf bookcase with overall dimensions of 10″ deep × 34″ wide × 48″ tall. While the depth of the case is directly tied to the 1 × 10 stock, you can vary the height. . . .

Materials List

Basic tools:	**Lumber:**
Portable circular saw	(1) 1/2 × 1/4″ parting strip
Block plane	(2) pieces of 1 × 4 pine
Combination square . . .	(5) pieces of 1×10 pine . . .

Cutting the Parts

1. Make a simple, self-aligning T-guide for your circular saw . . .
2. Butt the crossbar of the T-guide . . .

Assembly

1. Hook your tape measure to the top edge . . .
2. Extend the tap and place marks at the following locations . . .

Adding the Fascia

1. Mark the stock for crosscutting to fit the 1 × 2 fascia over the case front edges.
2. Make the vertical pieces 48″ long to match the sides . . .

1. Use the T-guide to mark crosscuts.

Adding the Finishing Touch

If you plan to paint your bookcase, first apply two coats of shellac over each knot to prevent the knots from bleeding through the final paint job. Then prime and paint the bookcase according to the manufacturer's instructions . . .

Exercise 8.1 *Internet assignment.   Use keywords such as "create Web site" to search for instructions on how to set up a Web site. Test them by setting up your own Web page. Evaluate the effectiveness of the instructions.*

Exercise 8.2 *Copy a set of instructions for a device. Critique the instructions by answering the following questions: What makes them effective? What can be improved? Do they pass the 6th-grader test?*

Exercise 8.3 *Write travel directions on how to get to your house from school, including a map.*

Exercise 8.4 *Write numbered instructions for one of the following procedures:*
- Hook up a VCR.
- Install a car radio.
- Set up a campsite.
- Set a digital watch.
- Prepare for a journey.

Exercise 8.5 *Expand the instructions from Exercise 8.4 into a formal set of instructions. Include the following elements:*
- An introduction (background, purpose, definitions)
- Materials list (all items, including quantities, needed for the procedure)
- Step-by-step instructions
- Conclusion (finishing tips or how to evaluate correct performance)

SPELLING: Adding *ly* and *ally*

One way to turn an adjective into an adverb or a noun into an adjective is to add *ly* to the end of the word. Sometimes writers are confused about whether to add *ly* or *ally*. A dictionary will always provide the answer, but there are some general rules that you can follow.

Rule: Add *ly* to the end of an adjective to make an adverb or to the end of a noun to make it an adjective.
   Add *ly* to the end of the following adjectives to turn them into adverbs.

        fortunate   + ly = _____
        serious   + ly = _____
        physical   + ly = _____
        friend   + ly = _____

   Most of the time, if a word ends with a final, silent *e*, just add *ly*. There are four common exceptions.

   Exception 1: In these words, drop the final *e* before adding *ly*.
        due   + ly = duly
        true   + ly = truly
        whole   + ly = wholly

Exception 2: If the word ends in *ble* or *ple*, drop the final *e* and just add the *y*.

terrible + ly = terribly
sensible + ly = sensibly
simple + ly = simply

Exercise 8.6 *Change the following adjectives into adverbs by adding* ly *or* y.

1. horrible _____ 2. awful _____

3. special _____ 4. precise _____

5. accurate _____ 6. true _____

7. probable _____ 8. whole _____

9. seasonal _____ 10. simple _____

Exception 3: If the word ends in a *y* that is pronounced like *e*, change the *y* to *i* before adding *ly*.

easy + ly = easily
heavy + ly = heavily

But if the final *y* sounds like *i* just add the *ly*.

dry + ly = dryly
sly + ly = slyly

Exception 4: If the word ends in *c*, add *ally*.

medic + ally = medically
critic + ally = critically
(In the case of *public*, add *ly* = publicly.)

Exercise 8.7 *Write the correct spelling of the adverb form of each word in parentheses.*

1. The mayor (public) _____ announced her intention to run for reelection.
2. The child played (happy) _____ in the yard.
3. Solving the problem was (simple) _____ a matter of rewording the question.
4. The new hairstyle changed his appearance (drastic) _____ .
5. We were (terrible) _____ shocked when Tim walked out of the interview.
6. The rain started, (fortunate) _____, before the crops died.
7. The passengers were (critic) _____ injured in the accident.
8. Steve finished the test (easy) _____ within the hour.
9. Graduation is (true) _____ an exciting experience.
10. With your positive attitude, you are (like) _____ to succeed.

VOCABULARY: *micro* and *macro*

Macro is a Greek prefix meaning "long" or "large."

Macroscopic means "something that is visible to the naked eye."

Micro is a Greek prefix meaning "small" or "little" (or a metric unit).

Microscopic means "something that is too small to see without using a microscope."

When these prefixes are combined with words, they are usually joined without a hyphen.

Exercise 8.8   *Use* macro *or* micro *to complete the words in the following sentences.*

1. A _____ circuit is a miniaturized integrated circuit found in a computer.

2. A _____ n is an extremely small unit of measurement.

3. A large, complex entity, such as the universe, is sometimes referred to as a _____ cosm.

4. A miniature universe, such as a pond, is sometimes referred to as a _____cosm.

5. A _____ meter is used to measure very small distances.

6. _____ economics is the branch dealing with the interrelationships of large sectors of the economy, such as employment and income.

7. A _____ scopic object is large enough to be observed by the naked eye.

8. _____ film is film that is reduced in size after being photographed.

9. A _____ processor is the controlling unit of a microcomputer laid out on a silicon chip.

10. A _____ wave is part of the electromagnetic spectrum of wavelengths ranging from 0.3 m to 10 mm.

WORD WATCH: *advice* and *advise*

Remembering how the pronunciation of a word affects its spelling will help you in difficult situations. *Advise* ends with the sound of "eyez," and it is a verb that means "to give an opinion or counsel." *Advise* is a regular verb (*advised, advising*), and the other forms are *advisory, advisement,* and *adviser.*

> Your *adviser* will *advise* you to join the *advisory* council.

Advice ends with, and sounds like, the word *ice*. *Advice* is a noun, meaning an "opinion" or "counsel that is given." It has no other forms. Although a plural form of *advice* is occasionally written as *advices,* it is usually considered an "uncountable" noun, meaning that we do not make it singular or plural.

> I will use the *advice* to guide my career.

Exercise 8.9   *Fill in each blank using the correct form,* advise *or* advice. *Drop the final* e *if necessary when adding an ending.*

1. Because I don't know anything about the subject, I can't offer you _____ .

2. Workers are _____d to wear safety glasses when entering the room.

3. I am too involved in the situation to have any objective _____ about resolving it.

4. The _____ory group was formed to handle the situation.

5. Personnel problems require special _____ment from qualified counselors.

6. He ignored the _____ of his staff and followed his natural instincts.

7. He learned to withhold _____ until it was asked for.

8. Sometimes just talking about a problem with a trained _____r solves the problem.

9. She spent so much time _____ing others that she didn't notice her own problems.

10. _____ is often easier to give than to follow.

CHAPTER 9

Comparison and Contrast

- Write a comparison and contrast outline of two similar products.
- Write a comparison and contrast essay of two similar products.
- Spell words with *ough* correctly.
- Use *retro, circum, intro, intra* and *inter* correctly.
- Use *effect* and *affect* correctly.

READING: Your PC is Listening

by Leslie Ayers

Say something. I'm listening. Go ahead—tell me, already!

If your computer could talk, and if you had speech-recognition software installed, that's probably what it would say. It would tell you to put the mouse down, pull your hands away from the keyboard, put your feet up, and simply talk. Dictate a letter, a memo, or a report. Format the document. Print it. Save it. Paste its contents into an e-mail message. Send it on its way.

You can do all this with continuous speech-recognition software. Still hunt-and-peck typing at 40 words per minute?

Voice software lets you feed text to your PC three times faster. The toughest part is getting used to talking out loud, especially if you work in Cubicle Land, and keeping your hands off the keyboard and mouse.

But once you get the hang of it, nothing will shut you up. With four speech programs to test, we certainly had a lot of talking to do. Two of them, Dragon Systems' Dragon NaturallySpeaking Preferred 3.0 and IBM's ViaVoice 98 Executive Edition, are upgrades. Lernout & Hauspie's Voice Xpress Plus and Philips Speech Processing's FreeSpeech 98 are brand-new.

FIGURE 9–1
Speech-Recognition Software

Speech-Recognition Software

	Dragon Naturally Speaking	IBM ViaVoice 98	L&H Voice Xpress Plus	Philips FreeSpeech 98
Control Menus in Windows Apps		✓	Limited	✓
Systemwide Voice Navigation		✓		✓
Supports Multiple Users	✓	✓	✓	
No Training Required to Start	✓	✓	✓	

Reprinted from *PC Computing,* September 1998, with permission. Copyright © 1998, ZD, Inc. All Rights Reserved.

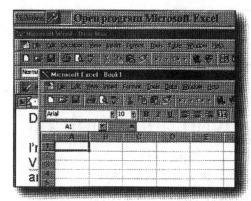

IBM ViaVoice 98 Executive Edition
★★★★★
Verdict: Best voice control across Windows and apps.
Pros: Dictate and format documents; control Windows and other applications.
Cons: Can't verbally deselect selected text.
System Requirements: 32MB of RAM for Win95/98, 48MB of RAM for WinNT, 180MB of hard disk space.

$149 est. street price / IBM Corp. / (800) 825-5866 / *Enter 668 at www.pccomputing.com/infolink*

FIGURE 9–2
IBM ViaVoice 98 Executive Edition

All four packages need lots of processing power. Minimum requirements range from 133MHz to 200MHz Pentium CPUs—but don't believe them. We tested the software on 266MHz Pentium II systems and still found ourselves staring at the Windows hourglass. Also be sure you've got memory to spare—at least 32MB, preferably more. You should have at least 200MB of free hard disk space before installation. You also need a 16-bit sound card, and a microphone is a must. For complete hands-free operation, get a microphone headset.

Can We Talk?

Of the four packages we tested, IBM ViaVoice 98 offers the most effortless combination of command and control and continuous speech dictation. It lets you rule your desktop in a way the other three programs don't. But ViaVoice was not the most accurate. That accolade goes hands-down to Dragon NaturallySpeaking.

ViaVoice lets you control Windows, however—a feature NaturallySpeaking lacks. Because ViaVoice lets you say what you see, you can run applications, open and close files, and activate menu functions. The only time you need to use the mouse is to click the microphone on, and if you configure ViaVoice to launch in sleep mode, just say "Wake up" to begin.

All the packages we tested let you dictate directly into Microsoft Word 97;

they also feature some degree of voice-activated formatting and editing capabilities. With ViaVoice and Naturally Speaking, you do all this in a modeless environment—the software picks up your intention based on what you say. Lernout & Hauspie's Voice Xpress Plus is almost modeless, but you do have to let it know when you're about to spell out a word. Philips' Free Speech makes you switch between command, dictation and spelling modes.

To dictate text for all the packages you speak normally without pausing between words. For formatting, punctuation, and other commands you must pause slightly before and after speaking. Other than that, the only trick is remembering how to phrase commands—like whether you say "*Go* to the end of the document" or "*Move* to the end."

L&H's Voice Xpress Plus deserves high praise for flexibility in understanding commands. There are several ways to say the same thing—which increases the odds that you'll get it right. For example, if you select a bolded word and say "Unbold that," "Clear bold," "Remove bold," or "Bold off," it will understand.

Voice Xpress Plus also offers formatting shortcuts. To change the font type and size in your entire document in other products, you have to say "Select document," then "Make font 12-point Arial."

With Voice Xpress Plus, just say "Make entire document 12-point Arial." IBM

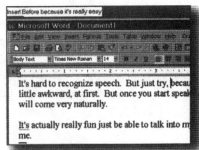

Dragon NaturallySpeaking Preferred 3.0

★★★☆

Verdict: The champ for heavy dictating.

Pros: Dead-on accuracy

Cons: Can't control Windows or other apps.

System Requirements: 32MB of RAM for Win95/98, 48MB of RAM for WinNT, 95MB of hard disk space.

$149 est. street price / Dragon Systems / (800) 437-2466, (617) 965-5200 / *Enter 669 at* www.pccomputing.com/infolink

FIGURE 9–3

Dragon NaturallySpeaking Preferred 3.0

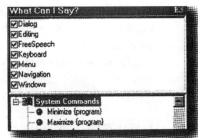

Philips FreeSpeech 98

★★★

Verdict: Great for light dictation.

Pros: Inexpensive; controls Windows and other apps.

Cons: Mode switching required.

System Requirements: 32MB of RAM for Win95/98, 48MB of RAM for WinNT, 200MB of hard disk space.

$39 est. street price / Phillips Speech Processing / (800) 851-8885 / *Enter 671 at* www.pccomputing.com/infolink

FIGURE 9–4

Philips FreeSpeech 98

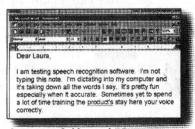

Lernout & Hauspie Voice Xpress Plus

★★★★

Verdict: You'll love it if you live in Word.

Pros: Most flexible command structure.

Cons: Can't control other Windows apps.

System Requirements: 40MB of RAM for Win95, 48MB of RAM for WinNT, 130MB of hard disk space.

$70 est. street price / Lernout & Hauspie / (888) 537-6688 / *Enter 670 at* www.pccomputing.com/infolink

FIGURE 9–5

Lernout & Hauspie Voice Xpress Plus

ViaVoice offers a similar shortcut. But Voice Xpress Plus wins for most intuitive undo feature. Instead of repeating "Undo that" a gazillion times like a babbling idiot, just say "Undo five times" to move back five levels of changes. If most of your work involves creating documents or repurposing existing ones in Word, Voice Xpress Plus is your best bet.

Take Note

Dragon NaturallySpeaking leads the pack in overall accuracy. Its vocabulary is vast—62,000 words—and it knows to capitalize certain words and phrases, like *Social Security Administration*. But NaturallySpeaking offers less control over Word than ViaVoice and Voice Xpress Plus do. It's a bit more flexible, however, about which word processors it hooks into, letting you use Corel WordPerfect 8 in addition to Word 97.

But the menu commands you can say when running NaturallySpeaking in Word are limited. You can preview and even print the document, but you can't open any of the main menus—which means you can't save, open a new file, or exit Word using your voice.

But NaturallySpeaking is the only one in the bunch that lets you move the cursor by issuing an "Insert" command. All the others make you say "Move up 17 words" or other absurd commands to get the cursor right where you want it.

Guide to PC Computing's Ratings

Excellent *Market leader that offers exceptional performance* .... ☆☆☆☆☆

Good *Excels in many areas; a good buy* .... ☆☆☆☆

Acceptable *Average for its class; a justifiable purchase* .... ☆☆☆

Poor *Out-of-date or substandard; offset by some positive features* .... ☆☆

Unacceptable *Missing necessary features; avoid* .... ☆

FIGURE 9–6
Guide to PC Computing's Ratings

Voice Control

Like ViaVoice, Philips FreeSpeech 98 lets you control Windows and other apps besides your word processor. Need to get at Windows settings? Just say "Start Menu, Settings, Control Panel"—and voilà! FreeSpeech's initial accuracy rivals that of ViaVoice and Voice Xpress Plus—but you must train it for 15 minutes first. Unlike the other three products in this roundup, FreeSpeech is under $50. At $39, you can afford to buy a copy for everyone in the office. (And you'll have to since it's single-user only.)

Despite the flexible control over Windows you get with FreeSpeech, there are a few drawbacks. The most glaring inconvenience is the inability to undo during dictation using a voice command. The other three products let you say "Undo that" or "Scratch that" during dictation to delete the words you just said if they were misrecognized or you changed your mind. But FreeSpeech won't take formatting or navigation commands during dictation.

The best all-around package for speech-recognition software is IBM's ViaVoice 98. It delivers accurate dictation and makes it easy to format and to control documents and applications. If accuracy is paramount, Dragon's Naturally Speaking is by far the most powerful and accurate. Lernout & Hauspie's Voice Xpress Plus offers the greatest flexibility in how you word commands and provides the most formatting options. For slow typists or an office on a tight budget, Philips' FreeSpeech 98 offers a low-cost alternative.

Reading Comprehension Questions

1. How can people use speech-recognition software programs? _____

_____

2. List three of the common equipment requirements of all four speech-recognition programs. _____

_____

_____

3. In the section "Can We Talk?" which program has the best overall dictation power?

_____

4. In the section "Take Notes," which program has the best dictation accuracy?

_____

5. In the section "Voice Control," which program has the best voice control of the Windows desktop and other software applications? _____

_____

6. What is the final recommendation of the author? _____

_____

WRITING: Comparison and Contrast

The purpose of the article "Your PC Is Listening" is to compare and contrast four similar speech-recognition software programs released at about the same time and to make a recommendation for the readers of the magazine.

Magazines that conduct product comparisons usually test each product under identical, controlled conditions and use the same testing procedures or scenarios—in the computer industry; this process is sometimes called a "shoot-out." Products being compared must be similar in purpose and general function. The writers then rate the products based on their experience with it, not on what the marketing literature claims. The article begins with a brief introduction to speech-recognition software for readers who are not familiar with this technology. The common elements of all four speech-recognition products are described briefly.

The article then describes how each program performed in three preselected categories:

- Continuous-speech dictation power
- Dictation accuracy
- Voice control of the Windows desktop and other applications

The final paragraph of the article summarizes the key strength of each product: the best all-around product, most accurate product, most flexible, and most cost-effective. The product summary charts display the comparative rating of each software program, as well as the pros, cons, system requirements, and product information.

Writing: Comparison and Contrast

When two or more objects are being compared, we often use a technique called *comparison* and *contrast*. Writers want to present facts and details in a meaningful, sometimes persuasive way, so that the reader can see the differences and similarities of the objects. The content and scope of the details will depend, naturally, on the purpose of the report: Is the writer simply informing the reader, or is the writer making an argument to persuade the reader?

In a **comparison,** we look for *correlation.* In a **contrast,** we look for *differences* in certain features. It is important to determine the standards of comparison before beginning a report. Many people are familiar with the phrase *comparison shopping.* When we shop, we find similar products and decide which one to purchase based on our standards, such as price, quantity, and quality. Depending on the product and how we intend to use it, one feature may be a priority. If all the products are equal in the priority feature, the decision will be made based on the remaining features. For other products, we try to get the best buy for our money with no single priority.

Certain magazines, such as *Consumer Reports,* specialize in comparing and contrasting products. Because they want readers to trust their opinions, it is vital for them to present complete and unbiased information. Technical magazines, such as *PC Computing,* occasionally compare and contrast new components and devices. People who are in the market for these products will read these magazines to gather pertinent information quickly. Sometimes, magazines cater to a certain audience, such as a group of people who use the products professionally. These magazines may not use cost as a feature of comparison, or may review only the most popular products, totally ignoring other worthy, but lesser-known competition. Readers need to evaluate whether an article is presenting completely neutral information or whether the magazine is biased toward a certain item or audience. As the saying goes, "Let the buyer beware!"

Elements of Comparison and Contrast Reports

Comparison and contrast reports contain three main parts: an introduction, a body of information, and a conclusion, which makes a recommendation.

The first part of a comparison report, the introduction, states the standards of comparison: a short description of the features that will be discussed and why those features were chosen.

The second part of the report, the body, is the comparison and contrast of the products and their features. It is important to write parallel descriptions of each product. The features described for one product are described in the same order for all the products. If a feature is missing on a product, it is noted.

In the following chart, two pieces of test equipment are contrasted. You already know that both the multimeter and the oscilloscope are devices used to measure ac/dc voltage, and they both have scaling adjustment knobs. Notice the criteria (features) on which the contrast will be based.

Features	A: Multimeter	B: Oscilloscope
Function	Measures resistance, ac/dc current and voltage	Measures ac/dc voltage, phase, and frequency
Display	Numerical	Graphical
Cost	$60–200	$200–800+
Complexity of use	Easy (push function buttons)	Difficult (adjust time and voltage scales)

Information in the body can be organized in two ways. The **point-by-point method** itemizes the features being examined: function, display, cost, and complexity of use in the example above. For example, if products A and B are being compared, the first paragraph might compare one feature of both products. Whenever something is said about A, the parallel information about B is presented. The next section will describe another feature of both products. The heading for each section is the feature being described.

The **block method** is organized by the products: the multimeter and the oscilloscope. First, all the features of product A are described in a block, and then the parallel features of product B are described in a block. The features of product B are described in the same order as they were described for product A. The heading for each block is the product being described.

EXAMPLES OF OUTLINES

Point-By-Point Method

A. Function
  1. Multimeter
  2. Oscilloscope
B. Display
  1. Multimeter
  2. Oscilloscope
C. Cost
  1. Multimeter
  2. Oscilloscope
D. Complexity of use
  1. Multimeter
  2. Oscilloscope

Block Method

A. Multimeter
  1. Function
  2. Display
  3. Cost
  4. Complexity of Use
B. Oscilloscope
  1. Function
  2. Display
  3. Cost
  4. Complexity of Use

The final section of a comparison and contrast report, the conclusion, restates only the major points that led the writer to make a final recommendation. The recommendation is a logical conclusion based on the evidence presented in the report. Sometimes the writer proposes several recommendations that take into account the possible priorities of the reader. For instance, a writer might conclude that if the reader's priority is one feature, she should buy product A; she should otherwise, buy product B. A "best buy" recommendation weighs all the features equally and makes the most cost-efficient choice.

Exercise 9.1 *Outline a comparison and contrast of two similar products that you are interested in purchasing. Pick three or more features to compare and contrast. Write the outline first in the point-by-point method. Then write the outline in the block method. You may choose from the following:*

- Sports equipment
- Stereo equipment
- Computer hardware or software

Exercise 9.2 *Write a comparison and contrast essay of the products from Exercise 9.1. Use either the point-by-point method or the block method. In paragraph 1, state the standards of comparison. In the middle paragraphs, discuss the two brands. In the last paragraph, discuss how and why you made (or would make) your decision between the two products.*

SPELLING: T*ough* Words

One of the most troublesome letter groups in English is the *ough* group. This group of letters has unpredictable pronunciations and is found in unrelated words.

tough	ought
rough	bought
slough	sought

These words, although difficult to spell at times, have consistent pronunciations. The following five words demonstrate all the different spellings and pronunciations:

1. *Through* is a preposition meaning "in one end and out the other" or "beyond." It is also an adjective meaning "finished." It is related to *throughout,* meaning "all the way through." *Through* is commonly abbreviated *thru* in notes, but not in formal writing. *Through* sounds the same as, but is not related to, the verb *threw,* the past tense form of *throw.*

    We went through the data throughout the experiment.
    We didn't stop until we were through.

Note: Although the spelling *thru* is generally reserved for informal writing, it is an accepted spelling for certain structures such as the *pass-thru* and *drive-thru.*

2. *Though* and *although* are condition-setting words meaning "yet" or "in spite of."

    Even though we failed the first test, we tried again.
    We tried again although we failed the first test.

3. *Thought* is either a past tense verb of *think* or a noun meaning "an idea." A related word is *thoughtful,* which is an adjective meaning "considerate" or "serious."

After he thought about it, the solution seemed obvious.
The thought of inventing a new device kept him working.

4. *Thorough* is an adjective meaning "complete" or "very exact." Related words are *thoroughbred* (of a high breed) and *thoroughfare* (a well-traveled road). The adverb form is *thoroughly*

We checked the data thoroughly.
The discovery was made due to his thorough research.

5. *Tough* is an adjective that usually means "strong but pliant." It rhymes with *rough* (do not spell it "ruff"). The verb form is *toughen.*

The case was made of a tough plastic.
The high temperature toughened the ceramic.

Exercise 9.3   *Using the following words, fill in the blanks.*

though   thought   through   thorough   tough

1. The researchers _____ the experiment was going well.
2. They had planned the tests _____ly.
3. Al_____ they spent several weeks planning the tests, the actual testing was finished in a few days.
4. The final test was an after_____, but it proved to be the most valuable test of all.
5. They sought the opinions of other scientists before they were _____ with the research.
6. The material used for electronic casing must be nonconductive and _____ .
7. The winner of the Kentucky Derby was a _____bred.
8. As Kathy walked _____ the doorway, she noticed that the light panel was flashing.
9. Choosing a career in technology was not a _____decision; nevertheless, I considered all the alternatives _____ly.
10. Everything was ready on stage before the run-_____.

Exercise 9.4   *Use each word in a sentence.*

1. (*Tough*) _____
2. (*Through*) _____
3. (*Torough*) _____
4. (*Though*) _____
5. (*Thought*) _____

VOCABULARY: *retro, circum, intro, intra* and *inter*

Retro is a root meaning "backward or behind." It is similar in meaning to the prefix *re*.

> *Retrorockets* are used to slow rockets down in space.
> *Retrogression* and *regression* are the opposite of *progression*.

Circum and *circu* are roots meaning "around."

> The *circumference* is the distance around.
> *Circuits* are the pathways through which current flows.

Intro and *intra* are roots meaning "into" or "within."

> *Intramural* teams are from within a single school.
> The *introduction* leads into the report.

Inter is a root word meaning "between" or "among."

> The scientists met to *interchange* ideas.
> The *intercom* connected all the rooms.

Exercise 9.5   *Add the correct root to complete each sentence.*

1. A pay increase that reimburses a worker for a period already worked is called _____active.
2. A/an _____venous injection goes into a vein.
3. Draw a line to _____sect the triangle.
4. People who are interested more in their own minds rather than in other people are called _____verts.
5. The details surrounding an event are called _____stances.
6. A/an _____jection is a comment interrupting a discussion.
7. An argument in which the premise is also the conclusion is called a/an _____lar argument.
8. Any unwanted signal that disturbs the reception or display of a wanted signal is called _____ference.
9. The formula for finding the _____ference of a circle is $c = \pi d$.
10. Taking a machine or device back to adapt it for a new procedure is called _____tooling.
11. An electric current that is interrupted at intervals but always flows in the same direction is called _____mittent current.
12. Paying someone back is called _____tribution.

Exercise 9.6   *Combine one of the roots above with the root* spect, *meaning "look" or "see," to complete the following sentences.*

A/an _____ view carefully considers all the information and circumstances before making a judgment or decision.

_____ion means analyzing one's own feelings, emotions, and behavior.

In _____ we can often find value resulting from past misfortunes.

WORD WATCH: *effect* and *affect*

Effect is usually a noun meaning "the result of an action or cause." It is usually found following an adjective such as *the* or *an,* and it can be made into a plural by adding *s*.

> One of the *effects* of stress is irritability.
> Stress can have many physical side *effects*.

Note: One trick for remembering this word is "Expect an Effect." Both major words begin with *e*. Remember that *an* is a noun marker. If the use of the word is a noun, use *effect*.

Affect is usually a verb meaning "to influence or produce change." It will have an *s* form and also a past, a future, and a participle form.

> Daily exercise *affected* my job performace.
> The stress on this project will *affect* my health.
> The pleasant surroundings have *affected* my attitude.

Note: A trick for remembering this word is "Affect is Active." Both major words begin with *a*. Action words are verbs. If the word is a verb, use *affect*.

Exercise 9.7   *Fill in the correct word,* effect *or* affect. *Add the final* s *or* ed *if necessary.*

1. One of the _____ of increasingly complex technology is the growing enrollment in technical colleges.
2. Technical curricula are _____ by the demands of industry.
3. Hands-on training has had a good _____.
4. A declining teenage population has not _____ enrollment in technical schools.
5. Technical programs have been _____ by a demand for more literacy skills in employees.
6. The growing number of workers who cannot read manuals and instructions will _____ future training programs.
7. The financial _____ of illiteracy in industry has been underestimated.
8. Public concern will have an influential _____ on the problem.
9. Technical students are _____ by the growing demand for "academic" skills.
10. The _____ of publicity has been increased awareness.

Forms of Technical Communication

CHAPTER 10

Technical Reports

- Write an outline for a report
- Write a technical report, a proposal, and a lab report
- Spell words ending with *ance* and *ence* correctly.
- Use *proto, trans* and *neo* correctly.
- Use *accept* and *except* correctly.

READING: Implementing ISO 9000: Three Perspectives

by Tony Fletcher, Joseph Hart, Jill Kitka and David Fitzwilliam

A service company, its registrar, and its consultant share the story of the company's ISO 9000 registration effort.

Since its release in 1987, the ISO 9000 quality management standard has steadily gained acceptance across many industries around the world. Until recently, the majority of organizations using the standard were from the manufacturing sector. Now, however, customers' quality expectations extend beyond manufacturing to include the service industry.

To address this new set of customer expectations, service providers such as hospitals, personnel placement offices and transportation organizations have begun to use the ISO 9000 standards to define and improve their quality management systems. Two such customers, Chrysler and General Motors, have even formalized their expectations by setting ISO 9000 registration deadlines for their transportation service providers of December 31, 1998, and July 31, 1998, respectively.

One service company's quality management process is examined here from three perspectives: the company's, the consultant's and the registrar's. The company, Ryan Transportation Group, is

Quality Digest, May 1998. Reprinted by permission of Tony Fletcher, Eagle Group.

a full-service transportation organization providing less-than-truckload and truckload services throughout the Midwest. As the consultant, Eagle Group USA Inc. offers full-service consulting and training, and specializes in the service industry. NSF International Strategic Registrations Ltd. serves as the project's registrar. Each stakeholder discusses its role, expectations and issues in the ongoing effort to make the project a winning proposition for all.

Company Perspective

When our major customers pointed out that we could solidify our standing as a transportation supplier by obtaining ISO 9000 registration, we at first wondered how this management system would make us a better, more efficient company. Being strictly a service company rather than one that produces parts or maintains a material inventory, we initially didn't feel that registration would be of much value to us. Nonetheless, with some trepidation and an attitude that lacked complete commitment, we began the process.

FIGURE 10-1
The Ryan Transportation Group teams up with Eagle Group Consultants to implement the ISO 9000 standards.

Now we're about 75 percent completed with the registration process, and our attitude is very positive. We're interested in ascertaining the value-added benefits of being registered, and we're beginning to understand how this will happen. Put simply, we're formalizing many facets of our business that previously were informal.

For a long time at our company, uncertainty surrounded even our most basic business processes. Naturally, senior managers met regularly, and staff members were informed of current business plans. We handled situations and put out fires as they occurred, but until we began the registration process, we had no formal system to review what we were doing and how we could do it better. We now believe that, with a properly documented management system and a desire to practice what we preach, most of this uncertainty will clear up.

For example, with our purchasing process, which entails buying and maintaining equipment as well as recruiting and retaining owner-operator drivers, we expect to realize a significant payback for the time and effort spent achieving registration. We'll have a purchasing system in place that will mandate a more consistent vendor review, including work performance quality and pricing. But I think we'll benefit most from using ISO 9000 as a catalyst to achieve systemwide customer service improvement goals. Our new management system will clearly indicate where the buck stops and with whom.

As far as the actual implementation process goes, Ryan senior management has expressed concern about how our current staff—which includes office clerks, dispatchers, maintenance personnel and midlevel managers—will handle the change to a management system that requires so much documentation and internal auditing. Ryan Transportation employs about 90 people, and we've trained seven staff members to serve as internal auditors.

At first, I thought that internal auditing might prove difficult for inexperienced staff members. They'd have to be responsible for auditing a business process they had very little prior exposure to. However, we recently completed our first round of internal audits, and I must admit I'm surprised at the confidence and enthusiasm our internal auditors have exhibited during this process.

Lisa Ann Throne, our deputy manager representative, is one of them. "Everyone here was very comfortable with the implementation process," she reports. "We mainly had to learn to write everything down, rather than simply going across the room and talking to someone. Now we have tracking devices to follow up our verbal quotes."

Concerning our consulting firm and registrar, our expectations were fairly simple. We wanted the consultants to be patient, thorough and committed as they taught the ISO 9000 system to Ryan Group staff. And they were. We're implementing a management system that will forever change the way we do business, and this radical change is not without its detractors. Having said that, I have full confidence that we will achieve all of our ISO 9000 goals.

With our registrar firm, NSF-ISR Ltd., we are expecting an audit that not only grades us on our ability to properly interpret the standard but also suggests ways we can continually improve our business processes. By the time we are ready for our certification audit this month, we anticipate having a functional system in place, but I predict it will take a couple of years for our company to have a well-oiled, excellent management system in process.

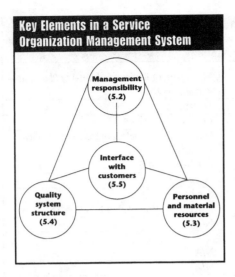

FIGURE 10–2
Key components of a service organization.

We hope NSF and Eagle Group will help us, through continuing education, achieve continued growth into the future.

Consultant Perspective

A service-based organization faces many challenges when it decides to set out on the path to ISO 9000 registration. Because the standard is geared toward the manufacturing industries, it can leave a service provider feeling lost and confused. The ISO 9000 series does provide several guidance documents, such as the ISO 9004 guidelines on quality management and systems elements, and the ISO 9004-2 guidelines for service industries, but making the connection without adjusting organizational processes or losing the standard's intent isn't always an easy task. As a service provider, Eagle Group had to come to terms with these issues before it could attain registration of its own management system. We addressed the unique opportunities presented to service organizations when we developed our quality management system. We wanted other service provider organizations—our clients, specifically—to be able to model from it. Eagle Group's registration success coupled with its expertise within the service arena provided Ryan Transportation with another case

study as well as a practical approach to ISO 9000 registration.

Our own management system became easier to define as we reviewed ISO 9004-2, keeping in mind the elements shown in Figure 10–2. ISO 9004-2 helps define key components of a service organization and gives direction to the organization for applying the manufacturing-based requirements listed in the ISO 9002 standard.

Figure 10-2 represents the key pieces of a service organization's management system: management, customers, policy and procedures, and resources. It's important when businesses begin to define their intent with ISO 9000 to tie the requirements back to these key elements.

Ryan Transportation is well-suited for a true service-based management system. It's looking to enhance its business processes and meet future customer needs. When Ryan and Eagle Group teamed up to implement an ISO 9002-registered management system, our goals were to develop a system based around Ryan's current practices rather than the ISO 9002 standard; keep the system lean, effective and value-added; and keep the company's processes responsive from a sales perspective.

Eagle Group's approach to implementing a service-based ISO 9000 system emphasizes flexibility and responsiveness. Service organizations, just like manufacturing companies, function best when they can offer quick response times to customer needs and order changes. However, the service provided often is the only "product" it has to measure.

Consequently, its management process must be a constantly evolving system based on whatever customers want, when they need it.

Our first step with Ryan was to perform a gap analysis comparing the systems currently in place against what the ISO 9002 standard requires. Then we mapped the standard elements to those systems and processes. By doing so, we laid the groundwork for determining what elements were applicable and how we could structure the content and quantity of procedures based upon existing

Level II Procedures for ISO 9002: Standard vs. Service	
ISO 9002—Procedures	**Service (Ryan)—Procedures**
Management Review—4.1	Management Review and Quality Planning—4.1, 4.2, 4.20
Quality Planning—4.2.3	See above
Contract Review—4.3	Contract Review—4.3
Document and Data Control—4.5	Document and Data Control, and Quality Records—4.5, 4.16
Purchasing—4.6	Purchasing—4.6
Control of Customer Supplied Product—4.7	See below
Product Identification and Traceability—4.8	See below
Process Control—4.9	Process Control—4.7, 4.8, 4.9, 4.15
Inspection and Testing—4.10	N/A
Control of Inspection, Measuring and Test Equipment—4.11	N/A
Inspection and Test Status—4.12	N/A
Control of Nonconforming Product—4.13	Control of Nonconforming Service—4.13
Corrective and Preventive Action—4.14	Corrective and Preventive Action—4.14
Handling, Storage, Packaging, Preservation and Delivery—4.15	See above
Control of Quality Records—4.16	See above
Internal Quality Audits—4.17	Internal Quality Audits—4.17
Training—4.18	Training—4.18
Servicing—4.19	N/A
Statistical Techniques—4.20	See above

FIGURE 10–3

Comparison of standard ISO 9000 procedures and the service procedures of Ryan Group.

business processes. This method differs from more typical procedures that structure business processes around ISO 9001's 20 elements or ISO 9002's 19 elements, as shown in Figure 10–3.

Many elements that apply to manufacturing organizations are easily left as procedures relative to a specific department. This isn't the case within a service industry. For example, when we think of process control procedures, we typically envision manufacturing lines with operators and equipment. For Ryan Transportation, process control actually means its interaction with a customer-supplied product, product identification and traceability, and handling, storage, packaging, preservation and delivery.

For a service provider, implementing a quality management system requires understanding nonmanufacturing environments in order to develop procedures that maintain ISO 9002's intent without overcomplicating the documentation.

Ryan Transportation also is typical of many organizations in that resources and time are valuable commodities. The system must be lean, functional and easy to maintain. Based on these factors, we developed a seven-page policy manual, less than 10 standard operating procedures tied to the business's various functional areas and a handful of key work instructions.

Standard operating procedures and work instructions are kept short and concise to ensure they are understood and used by all staff members. The streamlined process developed for changes, revisions and corrections works in conjunction with the company's continuous improvement process.

While developing procedures and supporting documents, we focused on improving process, communication and record keeping. This ensured that Ryan's system would help the company manage its growing business and keep it flexible in meeting customer needs as well as daily and hourly market changes. During this process, the management team interacted with staff members to mark the starting point for future process improvements. "This process means a lot of time and effort for a company our size, but it's going to help us define processes for our existing people as well as new people," noted one of the management team members.

Certain concerns arise when implementing ISO 9002 in a service organization, and Ryan experienced its share of them. Service providers always struggle with implementing elements 4.3—Contract Review and 4.6—Purchasing. Most service organizations' operations don't depend on lengthy contracts, with review and approval processes between the provider and customer.

Similarly, most service organizations don't have a formal process for subcontractor evaluation or purchasing. When they're not based on the practical business system of a service-based organization, these issues can become a burden. A system must allow for quick order turnarounds of one to four hours from start to finish as well as spontaneous purchasing within that time frame.

This type of implementation and process evaluation for unique solutions is one of Eagle Group's specialties. Although we have an extensive background in manufacturing and process-industry systems implementation, being registered to ISO 9001 ourselves gives us firsthand knowledge that we can pass along to our service-based clients.

Registrar Perspective

The ISO 9000 series of standards was written as generically as possible so that any type of organization can implement an ISO 9000 management system. As a registrar, NSF-ISR frequently is requested to certify the ISO 9000 management systems of service organizations, and overall our method is the same as for any registration. However, certain aspects of service organizations require a slightly different approach for both the client and registrar.

Perhaps the biggest difference service organizations find is that, for many aspects of their quality management system, they must first define their product (i.e., service) and how they inspect that product before they can move forward.

In many respects, the way a company approaches its implementation program will impact how easily it can define its service. If a company is counting on the ISO 9000 implementation process to enhance its operations and improve customer service, it will almost certainly evaluate what its customers really need and thus easily define what its service is.

However, an approach based more on minimizing efforts to obtain certification probably will skip some of the initial work of analyzing the service's true nature.

Certainly many of the elements—such as management review, document and data control, corrective and preventive action, control of quality records, internal quality audits and training—are relatively straightforward. However, mapping ISO 9002 or ISO 9001 into a service organization's relevant areas often proves more of a stretch. An important reference for any service company prior to program implementation is the ISO 9004-2 service guidelines.

As a registrar, we must first verify that the service has been defined realistically. The next issue is applicability. For example, ISO 9002's element 4.11—Control of Inspection, Measuring and Test Equipment—almost certainly won't be defined in terms a production environment would use. Key factors in interpreting and adapting the requirements would certainly include service specification, service delivery process and customer assessment.

Service specification deals with just what services a company will provide. We find that some service organizations already have developed company mission statements for their marketing proposals, while others have to start with a blank sheet of paper and create new specifications.

The service delivery process documents how the customer actually gets the service. A tanning salon, for example, typically might require its customers to make specific appointments before they come in. In the case of a trucking company like Ryan Transportation, there may be several different processes, depending, for example, on whether its customer is looking for local deliveries or a long haul.

The customers' assessment determines if customers received—sufficiently—what they expected; sometimes this information is available only when the service is delivered, other times it's available afterward. From a transportation perspective, this evaluation would include not only delivering correct and undamaged parts but doing so in the time frame required and with proper documentation.

Because service industries deal with interpretation and applicability more than their manufacturing counterparts, it's probably true that more service organizations use the various resources, such as consultants, available to assist their efforts. Typically, service providers will have more questions for their registrars. At NSF-ISR, we assign the lead auditor to our client immediately after they apply to us so that the client can call the auditor directly.

Many registrars will respond to their clients' questions relating to interpretation or applicability. However, unlike the consultant, the registrar's auditors can't specifically advise clients. The auditor will determine whether a client's interpretation or procedure under discussion meets the standard's requirements. If it doesn't, the lead auditor can't recommend a specific solution. Using a consulting firm can help provide that insight during the implementation process.

As for all registration audits, the audit team must have relevant experience in the industry. This is certainly an important factor for service providers because of the industry particulars previously discussed.

NSF-ISR uses a two-step approach for the on-site audit. Thus, after the document review has been completed and Ryan Transportation has had an opportunity to review the desk audit report, an on-site readiness review will take place. This will be followed some weeks later by the main audit. The readiness review provides an excellent opportunity to deal with those aspects of the ISO 9000 management system that are peculiar to a service organization. If any problem areas are detected, there is usually enough time to make any corrections before the main audit.

Ryan Transportation will soon undergo its certification audit, and as its registrar, NSF-ISR certainly can't prejudge the company's success. However, Ryan's sincere interest in the ISO 9002 standard and the benefits of a well-implemented system is evident from their involvement in this article, and we look forward to working with them.

Conclusion

For some service providers, a quality management system implementation might prove frustrating because the ISO 9000 standards are geared toward the manufacturing community. Their efforts therefore should focus on ways to integrate the ISO 9000 requirements into existing processes, making use of the ISO 9004-2 guideline and outside expertise for assistance.

All three stakeholders in the process described here are committed to attaining the same goal: a focused and effective quality management system that maintains and improves customer satisfaction. Teamwork is the key to reaching this common goal and succeeding in the ongoing effort to maintain the system in the future.

Reading Comprehension Questions

1. Who is the intended audience of the article? What will the audience do with the information in the article? _____

2. Based on your understanding of the article, describe the main function of the ISO 9000 quality management standard. _____

_____

3. What was the initial question the company had before it began the process? _____

_____

4. According to Ms. Throne, what was the biggest change in company processes due to implementing the ISO 9000 standards? _____

_____

5. Describe the role (function) of the consulting company. _____

_____

6. Describe the role (function) of the registrar. _____

_____

7. In one word, what do the authors feel is the key to a successful implementation?

_____

WRITING: Technical Reports

The reading article introduces formal report writing. The article mentions several formal documents required to become certified as ISO-9000 compliant in addition to the process control procedures required for certification, including the following:

- Gap analysis (highlights where improvements need to be made)
- Operating procedures
- Work instructions
- Process descriptions
- Company policy manual
- Certification documentation
- Audits

In addition, management might require an initial feasibility study to provide justification for funding the project, ongoing status and briefing reports during the implementation process, regular interdepartmental communication, and a final, comprehensive report for stockholders following implementation. In general, as the budget for a project increases, the expectation for formal documentation increases proportionally.

This chapter focuses on three basic report formats: the descriptive report, lab report, and proposal. You will find student examples of all three types in Appendix 3. Companies sometimes establish internal formats for reports. By practicing the three basic formats in this chapter, you can apply the guidelines to other formats for specific purposes. As you progress in your technical career, you will encounter a wide variety of reports to read, review and write, either singly or in collaboration with others.

Layout Guidelines

Recommended reading: The Non-Designer's Designer Book, by Robin Williams, 1994, Peachbit Press.

The appearance of a report adds to its visual interest and readability. Writers can use several techniques to make reports attractive and easy to read. Headings show the organization and flow of information; graphics add clarity to concepts, objects, or procedures; charts efficiently contrast or summarize details; and the white space of margins and between elements adds breathing room for readers.

As added insurance for eye-appeal, a writer designs the overall appearance of a report just as an artist plans the overall layout of the shapes and colors of a painting.

Design expert Robin Williams (not the comedian/actor) suggests the use of four basic guidelines to improve the design of any document, whether a brochure, letter, advertisement, or report. (You can find several excellent books by her on this subject for further reading.)

The following guidelines add consistency, unity, and cohesiveness to documents:

1. **Alignment.** Align elements to consistent margins (right, left, top, and bottom). For example, make sure all tables have a similar cell and row width. If you use indented information for quotes or examples, make sure the indentation is the same throughout the document. Align graphics with the margins of the text.

2. **Contrast.** Distinguish between differing elements on a page, such as headings from normal text, by varying the font, size, shape, and space. For example, make headings bold and a larger font size than normal text. Use one set of attributes (font and font size) for all main headings and another set of attributes for all subheadings, and so on. Add a consistent spacing between headings and the text.

3. **Proximity.** Group items that relate to each other close together. For example, reduce the spacing between bulleted or numbered items, such as this list, so that the list is viewed as a unit. Add a consistent spacing between paragraphs and between sections of the report.

4. **Repetition.** Repeat the visual elements of color, shape, texture, spacing, and line thickness throughout the document. For example, use headings and subheadings of consistent fonts and sizes. Use bullets with a consistent shape and indentation.

The following example shows a draft before basic design principles were applied.

EVALUATION CRITERIA

The research team investigated five cellular service providers and five types of cellular phones. Most phone providers offered a rebate for using their service and their phones. However, the team researched each phone and service separately.

The criteria for cellular phones are the following: *Features and Accessories, Options Packages, Warranty,* and *Cost.* The criteria for service providers are the following: *Cost, Coverage Areas,* and *Types of Phones.*

The following revision applies the principles of alignment (bullets and paragraphs), contrast (headings and subheadings), proximity (spacing and grouping of bulleted lists), and repetition (consistency of headings, subheadings, and bulleted lists).

EVALUATION CRITERIA

The research team investigated five cellular service providers and five types of cellular phones. Most phone providers offered a rebate for using their service and their phones. However, the team researched each phone and service separately.

Criteria for Cellular Phones:

- Features and Accessories
- Options Packages
- Warranty
- Cost

Criteria for Service Providers:

- Cost
- Coverage Areas
- Types of Phones

The application of the basic design principles takes experimentation and practice. And they are meant to be guidelines, not rules. Certain reports require other formats based on style guides or company policy.

Descriptive Reports

A descriptive report is a formal document with a specific audience, format, and purpose. The audience could be an instructor, supervisor, customer, or board of directors. As with any type of document, writers must analyze their audience before starting a report by asking the three basic questions: Who is my audience (technical, semitechnical, or non-technical)? What does my audience want to know about the subject? What does my audience intend to do with the information? The answers to these questions will influence the language, content, organization, and supporting elements of the report. Generally, the more technical the audience, the more complex and detailed the report.

The purpose of a descriptive report is to describe, define, explain, document, or teach, or any combination of these purposes. To achieve the purpose for the audience, writers must organize information clearly, emphasize main points, and provide the backup information needed by the audience. People usually do not have much time for reading in the workplace, so writers must get to the point quickly and stay focused. Internal reports might be more simplified and to the point than reports distributed to an external audience such as customers or clients.

A descriptive report relates the physical appearance of an object in words. It may be accompanied by pictures or line drawings that enhance the description, but pictures should never be considered a substitute for words. Most instruction manuals begin with a detailed description of the device, such as a computer or a multimeter. The description can include features, options, accessories, and specifications. You will be told what the various parts are called, how to maintain and replace them, and how to operate the device. Manuals for technical devices also include safety information.

In Chapter 5 you wrote a brief description of an object in a short paragraph. As you begin writing a descriptive report, which has many paragraphs, you will quickly see the need for careful organization. Plan the headings first, and then work on units within each heading. Pay attention to the relationship of one unit to another. As you will see in the example that follows, a report is arranged so that one description leads naturally to the next. Each part is described as it fits logically in the complete picture. Disjointed organization leaves the reader lost and confused.

Some basic guidelines for writing descriptions are as follows:

1. Start with an outline. Plan your report and report your plan.

2. Have the object in front of you as you write your outline. If you write about it from memory, you are more likely to omit or misrepresent important details.

3. Begin with an introduction. Lead in with general background information, a definition of the object, and a statement of the main sections of the report (a thesis sentence).

4. Use precise terms and measurements (use "toggle switch" rather than "switch," and "45 pounds" rather than "heavy"), but avoid highly technical terms and abbreviations when writing for a general audience (use "multiplex" rather than "MUX").

5. Describe the function of each part briefly, but do not confuse a descriptive report with an instruction manual (discussed in Chapter 8).

6. Include a line drawing or photograph if it reinforces your description, but do not rely on the picture.

7. Label each main section of the report. Use transition and sequence words as directions for the reader.

8. Describe specific parts moving in a consistent and logical direction, such as clockwise or top to bottom, when describing a large or multifeatured object.

9. End with a summary. Restate the purpose and the thesis of the report. Opinions are not appropriate in most descriptive reports.

10. Number pages (if the body of the report is over three pages) in the bottom margin, either centered or in the right corner (be consistent). Start numbering from the introduction to the report. Pages before the introduction are marked with lowercase Roman numerals.

   Some manuals and documentation are numbered using a decimal or hyphenated system. The chapter number is noted first, followed by the page within the chapter. At each new chapter, the page numbering begins with 1.

Examples of Page-Numbering Systems

	Sequential	Decimals	Hyphens
Chapter 1	1	1.1	1-1
	2	1.2	1-2
Chapter 2	3	2.1	2-1
	4	2.2	2-2

11. Staple the completed report in the upper left-hand corner. Do not include any blank sheets or use a plastic cover (unless required). If the report is longer than 10 pages, put it in a three-ring binder or folder.

Report Format

Use the following outline for report formats:

I. PRELIMINARY SECTION
   1. Letter of Transmittal or Preface
   2. Title Page
   3. Executive Summary or Abstract
   4. Table of Contents and List of Figures and Tables
II. MAIN SECTION
   5. Introduction and Thesis Sentence
   6. Body
   7. Summary and/or Conclusion
   8. Tables and Figures (if not included in the body)
III. DOCUMENTATION
   9. Notes (footnotes or endnotes, if needed)
   10. Bibliography
   11. Appendix

   1. *Letter of Transmittal or Preface (optional).* This page may be in memo or letter form (called a letter of transmittal), addressed to a specific reader. This page may also be in paragraph form (called a preface) for general readers. Use one or the other—not both. Provide background information: the reason for the report, the title of the report, special features of the report, and acknowledgement of any special assistance in the research/writing process.

2. *Title Page.* The title page, used for reports of three pages or more, is the first page of the report. Although styles vary, the information on this page includes the exact title, the author and position, the person to whom the report is submitted, and the date of submission. Arrange the information attractively and legibly.

3. *Executive Summary or Abstract (optional).* The executive summary or abstract is located at the beginning of a report and provides an overview of the contents. The *executive summary* lists all the main points, with emphasis on bottom-line information for people, such as those in upper management, who might have to make executive decisions based on the report. An *abstract,* on the other hand, provides a brief, but condensed overview of the contents for people such as researchers to determine whether the report contains information they want to read. Usually only one or the other is included, depending on the intended audience. Each should include the thesis (main purpose), the main ideas, and the conclusion.

4. *Table of Contents and List of Figures and Tables.* Each of these pages lists the order, topic headings, and page numbers for easy location by the reader. Include a table of contents for all reports over three pages. Include a list of figures and graphs if five or more formal graphics are included in the report. Arrange the information in neat columns. Use double-spaced dots to connect titles with page numbers.

5. *Introduction and Thesis Sentence.* The introduction is the first paragraph of the main section of the report, and it is always labeled. It provides a general lead-in to the subject of the report. This is the logical place to define a term, state the focus of a topic, or provide a brief background or history of the topic (assuming "History" is not a main heading in the report). The last sentence of the introduction is called the *thesis sentence.* The thesis sentence tells the reader the outline of the report.

> The purpose of this report is to discuss the history, function, and types of resistors.

The thesis sentence should match the sequence of the ideas that follow.

6. *Body.* This is the longest part of the report. It is made up of several individual sections with headings and includes many paragraphs. The order or sequence of the sections will depend on your purpose and topic. Each section must begin with a new paragraph. Do not begin a section or any new paragraph with a pronoun.

Labeling each major section of your report will provide a clear direction for the reader. Some common labels that state the purpose and the scope of sections are listed below.

History	Types	Operation
Applications	Construction	Features
Uses	Description	Function

To make the report easy to read, keep sentences under 25 words, and paragraphs under 10 lines of print (a suggestion, not a rule). To make the report interesting, vary your sentence length and style, using the methods you have practiced in grammar and mechanics exercises. Finally, proofread and edit carefully for clarity, grammar, spelling, and word choice. The revision process is often the most important step in effective writing.

7. *Summary and/or Conclusion.* The ending of the report must be clearly labeled and include at least one paragraph. A *summary,* like an abstract, is a brief review of the main ideas—no new information is provided. A *conclusion,* however, after restating the thesis, may include a personal opinion, analysis, or recommendation based on the research of the topic. At times, both endings are appropriate. The summary should come first, followed by the conclusion. Label each separately.

8. *Tables and Figures (optional).* Usually tables and figures belong in the body of the report, just following their reference in the text. However, if there are numerous or multipaged tables or figures, some writers include them at the end of the report, so the reader can continue reading and refer to all tables and figures in one place. Be sure to label each table and figure with a number and brief description and to refer to each table and figure in the text.

9. *Notes (optional).* Many authors find it necessary to "borrow" facts, statements, or figures from other authors. The careful and limited use of quotations is accepted as long as the writer does two things: place quotation marks around the exact words taken from another author (not necessary if the information is paraphrased), and cite the source of the quote (or paraphrased version) either by a footnote, endnote, or parenthetical note. Failing to do these things is considered *plagiarism,* passing someone else's work off as your own. It is a serious offense and a breach of ethics in technical writing.

The preferred format for citations, adopted by the Modern Language Association and the Society for Technical Communication, is the *parenthetical note.* This form simply cites the author's last name (or, if no name is provided, a key word from the title) and the page number in parentheses outside the final quotation mark, but inside the final punctuation of the sentence. Its efficiency makes it the best choice.

> **Samples of parenthetical notes:**
> "Teamwork is the key to reaching this common goal and succeeding in the ongoing effort to maintain the system in the future" (Fletcher, 123).
> The Ryan Group completed its ISO 9000 registration (Fletcher, 120).

The second entry is an indirect (paraphrased) quote. No quotation marks are needed, but the citation is included. Again, the full citation should be listed in the bibliography.

If you cannot use parenthetical notes, use endnotes (unless specifically told to use footnotes). When using endnotes, place a superscript number at the end of the quote or paraphrase. The full citation belongs at the end of the report on a new page titled "Notes," just preceding the bibliography. Number each entry with the corresponding number from the text, and add the page number where the quote is found. When using footnotes, insert the citation at the bottom of the same page of the quote or paraphrase.

Although the exact entry format for endnotes and footnotes differs slightly depending on the source, generally it starts with the author's name, the title, publishing information, date, and page(s). If more than one publishing location is listed, choose the one closest to you. If more than one publication date is listed, choose the most recent. Indent only the first line of the entry. Remember to italicize or underline the titles of books or magazines.

Refer to a style guide for the exact format required by your organization.

> **Samples of entries in endnotes and footnotes:**
> **Book:**
> [1]Joseph Gibaldi, *MLA Handbook for Writers of Research Papers* (New York: The Modern Language Association of America, 1999) 270.
> **Magazine article:**
> [2]Tony Fletcher, Joseph Hart, Jill Kitka, and David Fitzwilliam, "Implementing ISO 9000: Three Perspectives" (*Quality Digest,* May 1998) 120–124.
> **Manual:**
> [3]*Color Flatbed Scanner User's Guide.* (International Business Machines, Inc. 1998) 3.
> **Internet site:**
> [4]*Britannica Online,* Vers. 98.2, Apr. 1998, Encyclopaedia Britannica, 8 May 1998 <http://www.eb.com/>.

10. *Bibliography.* The bibliography is sometimes the last page of a report. Its purpose is to list all the references you used to research your report, whether you quoted from them or not. References can include books, magazines, encyclopedias, newspapers, manuals, documents, lecture notes, Internet sites, or personal interviews. Dictionaries are normally not considered references and are not listed in the bibliography. Any source from which a quote was used in the report must be listed in the bibliography.

If the entry is longer than one line of print, single-space and indent the second (third, etc.) line of the entry. Double or triple space between entries.

Entries in the bibliography contain the same information as footnotes, but in a slightly different format. They are arranged in alphabetical order by the first word of the entry, usually the author's last name. The first line of the entry is not indented, but each subsequent line is.

Samples of the entries in a bibliography:
Book:
Gibabli, Joseph. *MLA Handbook for Writers of Research Papers.* New York: The Modern Language Association of America, 1999.
Magazine article:
Fletcher, Tony, Joseph Hart, Jill Kitka, and David Fitzwilliam, "Implementing ISO 9000: Three Perspectives." *Quality Digest.* May 1998. 120–124.
Manual:
Color Flatbed Scanner User's Guide. International Business Machines, Inc. 1998.
Internet site:
Britannica Online, Vers. 98.2. Apr. 1998. Encyclopaedia Britannica. 8 May 1998 <http://www.eb.com/>.

11. *Appendix (if needed).* An appendix includes one or more sections containing supplemental information, such as detailed mathematical calculations, extended analyses, case histories or examples, or other backup information not essential to understanding the main report. Appendixes are usually lettered (Appendix A, Appendix B, and so on).

Before placing material in an appendix, however, consider its impact on the entire report. If the information is essential to understanding the major points of the report, it belongs in the body of the report, not an appendix. For example, if your audience is mixed, consisting of nontechnical and technical readers, you could place general information in the body of the report and place highly technical information in an appendix. That way, a nontechnical audience can follow the main report, and the detailed material is easily accessible to technical experts.

Figure 10–4 is an example of a short descriptive report. Examples of reports in Appendix 3 also demonstrate many of the writing conventions described in this section.

Two special types of descriptive reports are the white paper and specification.

White Paper   A white paper is an informal report, sometimes a draft, written on a technical topic by people in research or development. Companies publish white papers on Web sites to answer technical questions asked by peers or to solicit peer review and feedback. The format might be a proposal, description, or specification. Many companies encourage white papers as a way of generating and communicating ideas throughout the company or industry. Because the audience is primarily technical, the language and style include technical terms and acronyms and usually require specialized knowledge. The format is similar to that of the descriptive report.

Specification   A specification, informally called a "spec," is a highly structured (and confidential) report detailing the proposed features of a product or process. Engineers, programmers, and other technical specialists write specifications to detail how a product will be engineered, programmed, or built.

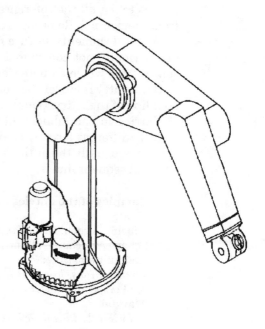

THE ROBOTIC ARM
by Robyn McKnight

A technologist who can easily identify the parts of a robotic arm saves time and effort when interfacing the arm with a computer. In this report the physical characteristics of a robotic arm will be discussed in detail to aid in locating specific parts of the arm.

A robotic arm is a mechanical device which, when interfaced with a computer, simulates the action of the human hand and arm. The major parts of the robotic arm described are the base, body, arm, and hand.

THE BASE The base of the robotic arm is made of blue, lightly textured metal. The base is a rectangular platform 8-1/2 inches long, 6 inches wide, and 1-1/2 inches high. The base supports the remaining parts of the arm. The body swivels relative to the base on a hollow shaft that is attached to the base.

THE BODY The body consists mainly of tan, lightweight plates of metal. Two main plates, each 6-1/2 by 8 inches, extend upward, housing gears and cables in the 3-inch space between the two plates. There are three motor casings on the outer side of each plate. The motors are numbered and attached to their corresponding function on the arm. The upper end of the body is connected to the arm at a shoulder joint.

THE ARM The arm has two parts: the upper arm and the forearm. The upper arm consists of two 10-inch plates, 2-1/2 inches apart, with drive cables housed inside the plates. An elbow connects the upper arm and the forearm.

The forearm bends downward from the elbow. Two plates, 8 1/2 inches long and 2 inches apart, house the cables that connect the forearm to the hand at a wrist joint.

THE HAND The orange hand is the last part of the robotic arm. It consists of two sets of links, 3 inches and 2-1/2 inches long, which function as fingers that are able to bend at a midjoint. The hand can swivel in a circular motion at the wrist joint. At the finger joint, the plates can either open or close.

The two fingers have a 4-inch spread when extended and meet when pulled together, resulting in a clamping action similar to pinching. Springs located in the hand provide the return force needed to open the hand.

CONCLUSION All the parts of the robotic arm discussed in this report are hollow, lightweight sheets of metal. The main parts are the base, body, arm, and hand. These parts form a robotic arm which can be interfaced with a computer and follow programmed instructions.

FIGURE 10–4
The Robotic Arm

Companies generally have their own standard format, but most specifications include the following basic elements:

- Executive summary, including a product overview and, if the product is based on an existing product, the revised or added features.
- Detailed descriptions of each component (including all subcomponents).
- Required resources, including equipment and staff.
- Dependencies, such as conditions or circumstances that must be in place or completed before this project can be completed.
- Detailed time line for development.

Budgets might be included, but usually a specification follows the acceptance of a formal proposal that includes the budget. As with white papers, the language and style are directed at a technical audience. A spec for a product revision or upgrade sometimes contains only the new features or changes from the original spec.

Exercise 10.1 *Rewrite the following list into a standard outline, using the format shown in the sample report in Figure 10–4.*

> THE PERSONAL COMPUTER
> Introduction: Definition of a microcomputer, brief history of computers (four generations).
> Logic Elements: Microprocessor (CPU), arithmetic-logic unit, input/output control, registers.
> Memory Storage Elements: Random access memory (RAM), read-only memory (ROM), hard disk drives, diskettes, zip disks, CD-ROMs.
> Conclusion: Advanced uses of the PC (interactive videodisc systems, touch screens, voice recognition, Internet browsers), growing function (memory and speed) for less cost.

Exercise 10.2 *Internet assignment: Using the keyword* ISO 9000, *find more information about its standards, including ISO 9000:2000 revisions, the relationship between ISO 9000 and SO 9000, and other case studies of companies making the transition to the quality management standards. Then write a descriptive report on the history, purpose, or changes in the standards.*

Exercise 10.3 *Use the descriptive paragraph you wrote in Chapter 5 as the basis for a formal descriptive report. Add the elements needed to produce a full report.*

Lab Reports

The formal lab report describes a lab test or experiment using industry conventions of methodology and reporting. A typical purpose is to report the results of an experiment to an audience of technical peers so they might replicate the experiment, or perhaps they might conduct further research based on your results. Because your results can have a far-reaching impact, it is critical to observe rigorous rules of procedure, objectivity, and reporting.

Scientists or engineers might distribute test findings in formally bound reports, abbreviated reports sent by memo or e-mail, or published journals. The elements of each type are the same.

1. *Title page.* The front sheet of the report states the exact title and/or number of the experiment. It also includes your name, section, and class, the date of submission, and the instructor's name (spelled correctly).

2. *Purpose.* This section states a one-sentence objective or purpose of the experiment. While you are a student, the objectives of lab experiments will often be given in your lab manual or by your instructor. Typical purposes include testing scientific

principles; examining the effects of certain procedures; and designing, modifying, and troubleshooting circuits or systems.

> *Poor example:*
> Objective: Compressive strength of concrete
> *Better example (list):*
> Objective: Determine the 28-day compressive strength of concrete (ASTM C39-84).
> *Better example (sentence):*
> The purpose of this experiment is to determine the 28-day compressive strength of concrete in accordance with ASTM C39-84.

3. *Theory.* State the basic formulas, theories, or assumptions that are used in the experiment. These can be written as a list of formulas or as a paragraph of background information.

> *Example of list:*
> Formulas: $V_c = V_{cc} - I_c R_c$ $V_b = (R_2/R_1 + R_2)V_{cc}$
> $V_e = V_b - V_{be}\,(0.7)$ $R_{in} = R_1 \times R_2/R_1 + R_2$
>   Gain $= v_{out}/V_{in}$

4. *Equipment and components.* List all the equipment and components and the number of these items that you used in the experiment. Remember that one purpose of the lab report is to direct someone who is trying to duplicate your experiment.

> *Poor example:*
> Equipment: Compression tester, gauge, caps, cylinders
> *Better example:*
> Equipment: Compression tester
>   Deformation gauge
>   Capping device
>   Cylinders of cured concrete
>   (3) 6″ dia. × 12″ long

5. *Procedures.* Explain in detail what you did and how you did it. Include any important illustrations, such as schematics and diagrams. Again, provide enough information so that the experiment could be repeated. Since you have already finished the experiment, use past-tense verbs. The steps can be written as a list or in paragraph form.

> *Poor example:*
> Procedures: 1. Set up circuit using different circuit values.
>   2. Calculate and measure normal operational values.
>   3. Solve all problems given in the lab manual, utilizing troubleshooting skills.

This example sounds like a slightly reworded lab manual. It uses present-tense verbs (set, calculate, solve) and directs the reader rather than reports the experiment.

> *Better example:*
> Procedure: 1. I constructed the circuit (see Figure 1) using the specified circuit values listed in Table 1.
>   2. I calculated and measured the normal operational values (see Table 2).
>   3. I used troubleshooting methods to solve the given problems (see Table 3).
> *Paragraph example:*
> I constructed the circuit shown in Figure 1. Using the given circuit values in Table 1, I calculated and then measured the operational values as shown in Table 2. Finally, I solved the given problems by troubleshooting the circuits as shown in Table 3.

The last two examples report the steps by using the active voice (*I* statements) and past-tense verbs. They also refer to specific figures and tables of information included in the experiment.

6. *Results.* List the raw data or results of the experiment. If charts, tables, or graphs are appropriate, create them with your computer, or draw them neatly using a template or ruler. Label each set of information as a figure (circuit drawings or graphs) or table (lists or charts), and number them sequentially throughout the report, starting with Figure 1 and Table 1.

7. *Discussion.* Analyze the results of the tests. Report the cause-and-effect relationships of the lab procedures. This section may be the longest part of the report. It is also the most important. In it, you will relate theory and scientific principles to the specific applications, equipment, and procedures that were used in the lab. State any points of doubt or error.

8. *Conclusions and recommendations.* Discuss the results and relate them to how successfully you achieved the objective. Suggest modifications in the procedure that you think would have improved the results. Restate what was learned or demonstrated during the experiment and after analyzing the results. In industry, this section could include recommendations for decisions such as necessary improvements, marketing strategies, or future enhancements.

The sample lab report, "Communications Lab #7: Low-Frequency Heterodyning," in Appendix 3 demonstrates many of these writing conventions.

Cause and Effect

One of the most important reading and writing skills in technical communication is the ability to identify cause-and-effect relationships. The entire study of science is built on this fundamental relationship. An *effect* is simply an observable situation resulting from the *cause,* or original event.

Service technicians must be aware of the 3 Cs: complaint, cause, and correction. Customers experience the effect, or symptom, of a problem and attempt to describe it to the technician in the complaint. So the effect and the complaint may be the same.

The technician will then isolate the cause and make the correction. For example, a flat tire (effect) may be the result of driving over a nail (cause) or of a worn-out tire (cause). It may be corrected by a patch or a new tire, depending on the exact cause. If the technician fails to identify and record the exact cause on the work order, the correction may be questioned later, causing more stress and paperwork.

Many words are used to signal a cause-and-effect relationship.

therefore	consequently	thereby
yields	because	for this reason
thus	as a result	the result
hence	the cause	as a consequence
so	that is why	since

Some students use a type of shorthand notation to indicate cause and effect. Attention to the correct relationship will help the student to understand difficult concepts. Following are some examples of cause-and-effect relationships. Notice the shorthand notation of cause and effect that follows.

Examples
The series relationship among resistors has a very important *consequence* for that circuit: the current is the same at every place in the circuit.

series circuit → current same

Generally, the individual resistances in a series circuit differ from each other. *Therefore,* the individual voltages also differ from each other.

series circuit → voltage differs

Both of these ideas could be noted together in shorthand.

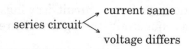

Exercise 10.4   *Underline the signal word and draw the cause/effect notation in each passage.*

1. Resistance produces heat. Therefore, the temperature rises.
2. An unreinforced footing may result in three types of movement: outward, downward, or inward.
3. Backfire is the result of combustion of the fuel mixture.
4. Random error is sure to affect the survey results; therefore, the skill and care of the surveyor are extremely important.
5. Because of the danger from sparks, hot metal, and rays, welders must wear safety glasses, leather welding gloves, and a hard hat.

Exercise 10.5   *Bring in an informal lab report that you have written. Rewrite it as a formal lab report.*

Technical Proposals

A proposal is a persuasive document that attempts to convince the reader to adopt or purchase a service or product. For example, the sales and marketing staff might write formal proposals to sell products or services to customers. The training staff might write a proposal to adopt a training program for employees. Technicians might write a proposal to purchase a new piece of equipment.

Proposals vary in size and length, usually becoming longer and more formal as the cost of the product or service increases. The decision-maker could be a committee or someone in upper management. The writer must take care to define technical terms and describe technical concepts.

Proposals can be:

- **External** (written to other companies), such as when proposing to sell your service or product to another company.
- **Internal** (written within your own company), such as when proposing a new service or product for your department to your manager.
- **Unsolicited** (sent without being requested), such as when your company wants to develop new customers.
- **Solicited** (requested by a department or company), such as when responding to customer inquiries, grants, or requests for proposals (discussed later in this chapter).

The size and formality of a proposal generally reflects the money at stake. For example, a proposal for a multimillion dollar, highly competitive sale might be written by professional proposal writers, bound, include colored graphics and be supplemented by on-site demonstrations and presentations. On the other hand, a proposal to a manager for a new laptop computer might be completed in an e-mail or memo as in Figure 10–5.

FIGURE 10–5
Example of informal
proposal.

Memorandum

To: Hannah Oberion, Manager

CC: Lyle Renquist

From: Carl Rugger

Date: 3/22/00

Re: Proposal for projector

Following our recent conversation about replacing the presentation projector, I researched several models that meet our requirements (portable, 1024 x 768 resolution). Lyle and I visited with several representatives of projection equipment, and we tested our existing sales program on each projector. Brands included InFocus, Compaq, Polaview, ExPro, and Proxima.

After comparing the quality, cost, and maintenance history, we found that the Proxima UltraLight LX1 LCD projector is the only model that met the basic requirements and our budget. By all standards, it is the best buy for our needs.

The Proxima UltraLight LX1 has also had good reviews in the trade journals which found the product durable, easy to use, lightweight (8.1 lbs.) and able to produce high-quality illumination in all types of lighted and darkened conference rooms.

We therefore recommend the purchase of the **Proxima UltraLight LX1 LCD projector.** The current price is $5079, available immediately through PC Connection. If this meets your approval, I will write the purchase order immediately.

TIP Longer formal proposals typically include the following elements:

- **Cover letter.** Address the letter to the primary decision-maker, such as the president of a company. This is a sales letter meant to catch the interest of the reader. It should be brief and to the point, such as emphasizing one main point of your service or product. Conclude with a date on which you will follow up with a phone call or visit.

- **Title page.** For competitive proposals, especially, use graphics programs or other software to create a title page that is eye-catching, modern, and professional.

- **Table of contents,** if the report is longer than a few pages, with a list of figures and tables.

- **Executive summary.** This is the most important part of the proposal, especially for longer proposals, and should be about one page. Upper management might read only the executive summary, without reading the body of the proposal. Include all the major benefits to the company, but not a discussion of them. Include bottom-line costs and implementation schedules. Use bulleted lists and short paragraphs to keep it easy to read. Use the active "you" voice, not "I" or "we." Write the executive summary after you have finished the actual proposal to ensure it is accurate and complete.

- **Discussion.** This section includes the rationale for the proposed service or product. If the proposal compares two or more products, the discussion includes a brief list of the products investigated for the proposal.

- **Proposed service or product, or comparison of products.** This is the main body of the report. It might include a technical section for technical experts and a management section for a more general audience. The management section describes overall how and when the work will be done, features, benefits, and impact on the company for people who will make the final decision. The technical section describes exactly the product, service, or idea you are selling for the staff who might implement the proposed service or product. Use headings for organization, and figures or graphics to increase clarity and interest. If products are compared, organize the description of each product in a parallel way, with consistent headings to show transitions.

- **Your company's background or experience** (for external proposals). This helps establish credibility and confidence in your company.
- **Budget.** Itemize the categories, such as components, labor, and overall totals, including unknown costs and how they will be determined.
- **Appendixes,** including supporting material and other product data (if needed).

The sample proposal, "Recommendation for the Purchase of a Cellular Phone and Phone Service," in Appendix 3 demonstrates many of these writing conventions.

Two particular types of proposals are the request for proposal and the feasibility study.

Request for Proposal, or RFP   An RFP is a formal solicitation for proposals. A company that is inviting competitive bids for a service or product might publish an RFP in newspapers, trade journals, or mailings. The RFP includes the agency's main needs, concerns, and constraints, as well as other information that is required from responders. RFPs sometimes include project-completion deadlines and budget limits.

Responding proposals must match the RFP requirements exactly, which requires careful reading. Responders should determine at the onset whether they can produce a competitive proposal—writing a proposal can be costly and time-consuming, and competition can be intense.

Responses are then compared and evaluated based on criteria, such as lowest cost or a combination of cost and quality of services. Proposals that fail to address a requirement or exceed the limits, costs, or deadlines are considered "nonresponsive" and eliminated from further review. The contract is awarded to the company that submitted the selected proposal.

Feasibility Study   A feasibility study investigates the practicality of one or more proposals. The people conducting a feasibility study are sometimes from an impartial, independent agency so they can conduct an objective comparison based on measurable criteria, such as cost, staff, and other requirements.

A typical feasibility study includes a methodical analysis of the various options and ends with a recommendation, which somewhat resembles a proposal itself. It is generally signed by the chief author, whose reputation is now tied to the honesty and credibility of the report.

Exercise 10.6   *Write a brief proposal to convince your instructor or manager to purchase a new piece of equipment. Include a cover letter, introduction, description of the proposed equipment, and budget.*

Exercise 10.7   *Write a formal proposal that sells a product or service. At a minimum, include a cover letter, title page, executive summary, description of the proposed product or service, and budget. Examples of topics include:*

- *A new process to be implemented at your school or workplace.*
- *A product that you want funding to develop.*
- *A service that you want to provide to your school or workplace.*

SPELLING: Suffixes *ance* and *ence*

The *ance* and *ence* noun-forming suffixes are used to add the meaning "an act of," "a quality or state of being," or "a thing that" to the root word.

> occurrence—the act of occurring
> resistance—the quality of being resistant
> conveyance—a means of conveying

Deciding between spelling a suffix *ance* or *ence* (or, similarly, *ant* or *ent*) can be a frustrating problem, seemingly decided long ago by someone who simply flipped a coin. There are no obvious patterns or rules for the correct ending. The unfortunate cause for this problem is that we use the Latin root word to determine the correct spelling. Those of us who have not had a formal study of Latin will have to rely on visual or artificial clues for remembering correct endings, with the exception of one minor pattern: When a verb ends in the letter *r* preceded by a single vowel and is accented on the last syllable, the ending is always spelled *ence*.

> pre-fer'—preference   oc-cur'—occurrence
> con-fer'—conference   con-cur'—concurrence

Although this rule is fairly reliable, it falls far short of the needs of most technical writers.

Some writers develop their own tricks for spelling certain words.

> attenDANCE—imagine people dancing into the room
> ambuLANCE—imagine a victim pierced by a lance
> baLANCE—imagine a ball balancing on a lance
> exisTENce—existing for ten million years
> obserVANce—imagine looking for a van

Other writers pay close attention to the words they use most frequently, and use the dictionary for lesser-used words. More words end in *ence* than in *ance*.

One bright note for electronics technicians, however, is that most electronics-related words end in *ance*.

Exercise 10.8   *Complete the following 20 words with either* ence *or* ance. *Try them first without using a dictionary.*

1. recurr_____   2. resist_____
3. deter_____   4. imped_____
5. differ_____   6. attend_____
7. observ_____   8. disson_____
9. bal_____   10. conduct_____
11. toler_____   12. persist_____
13. infer_____   14. capacit_____
15. excell_____   16. induct_____
17. react_____   18. exist_____
19. transfer_____   20. reson_____

Now check your words in the dictionary.

VOCABULARY: Prefixes *proto, trans,* and *neo*

Neo is a prefix meaning "new" or "recent."

> neo-Latin—modern Latin
> neoplasm—new, abnormal growth of tissue (tumor)

Proto and *pro* are prefixes meaning "first," "original," or "principal."

> protagonist—main character in a story
> protoplasm—essential living matter of plants and animals

Trans is a prefix meaning "across," "over," or "through."

> transcribe—write out in another form
> transfer—move from one place to another

Exercise 10.9   *Write the technical meaning of each word. Try to relate the meaning to the root word.*

1. *neo + natal* _____
2. *neo + phyte* _____
3. *neo + prene* _____
4. *proto + col* _____
5. *proto + n* _____
6. *proto + type* _____
7. *trans + former* _____
8. *trans + ient* _____
9. *trans + istor* _____
10. *trans + mission* _____

Exercise 10.10   *Write the meaning of these two confusing words.*

transparent   _____
transluscent   _____

Exercise 10.11   *Using the words from Exercise 10.9, fill in the appropriate word in each sentence.*

1. One of the chief parts of the atom is the _____.

2. A _____ is an amateur or a beginner.

3. Devices used for changing AC voltage and current levels are called _____.

4. The part of the vehicle that moves force from the engine to the wheels is the _____.

5. The _____ ward has special equipment for newborn infants.

6. The initial greeting, or "handshake," between two electronic communications devices is called _____.

7. A _____ voltage or current appears randomly without a fixed time interval between events.

8. _____ is a new, synthetic rubber that is highly resistant to heat, oil, and oxidation.

9. A transfer resistor, or _____, is a solid-state device having a collector, base, and emitter terminals.

10. The _____ is the original model of a particular type.

WORD WATCH: *accept* and *except*

These two words both have the same root, *cept,* which means "to take." The prefixes make them different.

> *Ac-* means "toward" (other forms: *a-, ad-, ap-, ar-, as-*)
> *Ex-* means "from" or "away from" (a negative prefix).

Write the literal meaning of *accept:* _____

Write the literal meaning of *except:* _____

Accept is a verb that means "to receive or approve." The noun form is *acceptance,* and the adjective form is *acceptable.*

> The public was not ready to *accept* robots.
> Industry's *acceptance* of standards took many years.

Except is usually a condition-setting (subordinate) conjunction or a preposition meaning "take out" or "unless." The noun form is *exception.* The adjective forms are *exceptionable* and *exceptional,* although the second word has come to mean "outstanding."

> Robots cannot function *except* by programmed commands.
> The *exception* proves the rule.

Note: *Expect* is a verb that means "to look for" or "to wait for." The root is another form of the root *spec,* meaning "to see." The noun form is *expectation.*

Exercise 10.12 *Fill in the correct form of* accept, except, *or* expect.

1. According to author Isaac Asimov, a robot must obey orders given to it by a human being _____ when those orders would violate the first law.

2. A robot must protect its own existence, _____ when that would violate the first or second laws.

3. Today's robots are not sophisticated enough to _____ and obey Asimov's rules of robotics.

4. Most industrial robots do nothing _____ carry, lift, or pull things.

5. Some workers find it difficult to _____ a robot as a coworker.

6. Modern robots use and _____ infrared and ultrasonic signals that tell when humans are around and where objects are.

7. Some people fear robots because they _____ to lose their jobs as robots invade the workplace.

8. Years ago, there were few robots around _____ for those built by hobbyists in their own basements.

9. We can _____ the robotlike toys to become more elaborate and more expensive.

10. Some programmable robots already can _____ and follow precise instructions.

CHAPTER 11

Forms, Memos, and E-Mail

- Fill out an accident report.
- Complete an insurance form.
- Fill out a purchase requisition.
- Write a status, personal, and negative memo and e-mail.
- Spell and use *ceed, cede,* and *sede* roots correctly.
- Use *past* and *passed* correctly.

READING: E-mail Etiquette: When and How to Communicate Electronically

Joseph M. Saul

With the advent of the "electronic revolution," many of us are turning to electronic mail for communications we used to handle by phone, letter, or face-to-face meetings. But effective e-mail communication involves far more than simply knowing how to use your e-mail program. This article discusses four major pitfalls of e-mail—missed signals, lack of context, permanence, and unfamiliarity—and offers tips for avoiding them.

How would you notify personnel in your department of a policy change? Set up a meeting with a colleague in another department? Wish a sister who lives in England happy birthday? Inform an employee of a change in job responsibilities?

Five years ago, most people would have said "official memo," "phone," "mail a card," and "face-to-face meeting," respectively.

These are all very different methods of communication, and they have different requirements.

Because the policy change is an official action, you would want people to see it on letterhead and have a file copy for reference. A quick phone call would suffice for the meeting, though. Phone calls to England are expensive, so you would probably mail your sister a card. Job actions, however, are sensitive and best handled face-to-face. Broadcasting e-mail messages to multiple e-mail lists and individuals regardless of their possible interest in the

FIGURE 11–1
With any written communication, all the non-verbal signals are lost.

Originally published in *Information Technology Digest,* University of Michigan, April 1996. Reprinted by permission of Joseph M. Saul, Communications Technology Consulting.

messages is called "spamming." Polite e-mail users do not send spam e-mail.

These days, many of us are turning to electronic mail for communications we used to handle by phone, letter, or even face-to-face meetings. E-mail is easy to send and does not depend on both parties being available at the same time. It even provides a written record.

E-mail may seem to be the perfect form of communication, but it does have some limitations, and it is not always the appropriate choice. It is important to understand its pitfalls and how to work around them. The four major pitfalls of e-mail are missed signals, lack of context, permanence, and unfamiliarity.

Missed Signals

You can't communicate as broad a range of information in e-mail as you can in a face-to-face meeting, or even in a telephone call. Your words come across, but all the nonverbal signals—facial expressions, eye contact, body language, tone of voice—are lost. We usually don't think about it, but we depend on those signals for information about the context of what is said; we need the signals to help us interpret the meaning beneath the words. Without them, we are often left to guess at the other person's intent.

These nonverbal signals are the main reason that most people prefer to handle sensitive issues (such as employment actions) in face-to-face meetings. When the situation is already potentially tense and you want your meaning to be absolutely clear, you want to have as much information as possible flowing back and forth.

Conversely, this is why e-mail conversations can become so heated. It's hard to say something "with a smile" in electronic mail, and it is all too easy to misinterpret an offhand, joking remark as a personal attack.

Once tempers flare, both parties—each operating without those important nonverbal cues to meaning—tend to read their worst fears into the written words and react in kind. This can happen even among friends, but when the parties involved don't know each other well, it can be worse.

As a result, experienced e-mail users have developed conventions for showing when they are joking—interjections such as "<grin>" or the use of "smileys," such as this: :-). (If you haven't seen one of these before, tip your head to the left to see the smile.) Unfortunately, these methods are not universally understood and communicate only a limited amount of meaning.

What is the best way to avoid misunderstandings due to missed signals? Give e-mail correspondents the benefit of the doubt and seek clarification (for example, "You sounded annoyed in that last reply. Am I reading you correctly?"). If there is a dispute, don't hesitate to call someone on the phone or talk to them in person.

Lack of Context

A note stuck to your door is informal; a signed memorandum on departmental letterhead is official. The way a message is sent tells the recipient a lot—people have learned to recognize the status of a message from its context and formatting cues. In e-mail, however, both kinds of message look the same. You can't send an e-mail message on letterhead or on scented stationery. As a result, your recipient not only lacks the nonverbal content of your speech, but he or she also lacks the traditional symbols that would show its status and context. If people in your department receive an e-mail message saying "Please get all grades in by the 25th," they don't necessarily know whether it is an official statement of policy or a plea for help from an overworked administrator.

As we start to use e-mail interchangeably with all of the other communication methods available to us, we have to develop ways of making the context of the message clear. Eventually, we may have "electronic letterhead" for verifiable official messages. Until then, the best solution is to explain your message's status and context right up front. You might, for example, state "This is a formal announcement from the office of the director," if indeed that's what it is.

Permanence

Unless your phone is bugged, a phone call leaves no permanent record. E-mail, however, does—and it can be forwarded again and again and come back to haunt you long after you have forgotten why you sent the original message. (This is especially true on mailing lists, where some list members may not see your message until weeks after you sent it.)

Because electronic mail is so easy to send and seems so ephemeral, people often forget just how permanent it is. You can achieve a kind of immortality through your e-mail well out of proportion to the amount of effort it takes to send it.

It can be a good idea to explain your intentions to the recipient of a message. If you do not want your message forwarded to anyone else, say so.

The convention on mailing lists and Usenet newsgroups is that private e-mail should not be publicly posted, but people are occasionally thoughtless or unaware of the convention. To be safe, think very carefully before sending a hostile or angry message; you can wind up defending your writings long after the feelings that motivated you to write them are past. And you can wind up defending them to people you never thought the message would reach.

Unfamiliarity

Most people learn to use the telephone and to write letters as small children. Appropriate phone or letter etiquette is second nature to most adults. Most people on this campus, however, have had electronic mail for a much shorter time—maybe one to five years. Many incoming students have their first experience with e-mail during college orientation. Electronic mail is a very new method of communication for most of the people worldwide who use it—and they're still learning the ropes.

E-MAIL ETIQUETTE

Do

- Do review messages before you send them out to make sure you are really saying what you want to say.
- Do be as polite as possible; terseness can be taken as hostility.
- Do make it clear to the recipient what type of message you are sending, especially if it is official.
- Do give correspondents the benefit of the doubt; try not to assume the worst.
- Do be patient with inexperienced e-mail users.
- Do, if possible, include the portion of the message you're replying to in your reply; people often forget the original context.
- Do include a subject line that accurately reflects the subject of your message.
- Do use e-mail for praise or for neutral messages (such as moving a meeting time).
- Do enjoy and use responsibly the e-mail resources available to you.

Don't

- Don't send a message when you're angry; cool down, look at the message again, and then decide whether you really want to send it.
- Don't copy an entire, large message in your response just to add a line or two of commentary.
- Don't reply to "all recipients" unless they *all* need to see your reply.
- Don't type in all capital letters; this is SHOUTING and is considered RUDE.
- Don't send off-topic messages to mailing lists, especially work-related lists.
- Don't send chain letters or messages recruiting participants in make-money-fast schemes.
- Don't edit quoted messages to change the overall meaning.
- Don't send criticism in e-mail; use the phone or, better yet, talk to the person face-to-face.

As a result, they make mistakes. This isn't surprising; e-mail etiquette is no more intuitive than phone etiquette, and everyone has heard children answer phones with "Who is this?" or simply with silence punctuated by giggles.

People do all kinds of things that offend experienced e-mail users—copying entire messages just to add "I agree," passing on chain letters, replying to entire mail groups instead of just the sender, typing in all capital (which is interpreted as shouting) or all lower-case letters. The list of "sins" goes on and on.

Never assume that another person is deliberately trying to be annoying over e-mail without supporting evidence; they simply may not know better. Most people, if told politely, will be happy to follow the conventions. They just need to know what the conventions are.

Use with Care

E-mail can be a wonderful communication tool when used with care. Avoid the pitfalls, think before you act, and remember that we are all learning the ropes together.

Reading Comprehension Questions

1. Why do people prefer using e-mail for communication?_____

   _____

2. How can you avoid the pitfall of missed signals? _____

   _____

3. How can you avoid the pitfall of lack of context? _____

   _____

4. How can you avoid the pitfall of permanence? _____

   _____

5. What are some typical mistakes made by inexperienced users? _____

   _____

WRITING: Forms, Memos, and E-mail

Most of the writing required of beginning technicians consists of in-house forms, memos, and e-mail. Forms are preprinted formats for recording specific pieces of information. Memos are brief, open-ended communications addressed to a specific person about a stated subject.

Electronic mail, or e-mail, provides an electronic medium for both types of communication—memos and forms. In addition, e-mail can transmit business letters and reports. Because of its simplicity, many people forget that all the rules of writing still apply, especially in the workplace. As the article suggests, inexperienced or new users can learn hard lessons from sending hastily written e-mails. Writers soon learn that effective e-mails require all the same concern for audience, clarity, grammar, and completeness as any other type of technical document.

This chapter focuses on forms, memos, and e-mails, sometimes considered informal or in-house types of communication. All three types are vulnerable to the pitfalls of haste, evasiveness, and insensitivity because of the lack of direct eye contact and immediate feedback from the audience. Most of us can recall personal experiences when events went haywire because of a "failure to communicate."

The article reminds us of the need to monitor the contents of e-mails, and we can apply those same reminders to other types of informal communication, including written, spoken, and nonverbal. This chapter includes some guidelines and practice for each format.

Forms

Each company and industry has its own standard and customized forms. Because formats are designed to furnish precise and efficient records, we must attend to the specific labels to determine what information to include in each line or box.

- Some forms include detailed instructions, such as tax forms. Most business forms, however, have brief labels or titles for sections, columns, and rows.

- Some forms are available online, and you can complete and transmit them on your computer. Others must be completed by hand or typewriter, and then mailed, faxed, or delivered.

- Some forms require an entry in every box or line—financial institutions are notorious for this and will return forms that are incomplete. But most service and retail businesses treat certain items on its forms as optional, depending on the particular use of the form or the entries you make in related areas on the form.

Examine and understand the purpose of the forms that you use in your career so you will fill them out correctly. Keep a copy of completed forms that you send by mail—it's always wise to have a copy for your records.

The form that is most frequently used by service technicians is the service record, also called a trouble log, customer service order, repair order, or status report. After the technician completes a service form, it may be used by the company for billing, inventory, and customer account records. Completed forms are usually kept in a central location and often entered into a computer databank. Since many departments may review the record for a variety of reasons, it is vital to record the information neatly and accurately.

Some companies review all records and return incomplete records to the technician for more information. Since the review could take place several days after the service was performed, it would be understandably difficult to recall and supply the missing information.

Figure 11–2 shows two service records from an automobile dealership. Figure 11–2a is incomplete and was sent back to the technician. Before reading on, examine it and see if you can determine which critical items were left blank. The record is missing the customer's signature, which authorizes the repairs; the vehicle identification number, which may be needed for legal matters; and the condition and cause of each repair. Figure 11–2b is complete for the dealership's needs.

Other common forms are the purchase requisition (Figure 11–3) and the monthly vehicle report (Figure 11–4). These forms also ask for specific, precise information that is necessary for efficient accounting, ordering, billing, or reimbursement.

TIP   Some points to remember in filling out forms are the following:

1. If you are writing by hand, write or print neatly. Many people may have to decipher your handwriting. A misunderstanding may have negative consequences for the customer or for you. Correct spelling and clear wording help.

2. Check out any information that you cannot easily supply. This includes the date and time, part numbers or descriptions, and the customer's full name.

3. Do not overlook supplying information you take for granted, such as the customer's complaint. Anticipate the question that your company may have concerning the service.

4. Record your procedures as soon as possible after you perform them. Some technicians keep a personal work record in addition to the company forms they submit. If questions arise later, or a similar problem comes up, the personal record provides details of past work.

5. Use standard abbreviations and symbols on in-house forms. If the form is given to the customer, use full words. Some customers are reluctant to pay for services they cannot interpret.

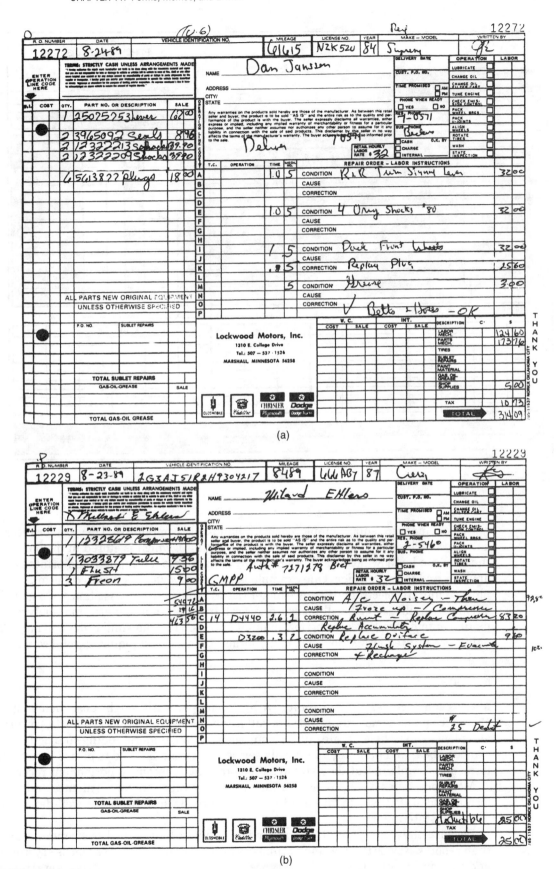

FIGURE 11–2

a) Incomplete service record. b) Complete service record.

Courtesy of Richard Ottum.

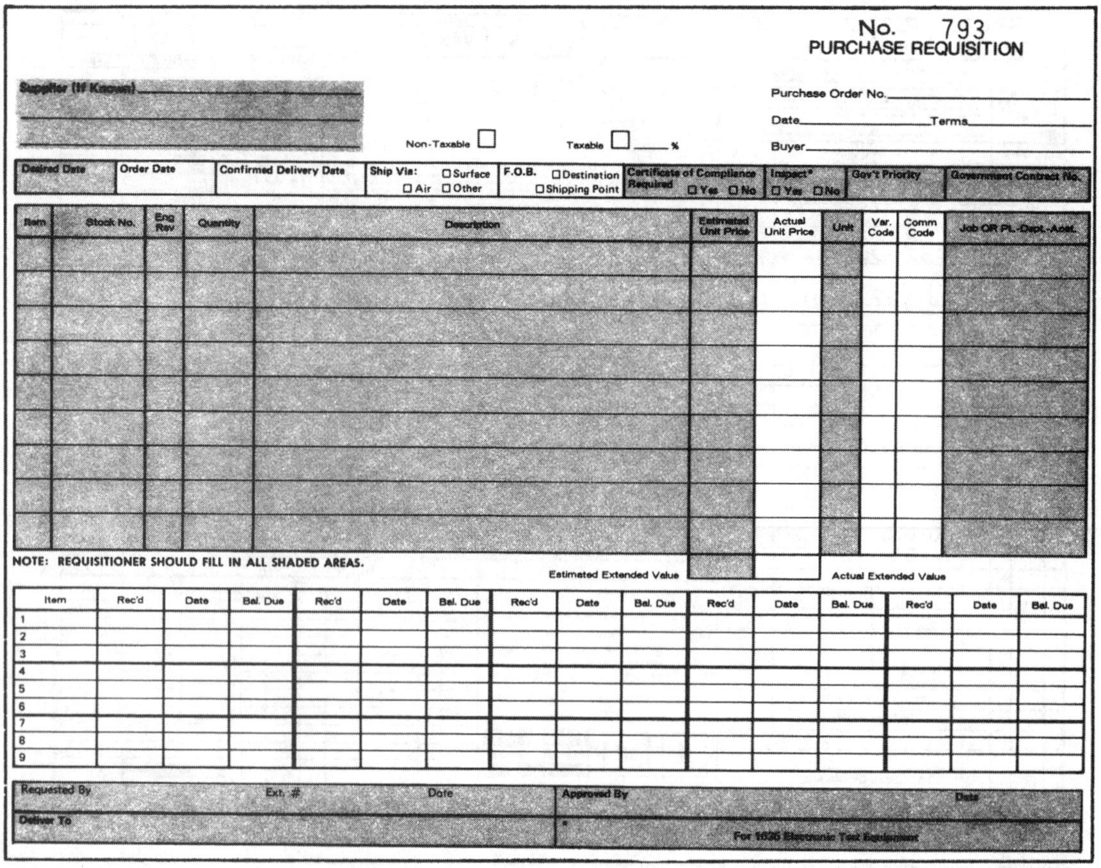

No. 793
PURCHASE REQUISITION

Supplier (If Known) ____

Non-Taxable ☐   Taxable ☐ ____ %

Purchase Order No. ____
Date ____ Terms ____
Buyer ____

Desired Date	Order Date	Confirmed Delivery Date	Ship Via: ☐ Surface ☐ Air ☐ Other	F.O.B. ☐ Destination ☐ Shipping Point	Certificate of Compliance Required ☐ Yes ☐ No	Inspect* ☐ Yes ☐ No	Gov't Priority	Government Contract No.

Item	Stock No.	Eng Rev	Quantity	Description	Estimated Unit Price	Actual Unit Price	Unit	Var. Code	Comm Code	Job OR PL.-Dept.-Acct.

NOTE: REQUISITIONER SHOULD FILL IN ALL SHADED AREAS.

Estimated Extended Value   Actual Extended Value

Item	Rec'd	Date	Bal. Due	Rec'd	Date	Bal. Due	Rec'd	Date	Bal. Due	Rec'd	Date	Bal. Due	Rec'd	Date	Bal. Due
1															
2															
3															
4															
5															
6															
7															
8															
9															

Requested By   Ext. #   Date   Approved By   Date
Deliver To   *For 1626 Electronic Test Equipment

FIGURE 11–3
Purchase requisition.

Some companies supply employees with hand-held computerized devices for recording data. Technicians "key in data" (type the information), which is stored in memory. Later the data can be "downloaded" (automatically loaded) into a computer and printed out for analysis and interpretation. One example is the surveyor's sealed data collector, which is plugged into the electronic distance meter (EDM) and, with the assistance of a global positioning satellite (GPS), will measure, record, transmit, manipulate, and analyze information automatically, requiring few manual entries. This and similar devices may eliminate some printed forms; however, many forms still exist and require careful completion.

Memos

Memos can increase your visibility within a company. Although most memos have a short life span, memos can be used to call attention to your projects, efforts, coworkers, and plans. They can generate enthusiasm, cooperation, understanding, and action. They can be upbeat and humorous, or deadly serious. They can be used to praise, question, inform, and complain. But they all have the potential to become a permanent record.

Most companies have protocol for memos. Study your company's memos and use the format and style that are familiar to your readers. Be sure to have a well-defined purpose and state it early.

Because they are generally short, memos should be planned even more carefully than reports. Revise them as many times as necessary until the message is crystal clear and the tone is appropriate. Three types worth practicing are **status, negative,** and **personal memos.**

FIGURE 11–4
Monthly vehicle report.

MONTHLY VEHICLE REPORT — ________

EMPLOYEE NAME (PRINT) | COMPANY CAR # | ENDING ODOMETER READING
EMPLOYEE # 6 | YEAR/MAKE/MODEL | BEGIN. ODOMETER READING
COST CTR # | OFFICE # | RENTAL CONTRACT #

TOTAL MONTHLY MILEAGE
LESS BUSINESS MILEAGE
TOTAL PERSONAL MILES
RAN # (IF NEEDED)

CASH AND PERSONAL CREDIT CARD PURCHASES ONLY
(ATTACH RECEIPTS FOR ANY EXPENSE OVER $10.00)

		ACCOUNT 371				ACCOUNT 372	ACCOUNT 373	TOTAL
		GASOLINE		OIL		MAINTENANCE & TIRES	MISCELLANEOUS	REIMBURSEMENT
DATE	BUSINESS MILEAGE	GALS. LTRS.	$	QTS.	$			
TOTAL			.		.	.	.	.

SIGNATURE AND DATE

APPROVAL SIGNATURE AND DATE

TITLE

MILEAGE REIMBURSEMENT: A. FLAT RATE	.
ACCOUNT 377 B. PER MILE RATE	.
RENTAL EXPENSE - ACCOUNT 379	.
GROSS REIMBURSEMENT	.
PETTY CASH REIMBURSEMENT	.
NET REIMBURSEMENT DUE	.

M-1176-1 FLEET - WHITE COPY; ACCOUNTS PAYABLE - CANARY COPY; OFFICE - PINK COPY; DRIVER - GOLD COPY.

A **status memo** informs others of the state of a project or situation. Stay positive and action-oriented, yet realistic. When you promise something in writing, your integrity is at stake. If you fail to follow through, people will begin to doubt your ability and commitment. Share credit when it is due—hogging the spotlight is not a sign of a team player, and American businesses value teamwork.

Figure 11–5 shows an example of poorly written and well-written memos about a work-related accident. Notice the differences between the two. Figure 11–5a does not answer all the questions a supervisor needs to have answered, and the writer comes across as unprofessional and vague. Figure 11–5b not only looks professional, but also provides an account of the accident, the results of the accident, and what further action is required.

If a table or graph will express your message more succinctly, include it. When covering several points, number them to simplify the organization. Start with the most important points since some readers may not finish reading a memo that starts with dull information.

Negative memos are written to reject, disagree, complain, or admonish. When writing a negative memo, remember that it cannot be erased later. Be sure that the message you have written will not be damaging to you or others. Time will not heal some wounds, and you may regret harsh words. Because negative memos cannot be softened with body language or voice tone, your written words may be stronger than you intend. Try to sound businesslike and factual. Attempt to identify and resolve underlying issues ("We need to change your performance plan") rather than focusing on personalities or emotions ("Your attitude has to change"). This will come more easily

FIGURE 11–5
Status memos: a) Poorly-written memo. b) Well-written memo.

To: Ed   **Date:** Tues.
From: Chris
SUBJECT: Scott Van Hoffman

Scott tripped over a toolbox and had to go to the hospital. He was treated in the emergency room and released. Figures he'll be out about 2 weeks. No problem to reschedule.

(a)

To: Ed Cramer, Human Resources   DATE: 1/31/87
CC: Scott Van Hoffman   *CA*
FROM: Christ Anderson, Foreman
SUBJECT: Incident Report—Scott Van Hoffman
                                ID# 30045
                                Hire Date: 5/15/86

$\left[\begin{array}{c}\text{What}\\\text{happened}\end{array}\right]$ On 1/30/87, at 10:15 a.m., Scott Van Hoffman backed up from his workstation and tripped on his toolbox, which had not been returned to its shelf. The accident was apparently caused by his own negligence, but I will gather more information from his coworkers and from Scott by this Friday.

$\left[\begin{array}{c}\text{Result}\end{array}\right]$ I drove Scott immediately to General Hospital, where he was treated in the emergency room (X rays and a prescription for pain) for a sprained wrist and multiple bruises on his tailbone. The doctor released Scott and recommended bedrest for three days, a check-up next Monday, and limited use of his right hand for two weeks. The standard hospital insurance forms have been filled out and submitted.

$\left[\begin{array}{c}\text{Further}\\\text{action}\end{array}\right]$ Scott's absence will require only minor scheduling revisions, which are already in process. Scott will be contacting you concerning his disability insurance and workers' compensation benefits.

(b)

Poorly written negative memo

To: Pete Ginini

From: Lee Chin

Date: 1/31/00

Subject: Bugs

What happened on your latest build? You said you would have the bugs worked out, and we could not get it running at all. It's still full of bugs, mostly in Hank's component. We've wasted many hours tracking down the problems ourselves, but this is your code and your responsibility. We are not going to spend any more time on it until you can get the code working and verify it on your own test machines.

If we're going to continue to work together, you'll have to change your attitude toward testing and get your staff motivated to turn in a quality, bug-free code. Let me know what your team is doing to ensure that these types of problems don't happen again.

(a)

Revised negative memo

To: Pete Ginini, Development Team Lead

From: Lee Chin, System Test Lead

Date: 1/31/00

Subject: Completing your system test

We encountered a major problem with our latest build (104_0125) of the three remaining components for the beta software.   [State the problem.]

We spent two days tracking down a start-up problem that turned out to be caused by two errors in the code checked in by Hank. We were able to isolate the errors, and we have sent a description of them to Hank. He responded that he will fix them ASAP. I would appreciate your assistance in making this a priority item for him.   [Describe the problem.]

The other components have only minor problems (see attached description). I want to stick to our original schedule to pass the beta software out of system testing by the end of the week. With your help, we can still achieve this goal.

If you have not done it already, please remind the developers to verify that their code works on your own test machines before they send it to us. I realize that Hank is the newest developer on your team, and perhaps he is still unfamiliar with our testing process. Let's meet for lunch to discuss what each of us can do to keep on schedule. Wednesday and Thursday noon are open for me.   [Request action and followup.]

Lee

(b)

FIGURE 11–6

Negative memos: a) Poorly-written negative memo. b) Revised negative memo.

with experience, but remember that the people who are promoted are those who can resolve conflicts constructively and make everyone feel like a winner.

Figure 11–6 shows an example of a poorly written negative memo and a well-written (revised) negative memo. Notice that figure 11–6a contains harsh, blaming, and vague language that will certainly provoke further negative interaction and bad feelings between the two team leads. The writer does not invite any personal discussion of the problem, but instead, seems to retreat behind a parentlike admonishment. Figure 11–6b contains neutral language, states the problem clearly, stresses teamwork to achieve their mutual goals, and invites a face-to-face conversation in a neutral location.

Personal memos are written to convey private information. When writing a personal memo, beware of including confidential information—if read by the wrong person, the message could quickly find its way into the company grapevine. Many news stories illustrate the devastating consequences of "leaked" personal memos. If your message could be embarrassing to you or someone else, find another way to deliver it.

A memo might be a preprinted form, but it is an open-ended form. Some companies supply workers with preprinted memo forms or computer templates. Follow your company's policy.

The standard labels of a memo are "To" (the reader), "From" (the writer), "Date" (day, month, and year), and "Subject" (the specific topic), which are all followed by colons and the appropriate information.

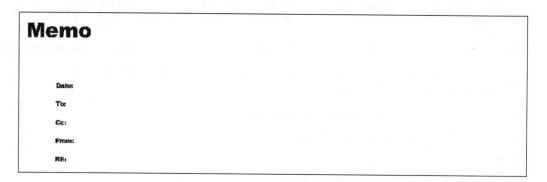

Some points to remember when writing memos are the following:

1. Single-space memos, and double-space between paragraphs.

2. Use standard spelling, punctuation, and grammar. Write complete sentences.

3. Include the names of those to whom copies of the memo will be sent. They are listed following the abbreviation "CC" (carbon copy) either in the "To" section or at the bottom of the memo. Although actual carbon is seldom used anymore, the abbreviation has become standard.

4. If you print your memo, write your initials beside your name. This serves as a signature. Or sign your name at the end of the memo.

5. Keep a copy of all the memos you send because you may have to follow up on them. Always assume that the receivers will keep your memos, too.

E-Mail

Most businesses routinely provide Internet access to employees. In addition, anyone who owns a computer with a modem can quickly set up an Internet account using an Internet provider company such as America Online, Mindspring, Yahoo, or Cybergate. Most accounts are fee-based, as in a flat monthly or annual fee plus a long-distance charge, if applicable, for minutes of use.

To receive and send e-mail, users must then set up an e-mail program. Some providers have their own e-mail programs and will instruct subscribers on how to create a password and e-mail address. Some advanced e-mail programs, such as Lotus Notes, include features such as corporatewide calendar programs and databases that allow collaborative computing—individuals in different sites working on a project together.

Because e-mail is so quick and easy to use, writers sometimes get careless and forget the basic rules of business communication. The rules of grammar and spelling are just as important in e-mails as in any other type of writing.

TIPS   Most e-mail programs provide a template, similar to a memo format, at the top of the message. Use these general guidelines for memos, where applicable, and the reminders in the article at the beginning of the chapter, as well as these guidelines specifically for e-mails:

1. The program inserts the date and the "From" address (your own) automatically. You cannot change them.

2. Enter a "To" address. You usually can add other addresses as *CCs* (copies) or *BCCs:* (blind copies which means that none of the other receivers will see this address as a recipient). If your program includes an address book, you can save time in the long run by entering the e-mail addresses of all your business associates right away. Then, depending on your program, when you start an e-mail, you simply select the person's name or the predefined group from an alphabetical listing, and the program automatically inserts the e-mail address.

3. Attach documents and files with the e-mail, if needed. Describe the attachments in the text of the e-mail, including the size of the file (especially for large files) and whether it requires a specific application, such as Microsoft Excel or Word 2000, to open. Each e-mail program has its own way of attaching (and opening) attachments. Some e-mail companies have a file-size limit for attachments, such as 1 megabyte. E-mails with attachments larger than the limit will either be returned to the sender as undelivered or will arrive without the attachment. When sending attachments, check with your recipients to see if they are able to open them.

4. Fill in the "Subject" line with a brief topic. Although e-mails will transmit without a subject, it helps the recipient gear up for that topic before opening the e-mail. And it helps the recipient find the message later.

5. State the purpose or scope of your e-mail in the first paragraph. Include whether a response is necessary, and if so, by when.

6. The language of your e-mail depends on your audience and purpose. If the recipient is a friend, the language can be somewhat casual, but keep the pitfalls of e-mail always in mind—even with friends. If the recipient is a customer, client, or upper management, be *very* polite. Use neutral language and state your points clearly. Keep paragraphs short so readers have some white space while reading—long sections of text can be difficult to read on a computer screen.

7. If you are requesting an action from the recipient, be specific about what you expect and how to reach a resolution. For example, if you are unsatisfied with a service, you might make the initial complaint in an e-mail to the company. In it, you would provide the supporting facts surrounding the service, what you expect the recipient to do (such as complete the service or refund the fee), and what you will do to resolve the situation (such as make an appointment to examine or discuss the faulty service).

8. Keep humor to a minimum—without body language or facial expressions, humor can be lost. What may seem to be an obvious joke to you might come across as strange, crass, or sarcastic to your recipient. In informal e-mails, you can add "emotions," such as a :-) to lighten the mood, but avoid them in formal e-mails.

9. Include your name, title, mailing address, phone and fax numbers, and e-mail address in your closing. This provides alternatives for return contact from the recipient. Some programs allow you to customize your e-mails with automatic headings and closing information.

10. Take the time to re-read your e-mail before sending it. Step away from the e-mail for a few moments before re-reading it. Check for complete sentences, spelling, and grammar—use the spelling check if available. Look for gaps in information, logic, and details. Revise as needed.

Remember that any criticism can seem excessively harsh in writing, without voice tone or facial expressions to soften it. In touchy situations, especially if you send disappointing or negative information, follow up with a personal phone call. It might give you an opportunity to straighten out any misunderstandings and salvage a working relationship before things escalate.

If you receive an e-mail that seems negative, resist the urge to shoot back a scathing response. Instead, ask the sender for clarification, just in case the sender didn't mean the e-mail to be negative at all. And if the criticism is intentional, do your best to reestablish a professional relationship with the sender—respond with neutral language, state your mutual goals, keep the emphasis on actions rather than personality, and attempt to discuss the issues face-to-face.

Above all, never assume that e-mail in the workplace is private or deleted. As the article at the beginning of this chapter reminds us, your e-mail might get forwarded to someone else without your knowing it. And even if you delete e-mail from your system, your recipient might not, or your company might have a permanent database of all e-mail as part of its routine back-up procedure. All of these could result in embarrassment, or even, in extreme cases, a civil lawsuit at a later time.

Figure 11–7 shows examples of a drafted and revised e-mail. The revision adds many details that make the message clear, complete, and polite.

The following article illustrates some humorous examples of careless writing.

---

claims work has light side

State Farm Year polled claims people in all regions for examples of humorous or unusual claims they have handled in the last few years. The response was tremendous. Here are some of the best ones:

"I was driving my truck under a bridge and it didn't fit."

"I was traveling down the road at approximately 35 m.p.h. As I rounded a curve I felt a sudden rush of air. When I looked over, my wife was gone!"

"I wanted somebody to come and talk to me. I was in a wreck more than two years ago and I have been under a doctor's car ever since."

"I let (name) try out my motorcycle. He was climbing a hill and didn't know hill went down other side and crashed."

"A drunk walked up to the side of my car as I was waiting at a stop light. He yelled 'I hate Ramblers,' and kicked in the right door."

From a claimant whose car was struck from behind: "I heard his skid marks coming and he collapsed into my rear whip, snapping my neck."

Policyholder explaining why he backed 330 feet across a parking lot and into a store: "Car sat in sun too long and absorbed too much solar energy. When I started car, I could not control it."

Telling how she was injured in a supermarket fall, policyholder wrote that she "slept on a string bean." Describing her injury, she said: "Right leg bruised, hard to walk on lower back."

Courtesy of State Farm; from STATE FARM YEAR, 1976.

Exercise 11.1 *Complete an accident report that will be sent to the State Board of Workers' Compensation. Assume that you had an injury at work. Fill in the details and information to make the report realistic. Use Figure 11–8.*

Exercise 11.2 *Complete an insurance form to file concerning a car accident that you may have had. Use Figure 11–9.*

Subject: Hi Kyle

Kyle,

I am attaching the agenda of sessions for the conference. Sessions without a confirmed presenter are marked TBD.

Please increase your efforts to confirm a presenter for these sessions. Please send any further information and status reports directly to me, instead of through Nick.

Keep me posted.

Andrea

(a)

Subject: Unconfirmed presenters for 0304 seminar

Kyle,

I am attaching a preliminary agenda of sessions for the March 3rd conference (attachment is an Excel spreadsheet, 2 pages). It includes all the confirmed presenters, as of this date. Each has submitted an abstract that can be published with the program.

Sessions without a confirmed presenter are marked TBD. They include the following:

        10:00   Evaluation techniques
        1:30   Trends in employment testing
        2:00   Using the Internet for recruiting

Please increase your efforts to confirm a presenter for these sessions so I can send the agenda to the printer by February 11. I realize that you might have made some confirmations late last week that I haven't received from Nick Stillers, your committee chair. From this point on, please send the information and status reports directly to me, instead of through Nick.

Generally, the other arrangements for the seminar (luncheon, reservations, vendor displays, and A/V equipment) are completed, and all the committees are confident that this year's seminar will go off without a hitch.

Please let me know your progress. If I can be of any assistance, just let me know.

Andrea

(b)

FIGURE 11–7

E-mail Examples: a) First draft of an e-mail. b) Revised draft of an e-mail.

A

STATE OF GEORGIA
EMPLOYER'S FIRST REPORT OF INJURY OR OCCUPATIONAL DISEASE
(See Instructions on Reverse Side)

Ga. Form WC 1 (Rev. 7-82)

OSHA File Number

Insurer File Number

Employer	Employer Phone No.	**DO NOT WRITE IN THIS COLUMN**
Address	Employee Soc. Sec. No.	←
City State Zip	Date of Injury	←

2. Employee/Claimant Name (Last) (First) (Middle)

Address

City   State   Zip

3. Insurer

→ **COMPLETE ORIGINAL AND ONE COPY AND** ←
SEND IMMEDIATELY TO

☐ LUMBERMENS MUT. CAS. CO.   ☐ AMERICAN MOTORISTS INS. CO.
☐ FEDERAL KEMPER INS. CO.   ☐ AMERICAN MFRS. MUT. INS. CO.

1401 PEACHTREE STREET, N.E.
ATLANTA, GA 30309

Carrier No.

Sic

4. Nature of Business (Mfg., Trade, Transportation, Etc.)   Specific Products

5. Employee's Home Telephone   Area Code   Number

Marital Status   Single ( )   Married ( )   Divorced ( )   Separated ( )   Widowed ( )

Date of Birth   Age   Sex   Male ( )   Female ( )

Age

Sex

6. Regular Occupation   Department in Which Regularly Employed

Occupation

7. Place of Accident or Exposure (Address or Location)   On Employer's Premises?   Yes ( )   No ( )

Location

8. County of Injury   Time of Injury   A.M. ( ) P.M. ( )   Time Workday Began on Day of Injury   A.M. ( ) P.M. ( )   First Date Employer Aware

Employer Aware

9. Describe the injury or occupational disease in detail and indicate the part of body affected. (e.g., amputation of right index finger at second joint; fracture of ribs; lead poisoning; dermatitis of left hand, etc.)

Nature

10. If Fatal: Give Date of Death   Number of Dependents

Body Part

11. What was employee doing when injured? (Be specific. If employee was using tools or equipment or handling materials, name them and tell what employee was doing with them.)

Type

12. How did accident or exposure occur? (Describe fully the events which resulted in injury or occupational disease. Tell what happened and how it happened. Name any objects or substances involved and tell how they were involved. Give full details on all factors which led or contributed to the accident or exposure.)

13. What thing directly injured the employee or made employee ill? (Name object struck against or struck by: vapor, poison, chemical or radiation, if strain or hernia, the thing being lifted, pulled, etc.: if injury resulted solely from bodily motion, the stretching, twisting, etc., which resulted in injury.)

Source

14. Name and Address of Treating Practitioner (Include Zip Code)   Name and Address of Hospital If Hospitalized (Include Zip Code)

15. Did employee work the next day following injury? Yes ( ) No ( )   First Date Employee Failed to Work A Full Day

16. Time Discontinued Work: A.M. ( ) P.M. ( )   If Returned to Work, Give Date   Returned at What Wage? $ Per Week

17. Length of Time In Your Employ   years   months   Did employee receive full pay for date of injury? Yes ( ) No ( )   Wage Rate at Time of Injury or Disease $ Per hour ( )   Week ( )   Day ( )   Month ( )

18. Hours Worked   Per Day ( ) Week ( )   Number of Days Worked Per Week   List Normally Scheduled Off Days

19. If employee is paid on commission or piece work basis, enter average weekly amount. $ If board, lodging or other advantages were furnished, enter average weekly amount. $

20. Report Prepared by (Print or Type Name)   Position   Date of Report

EMPLOYER'S FAILURE TO SUBMIT THIS REPORT TO INSURER IMMEDIATELY MAY RESULT IN PENALTY.

FIGURE 11–8
Accident report for worker's compensation.

APPLICABLE ONLY IN CALIFORNIA — FOR YOUR PROTECTION CALIFORNIA LAW REQUIRES THE FOLLOWING TO APPEAR ON THIS FORM: IT IS UNLAWFUL TO (A) KNOWINGLY PRESENT OR CAUSE TO BE PRESENTED ANY FALSE OR FRAUDULENT CLAIM FOR THE PAYMENT OF A LOSS UNDER A CONTRACT OF INSURANCE, (B) KNOWINGLY FILE MULTIPLE CLAIMS FOR THE SAME LOSS OR INJURY WITH MORE THAN ONE INSURER WITH AN INTENT TO DEFRAUD THE INSURER, (C) KNOWINGLY PREPARE, MAKE, OR SUBSCRIBE ANY WRITING, WITH INTENT TO PRESENT OR USE THE SAME, OR TO ALLOW IT TO BE PRESENTED OR USED IN SUPPORT OF ANY SUCH CLAIM, (D) EVERY PERSON WHO VIOLATES ANY PROVISION OF THIS SECTION IS PUNISHABLE BY IMPRISONMENT IN THE STATE PRISON, FOR TWO, THREE, OR FOUR YEARS, OR BY FINE NOT EXCEEDING TEN THOUSAND DOLLARS ($10,000) OR BY BOTH.

STATE FARM INSURANCE

AUTOMOBILE CLAIM REPORT

REPORTING AGENT | AGT. CODE | MGR. CODE | PHONE | CLAIM NUMBER

POLICY NO. | ST./PROV. CODE | CHG. CODE | CAR NO | INSURED'S/BUSINESS NAME | LAST | FIRST | MIDDLE

ADDRESS | STREET | APT./SUITE | CITY | STATE/PROV. | ZIP/POSTAL CODE

DATE OF BIRTH | HOME PHONE | WORK PHONE | EXT

DATE OF ACCIDENT/LOSS | TIME | AM | PM | LOCATION OF ACCIDENT/LOSS | STREET | CITY | STATE/PROV. | ZIP/POSTAL CODE

VEH. 1

INSURED'S VEHICLE-YEAR-MAKE-MODEL-BODY STYLE | VEH. IDENTIFICATION NO. | VEH. LICENSE NO.

PRIOR DAMAGE? IF YES, DESCRIBE | YES | NO | IS THIS THE VEHICLE INVOLVED? IF NO, EXPLAIN | YES | NO | DRIVER: SAME | PARKED | UNOCCUPIED | UNKNOWN

DRIVER'S NAME | LAST | FIRST | MI | ADDRESS | STREET | APT./SUITE | CITY | STATE/PROV. | ZIP/POSTAL CODE

EMPLOYER | AGE | HOME PHONE: | WORK PHONE: | EXT. | RELATIONSHIP OF DRIVER TO INSURED? | WHO GAVE DRIVER PERMISSION?

WAS DRIVER ON MISSION FOR OWNER? YES NO | PURPOSE OF TRIP? | PRINCIPAL DAMAGE

IS VEHICLE DRIVEABLE? YES NO | DRIVE-IN SERVICE? IF YES, OFFICE YES NO | LOCATION OF INSURED VEHICLE, IF NOT DRIVEABLE

VEH. 2

OTHER VEHICLE-YEAR-MAKE-MODEL-BODY STYLE | VEHICLE LICENSE NUMBER | DRIVER: SAME | PARKED | UNOCCUPIED | UNKNOWN

OWNER | LAST | FIRST | MI | ADDRESS | STREET | CITY | ST./PROV. | ZIP/POSTAL CODE | HOME PHONE | WORK PHONE

DRIVER: | LAST | FIRST | ADDRESS | STREET | CITY | ST./PROV. | ZIP/POSTAL CODE | AGE | HOME PHONE | WORK PHONE

PRINCIPAL DAMAGE | IS VEHICLE DRIVEABLE? YES NO | DRIVE-IN SERVICE? IF YES, OFFICE YES NO | WHERE IS VEHICLE

INSURANCE COMPANY | POLICY NUMBER | ADDRESS | STREET | CITY | STATE/PROV. | ZIP/POSTAL CODE

FACTS OF ACCIDENT OR THEFT?

SPEED LIMIT? | SIGNALS GIVEN VEH 1 YES NO VEH 2 YES NO | HEADLIGHTS ON? VEH 1 YES NO VEH 2 YES NO | ROAD CONDITIONS? | TRAFFIC CONTROL? IF YES, STATE TYPE YES NO

POLICE THERE? YES NO | WHERE REPORTED? | REPORT NO.? | DATE REPORTED? | TIME? AM PM | SCENE INVESTIGATION? YES NO

WHO RECEIVED TRAFFIC CITATION? | TYPE OF VIOLATION? | DATE/TIME THEFT REPORTED? AM PM | RECOVERED? YES NO DATE | WHERE RECOVERED?

WHO RECOVERED? | ITEMS FOR DEPRECIATION (SHOW AGE, CONDITION, COST AND VALUE)

INJURIES

VEH. NO. | ENTER NAME, AGE, PHONE, ADDRESS, INJURIES | DRIVER/PASS/PED

VEH. NO. | | DRIVER/PASS/PED

VEH. NO. | | DRIVER/PASS/PED

WITNESSES AND OTHER PARTIES TO THE LOSS? IF YES, ENTER NAME, RELATIONSHIP TO LOSS, ADDRESS, PHONE YES NO

I HEREBY DECLARE THAT THE FACTS IN THIS REPORT ARE TRUE AND ACCURATE. | I HEREBY DECLARE THAT THE FACTS IN THIS REPORT ARE TRUE AND ACCURATE.

INSURED SIGN HERE X | DATE | DRIVER SIGN HERE X | DATE

COVERAGES IN FORCE? | AGT. CONFIRMED? YES NO | DATE REPORTED TO AGENT? | OTHER INSURANCE, UMBRELLA, ETC? EXPLAIN: YES NO

F.R. REPORT FILED? YES NO | HANDLER? | UNIT NO.? | HANDLER? | UNIT NO.? | DATE AND PERSON TAKING LOSS?

(160) G 4684m.16 Printed in U.S.A. Rev. 6-86 USE REVERSE SIDE IF NECESSARY

FIGURE 11–9

```
Memo

Date:

To:

Cc:

From:

RE:
```

FIGURE 11–10

Exercise 11.3   *Write a status memo to your supervisor, Ms. Franklin, about the accident from either Exercise 11.1 or 11.2. Include what happened, the results of the accident, and what further action is required of you or her. Use the format shown in Figure 11–10.*

Exercise 11.4   *Keep a vehicle report for one week of your travels to and from work and school as though you will be reimbursed for your travel time. Submit a vehicle report for the final record. Use Figure 11–11.*

Exercise 11.5   *Write a personal memo or e-mail to someone in your class to congratulate that person for an achievement. Copy the memo or e-mail to the person's supervisor (your instructor).*

Exercise 11.6   *Write a negative memo or e-mail to a department manager or dean at your school to identify and resolve a conflict, such as a disagreement over a fee, grade, or policy. Include a factual account of the problem, what you expect the other person to do, and what you will do to resolve the issue.*

SPELLING/VOCABULARY: Roots *sede, cede* and *ceed*

In this chapter we combine the spelling and vocabulary exercises because of the unusual nature of this topic.

The root word we pronounce as "seed" is spelled in three different ways: *ceed, cede,* and *sede.* Each spelling follows certain prefixes, many of which you have already studied.

The reason for the three spellings is that they are derived from two Latin roots with slightly different meanings, and over the centuries, as the words crossed languages, certain spellings for certain words became standard. Knowing this does not make remembering the correct spellings any easier; however, we spell these roots in only a dozen verbs. Related words also use forms of the roots, and knowing the roots gives us clues to the meaning of these words.

BUSINESS AND TRAVEL EXPENSE REPORT   ACCOUNTING COPY

NAME (LAST, FIRST, INITIAL)						DEPT. NO.	ACCT. NO.		WEEK ENDING	

STREET			CITY	STATE	ZIP CODE	REP # 1-5	REGION 9 12

DAY OF WEEK	MONDAY	TUESDAY	WEDNESDAY	THURSDAY	FRIDAY	SATURDAY	SUNDAY	
DAY OF MONTH (WRITE IN)								
TRAVEL FROM (CITY)								
TRAVEL TO (CITY)								
PURPOSE OF TRAVEL								TOTALS FOR WEEK
BUSINESS MILEAGE								14

MILEAGE ALLOWANCE								
GAS/OIL								01
PARKING & TOLLS								02
OTHER TRANSPORTATION* (PAID BY EMPLOYEE)								10
MEALS & ENTERTAINMENT*								04
LODGING								03
POSTAL, TELEPHONE								08
MISCELLANEOUS*								05
AIR FARE								09
TOTAL (PAID BY EMPLOYEE)								

LESS TRAVEL ADVANCES RECEIVED	DATE	AMOUNT	TOTAL TRAVEL ADVANCES ( )

TOTAL AMOUNT TO BE PAID TO EMPLOYEE (REFUNDED TO DEVRY)

AIRFARE BILLED TO CO. (ATTACH STUB & INVOICE)								09
APPROVED DIRECT BILLINGS TO CO. (ATTACH RECEIPTS)								

MISCELLANEOUS EXPENSES PAID BY YOU* (ATTACH RECEIPTS)

DATE	EXPLANATION	AMOUNT

OTHER TRANSPORTATION* (ATTACH RECEIPTS)

DATE	EXPLANATION	AMOUNT

DESCRIPTION OF MEALS & ENTERTAINMENT EXPENSE* (ATTACH RECEIPTS)

DATE	NAMES & TITLES OF PERSONS	RESTAURANT/ENTERTAINMENT SITE & CITY	BUSINESS DISCUSSED (IF NOT ALONE)	TYPE OF EXPENSE	AMOUNT

EMPLOYEE SIGNATURE		DATE	A/P CHECK	ODOMETER READINGS THIS WEEK	
			12	ENDING NO.	
APPROVED BY	DATE	AUDITED BY	ADJUSTMENTS	BEGINNING NO.	
			11	TOTAL MILES DRIVEN	

909235 REV BTER 5510

FIGURE 11–11

Business and travel expense report.

Sede is derived from the Latin word *sedere,* meaning "to sit." We use this spelling with only one prefix.

Supersede = super (over) *sede* (to sit), to replace or to cause another to become obsolete
The transistor *supersedes* the vacuum tube in solid-state electronics.

Related words: sedate (to sit calmly)
sedentary (moving little, as in a sedentary lifestyle)
presides (sits before others as the head of a meeting)
session (a sitting or assembly of many people)

Cede is derived from the Latin word *cedere,* meaning "to yield," "to go," or "to leave."

Cede = to surrender formally
Spain *ceded* Florida to the United States in 1819.

Secede = se (apart) *cede* (to go), to formally withdraw or separate
Florida *seceded* from the Union in 1861.
Related word: secession (a formal separation)

Intercede = inter (between) *cede* (to go), to mediate or make a request on behalf of
another.
An attorney can *intercede* in legal disputes.
Related words: intercession (the mediation or pleading for another)
intercessor (a person who intercedes)

Antecede = ante (before) *cede* (to go), to go before in rank, time, or place
The Univac computer *anteceded* microcomputers.
Related words: antecedent (going before in time, logic, or order: prior)
ancestor (one who lived earlier in a family line)

Precede = pre (before) *cede* (to go), to go before in rank, time, place, or importance
The development of solid-state electronics *preceded* the use of integrated circuits.
Related word: precedent (a fact or procedure established before)

Concede = con (with) *cede* (to go), to admit as true or acknowledge
After the votes were counted, the loser *conceded* the election to his opponent.
Related word: concession (a privilege, right, or lease)

Recede = re (back) *cede* (to go), to go or move back
After the floodwaters *receded,* people returned to their homes.
Related words: recess (a temporary halt or withdrawal)
recession (a departing processional or inactivity in the economy)
recessive (a nondominant gene)

Accede = ac (to) *cede* (to yield), to consent or enter into duties—rarely used except in sophisticated, formal language.
Both nations *acceded* to the treaty.
Related words: access (approach or come near)
accessory (helping in a subordinate way)
accessible (can be approached easily)

Ceed is also derived from the Latin word *cedere,* meaning "to go." This spelling is used with only three prefixes.

Exceed = ex (beyond) *ceed* (to go), to surpass or outdo
The driver was fined for *exceeding* the speed limit.
Related words: exceeding (extraordinary)
exceedingly (extremely)
excess (an amount more than is needed)
excessive (being too much or too great)

Proceed = pro (forward) *ceed* (to go), to advance, or to go on after stopping

After a brief recess, the lawyers were asked to *proceed* with the trial.

Related words: process (a method of development)

procession (moving forward, a parade)

procedure (a step in a process)

procedural (having to do with a step)

Succeed = suc (under) *ceed* (to go), follow or come after; to happen or turn out as planned; have a favorable outcome.

The election will determine who will *succeed* the outgoing president.

Related words: succession (series or sequence)

success (favorable outcome or result)

successful (having achieved success)

Spelling Review

sede	cede	ceed
supersede	accede	exceed
	antecede	proceed
	cede	succeed
	concede	
	intercede	
	precede	
	recede	
	secede	

Remembering the correct spelling for these words will take some memorization and frequent use. Use a dictionary or spelling check when you are in doubt. Since there is only one word using *sede*, you should not have too much trouble remembering it.

A memory aid for remembering the three words using *ceed* is to think of the word speed. Speed ends with the *eed* ending, and the first letters of *succeed, proceed,* and *exceed* form the first three letters of *speed*. All the other words use the *cede* spelling.

Exercise 11.7   *Fill in the missing letters to form words derived from* sede, cede, *or* ceed *roots.*

1. The pro_____dures of the meeting were determined by the presider.
2. The issue of retroactive pay raises super_____ded all other issues.
3. The presider allowed arguments concerning the high costs of living to pre_____de other items on the agenda.
4. The economy was not expected to re_____de in the suc_____ing months, according to the market forecaster.
5. Inflation was making meeting day-to-day expenses ex_____dingly difficult.
6. The accountant was asked to inter_____de for the employees at the budget negotiations meetings.
7. The accountant was given ac_____ss to all company financial records.

8. The CEO con_____ded that harmony within the company was the primary goal of management.

9. As the negotiations pro_____ded, the workers became more optimistic that they would suc_____.

10. The day the new policy went into effect, a valuable pre_____nt was established that ex_____ded the employees' expectations.

WORD WATCH: *past* and *passed*

The words *past* and *passed* are pronounced the same. The simplest uses of these words are easily categorized.

Passed is a verb, the past tense of *pass*. Use *passed* if the word is a verb or part of the verb in a sentence.

> He *passed* the test.
> She has *passed* back the papers.

Past can be used as a noun, adjective, adverb, or preposition. Use *past* for any use other than a verb.

> His *past* employers speak highly of him. (*adjective*)
> In the *past,* he has worked hard. (*noun*)
> I wouldn't put it *past* him to work overtime. (*adverb*)
> He has gotten *past* the first interview. (*preposition*)

Exercise 11.8   *Write the correct form of* passed/past *for each sentence.*

1. Customers who have had poor service in the _____ are likely to tell their friends about it.

2. Satisfied _____ customers are the best advertisement for new customers.

3. Bad news gets _____ on faster than good news.

4. Customers who have had good, friendly service in the _____ usually remain customers.

5. It is twice as expensive to recruit new customers as it is to keep _____ customers.

6. Companies with cold representatives and uncaring technical staff will often be _____ over for a friendlier company.

7. The business that _____ on helpful information to the customer in the _____ was remembered.

8. The field technician is usually the representative closest to the customer once the customer is _____ the sales calls.

9. Field technicians who are friendly, competent, and efficient are valued by their company for retaining _____ customers.

10. The technician who couldn't relate to his customers, even though he was technically competent, was _____ over for promotion.

Business Letters

- Correct ineffective wording.
- Write a correct, formal business letter and envelope.
- Write a formal letter requesting information.
- Write a letter to an elected official.
- Write a letter of goodwill: sympathy, congratulations, or thanks.
- Write a negative letter refusing information.
- Spell *ible* and *able* endings correctly.
- Use *grad* and *gress* roots correctly.
- Use *stationary, stationery* and *compliment, complement* correctly.

READING: Turning Confrontation into Communication

---

by Arch Lustberg

Disputes are inevitable. The first step toward winning: Control yourself.

In today's competitive world, confrontation is inevitable. The possibility of a face-off exists in every facet of your life—professional as well as personal.

It exists at public meetings and in private arguments, at the office and at home, with your senior partner and with your son, in practically every situation involving more than one person. You cannot pick the time or the place for a confrontation, but you can control your response and even the terms of the debate—and thus convey your ideas effectively.

First, concentrate on your attitude. When someone hits you, you want to strike back. When someone shouts at you, you want to shout back. Though a natural tendency, it is not a winning one. In fact, it is

a waste of energy. Do not shout denials. Calmly explain your stand and be positive.

Next, concentrate on breathing and relaxation. Proper breathing creates an almost instant feeling of well-being. You will find you can immediately banish tension and stress. The relaxed, self-assured person can think—and the thinking person can take control.

The key to proper breathing involves moving the diaphragm (the muscle just below the rib cage) to make room for your lungs to fill with air. Difficulty arises when we incorrectly move the diaphragm *in* on inhalation and *out* on exhalation. We fall into that bad habit when we are nervous, thus reducing our oxygen supply just when we need it most. The trained singer

---

practices breathing exercises as assiduously as scales. You must do the same.

Next, know exactly what you want to do. If you are certain that a meeting has potential for confrontation, take time to plan your strategy. Determine what you want to accomplish and how you can best make your point. Never lose sight of your objective.

Do not spoil your planning by getting irritated and snapping at the person you are talking to. You will undoubtedly say something you will regret. And if that happens when you are talking with the media, your unfortunate comment may be used to conclude a broadcast interview— or it may be the only thing you say that is broadcast.

Expert communicators direct their thoughts to those people who do not have an opinion on the subject at issue. (This "audience" may even be imaginary, as when you and your adversary are the only people present at a meeting.) An upbeat statement directed toward them will serve your purposes much better than a defensive quip aimed at your adversary.

Speak simply, clearly, concisely. We have become a people who think that we have to use big words or jargon to appear "in the know." I recommend that you replicate, interface and offload only in the privacy of your own home. Speak English in public.

All of us have four weapons that will help us communicate: the mind, the face, the body and the voice. Though we generally use them correctly in animated conversation, we almost never use them correctly when we are tense, afraid and intimidated.

How can you use your mind creatively when face to face with an adversary? The best tool is the pause. It gives you time to think of a positive response, time to eliminate negative comments.

Remember, however, that the pause will work only if it looks comfortable. The key is to remain silent and maintain eye contact with the person you are talking to.

This means that you should avoid such audible pauses as "uh . . . uh . . . ," "like," "I mean" and "you know" and that you should not give the impression that you are afraid to look your adversary in the eye. Eye contact does not necessarily mean eye to eye. If you find that uncomfortable, try focusing on a certain part of the person's face. Most people, in fact, look at the lips.

Television provides a wonderful opportunity to study the pause. The next time

FIGURE 12.1

In a confrontational situation, never lose sight of your objective.

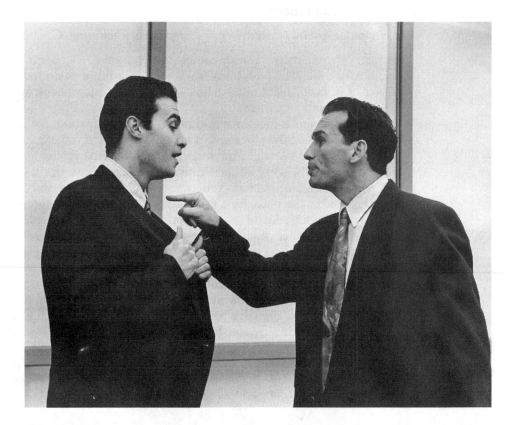

you are watching an interview, notice how the pause—when used correctly—makes a person appear more confident.

What about effective use of the face? The smiling, animated face is one of your strongest communications tools. And it is one you will probably use most often after you have mastered certain basics.

All of us usually "open" our faces when we are talking with someone. The problem is that in confrontational situations we often close the facial muscles, creating a frown line that we think makes us look "professional."

Try this exercise in front of the mirror to get your facial muscles to work for you: First, tighten the brow as much as you can. Hold that position and count aloud from 1 to 5, then relax. Repeat the exercise, counting a bit louder. Now open your brow. Make the lines in your forehead that come when your eyes are wide open and your eyebrows are arched. Hold the position and count aloud to 5.

Repeat the exercise. This time count as quietly as possible. You are not only opening your face, you are also speaking in a warmer, friendlier, more communicative way.

Get your body into the act of communicating, too. Probably the most animated forms of communication are gossiping and telling secrets. Watch two people deep in conversation. They are not just talking, they are painting pictures with their faces. Their entire bodies are alive.

You can do the same. Talk with your body as well as with your face.

The hands are the most used—and abused—part of the body when it comes to communication. You need to make them work in concert with the face—and make them less conspicuous when that is appropriate.

All of us tend to hide our hands, just as we do our faces. You probably clutch one in the other, hold them behind your back, put them in the folds of your arms or put one in a pocket and grab your wrist or forearm with the other. You thus end up calling attention to your hands—not communicating.

Try standing with your feet about as far apart as the width of your shoulders. Shake your shoulders a few times. Then let your hands fall naturally. Now you are in position to use your hands in animated conversation. When you begin talking

---

FEAR OF SPEAKING

For thousands of Americans, fear of public speaking has serious consequences because their jobs require them to give testimony before local, state or federal legislators.

Their fear—combined with a lack of the proper communications skills—often keeps them from making the impact they want.

Among the types guaranteed to turn off an audience:

- The droner who delivers a boring text without vocal inflection, facial expression or eye contact.
- The ill-prepared scaredy-cat who stumbles through his text with an assortment of nervous tics.
- The long-winded bore who thinks more is better.
- The greeting reader who keeps his eyes glued to the page, even reading "good morning."
- The statistician who overwhelms his audience with numbers that no one can relate to.

You can avoid falling into those deadly categories by mastering some basic principles:

- Strive for an easy, open facial expression.
- Use body language to emphasize words.
- Pause to allow time to breathe and throw off stress.
- Control your voice, varying your inflection and emphasis.
- Maintain eye contact with your audience.

You can also make your testimony more effective by keeping it simple and short.

Remember, you want to communicate ideas, not just read words. Testimony that is well delivered will win the day.

Adapted from Testifying With Impact, *by Arch Lustberg. © 1983 by the Association Department, U.S. Chamber of Commerce.*

with someone, keep your hands and fingers still. Make gestures when you feel they will complement the points you are making with your face and voice.

Your voice tends to follow the personality created by your face and body. When one is warm and friendly, so is the other. When one is cold and hostile, the other follows suit.

You can add interest and drama to your voice by learning when and how to use pitch, rate and volume. Pitch is the position of the sound on the musical scale. Rate is the duration of sound. Volume is the decibel level.

Volume is the least effective vocal tool. It is useful only when your purpose is to discipline a child or a pet. Loud sounds are irritating.

To communicate, you must express yourself. The best expression comes with uninhibited and unselfconscious use of pitch and rate. For example:

- "I had a wonderful time." (Unless you do something wonderful with *wonderful,* your host will think you are lying.)
- "It was a magnificent day." (Make *magnificent* truly magnificent.)
- "That garbage gave off the most foul smell." (It couldn't have been worse!)
- "Give me that knife." (If you don't, you'll be sorry.)

That is vocal flexibility: the willingness and ability to use these vocal tools.

So there you have the steps you can take to become an effective communicator—to win at confrontational situations. You can adapt them to any situation. If you are prepared, you will have an excellent opportunity to win.

You have something valuable to say. Learn to say it well.

Reading Comprehension Questions

1. How will concentrating on your attitude help in a confrontational situation? _____

2. What is the key to proper breathing? _____

3. How can you use your mind as a "weapon" in a confrontational situation? _____

4. How can you use your face and body as weapons in a confrontational situation? __

5. How can you use your voice to gain control of a confrontational situation? _____

6. Describe a personal experience in which you handled a confrontational situation maturely and effectively. _____

WRITING: Business Letters

The reading article suggests methods to reduce your anxiety during confrontations or stressful situations. In the business world and in everyday life, we encounter situations that require a cool head and a steady voice. Practicing strategies to reduce anxiety gives us the tools we need to deal with people effectively in all situations.

Many people are just as nervous about writing a formal letter as they are about giving a formal presentation, and with good reason. Poorly worded letters can have serious consequences because it takes time to detect, pinpoint, and resolve any miscommunication.

This chapter focuses on the different types of formal business letters: letters of inquiry, follow-up letters, and positive and negative letters. As you practice writing different types of letters, keep in mind that you already have many tools to get started. Your knowledge of spelling, mechanics, and grammar can help you prevent or revise distracting errors. Your technical vocabulary can add the precision needed for clear messages. And the skills you practiced in other types of writing transfer into letter-writing, as well.

The final skill is the ability to choose words that convey an appropriate tone for the purpose of the letter. As a general rule, written correspondence requires polite language, a neutral tone, clear requests, and careful editing to eliminate any chance of misunderstanding.

Personal vs. Business Letters

Letters usually fall into two groups: personal and business. You can write personal letters in an informal, breezy style that reflects the way you speak—with slang, familiar expressions, and a variety of loosely organized topics. Many people now use e-mail as a substitute for personal mail, and the styles are similar.

When you write business letters, however, you must use formal language, standard English—the kind of English used in schools and business—and a conventional style. Like all technical writing, business letters should be clear, concise, and well edited—you might want someone else to review your first draft and offer suggestions.

Business letters have a clear purpose, usually stated in the first paragraph. They might cover more than one topic, and if so, the ideas and paragraphs are organized and presented in a logical, deliberate order.

Many inexperienced writers are reluctant to write business letters because, despite their original purpose, the words used in a letter not only send a message, but they also convey the attitude and image of the writer. Words can portray the writer as mature or childish, sincere or phony, direct or evasive. Also, letters are often the first and last contact with customers and business associates, which means they provide the first and last impressions of the writer.

The format helps convey a businesslike tone. Today's letters have either a blocked or modified-block style, with single spacing within paragraphs and double spaces (one empty line) between paragraphs and letter parts. Examples in this chapter illustrate these styles.

Envelopes, too, have a conventional style recommended by the U.S. Postal Service to facilitate quick and accurate delivery.

Letter Formats

The two standard formats for business letters are the blocked (Examples 1 and 3), and the modified-block with indented paragraphs (Examples 2 and 4).

The **blocked style** is the simplest: Start every part of the letter and every paragraph in the body of the letter at the left margin, and skip one line between parts and paragraphs.

In the **modified block with indented paragraphs,** indent your address (unless using letterhead stationery) and your closing and signature to the same number of spaces, usually near the center. You can also indent the first line of each paragraph a consistent number of spaces (5–10) or not indent any paragraphs. Choose your preference and be consistent.

TIP Use these following general formatting guidelines:

- Use the computer (or typewriter) to type all business letters. Do not send handwritten business correspondence.

- Single-space the text of a letter, but skip one line between the parts and paragraphs of a letter.

- Use the white space (margins) to create a "frame" around the letter. Adjust the margins so they are similar.

- If the letter is more than one page, number the pages, starting with page 2, in a consistent place, such as the upper-right or lower-right corner.

- Indicate at the end of the letter if you are sending enclosures and if you have copied the letter to someone else.

- Type envelopes using the style recommended by the U.S. Postal Service (USPS) (see the section "Envelopes" later in this chapter). Make sure the envelope is wide enough for the paper.

Elements of a Business Letter

TIP   Business letters have the following parts, separated by a double space:

1. **Heading.** Type the sender's complete mailing address, sometimes included in the letterhead. (You can add phone and fax numbers and e-mail addresses here or within the body of the letter.) Include the following:

   - Street address, apartment number, or office number
   - City (followed by a comma), the two-letter state abbreviation (see Appendix 1 for a list of state abbreviations), and zip code (no comma between state and zip code)

   Short Circuit Engineering
   2110 Hotwire Drive
   Minneapolis, MN 55455

2. **Date.** Type the date of the letter directly below the heading address. Use the American style (August 15, 2000) or the international style (15 August 2000).

3. **Receiver's name and title (if known) and address.** Skip one line after the date. Align the receiver's information with the left margin. Include the following:

   - Full name
   - Title, department (if known)
   - Company name
   - Street address, apartment number, or office number (see Appendix 1 for a USPS list of address abbreviations)
   - City, two-letter state abbreviation (see the USPS list of state abbreviations in Appendix 1), and zip code

   Forrest Henry
   Century Products, Incorporated
   1800 Overlook Lane
   Bismarck, ND 58501

   Take care to spell the person's name and company name correctly. If the person's title is one or two words, type the title following the name:

   Forrest Henry, Manager

   If the title is more than two words, type the title below the name:

   Forrest Henry
   Application and Design Engineering Manager

Use the phone book or call the company's receptionist to confirm spellings and titles.

4. **Subject** (optional). Skip one line after the address. If your letter has a direct purpose, add a brief description to focus the reader on the topic.

   Subject: Schedule a sales call

5. **Greeting.** Skip one line after the address. Align the greeting with the left margin. The most common greeting is "Dear" followed by the formal name and a colon. If the person has a distinctive title, address the person by the title. If you do not know the name of the person, address the letter to a position or title. If you are already on a first-name basis with the recipient, use the person's first name, and then follow it with a comma.

   Dear Mr. Henry:
   Dear Dr. Henry:
   Dear Human Resources Manager:
   Dear Forrest,

6. **Body of the letter.** Skip one line after the greeting. The body of the letter has three main divisions. Skip lines between each paragraph.

   - First paragraph: Introduction and statement of the purpose. State the purpose clearly.
   - Middle paragraphs: Details and other information surrounding the purpose.
   - Final paragraph: Polite closing, expression of appreciation for assistance (even if in advance), and/or statement of follow-up intentions.

7. **Closing.** Skip one line after the last paragraph. Either align it with the left margin (blocked style), or align it with your address at the top of the letter (modified block style). Use a neutral closing, such as "Sincerely" or "Cordially," followed by a comma.

8. **Handwritten signature of the sender.** Leave four blank lines for your signature. Use a pen with black or blue ink to sign the letter. In most cases, sign your first and last names. If you are already on a first-name basis with the recipient, sign just your first name.

9. **Typed name and title of the sender.** Type your name (first and last), followed by your title, if appropriate, below the space for your signature, aligned with the closing.

   Eric V. Christianson
   President

10. **Enclosure or copy information** (if needed). Skip one line after your typed name and title. If you plan to add materials in the envelope, such as a resume, type "Enclosure:" followed by a space, and name the enclosed document(s). If you plan to send a copy of the letter to other people, type "cc:" followed by a space, and type the names and titles. Align both with the left margin.

    Enclosure: Sample specification
    cc: Alice Rostovich, Purchasing, Short Circuit Associates
           Ken Byrnes, Design Engineer

The following examples show the common elements in business letters.

[Heading]	**Short Circuit Engineering** **2110 Hotwire Drive** **Minneapolis, MN 55455**
[Date]	June 12, 2000
[Address]	Forrest Henry Century Products, Incorporated 1800 Overlook Lane Bismarck, ND 58501
[Greeting]	Subject: Schedule a sales call Dear Mr. Henry:
[Introduction & Purpose]	Recently, a mutual friend, Cal Johnson, mentioned your company and that you manufacture customized connectors using innovative designs. I would like to meet with one of your representatives to see examples of your products and discuss whether your company can help me meet the needs of my customers.
[Details]	In particular, I am looking for a connector for ink-handling applications. Currently we are purchasing connectors from Ace Engineering, but I have recently received several complaints from customers about ink leaks.
[Follow-up]	Please have a representative call me at (651) 555-4364 early next week to schedule a meeting here with my installation managers. We would like to see product samples at that time. I am available any weekday, and I would prefer to receive your call after 1 P.M. Thank you for your assistance.
[Closing]	Cordially,
[Signature] [Typed name & title]	*Eric V. Christianson* Eric V. Christianson President
[Enclosure]	Enclosure: Schematics for various ink-handling products

EXAMPLE 1:
Letter of Inquiry, blocked style

Century Products, Incorporated
1800 Overlook Lane
Bismarck, ND 58501

June 15, 2000

Eric Christianson, President
Short Circuit Engineering
2110 Hotwire Drive
Minneapolis, MN 55455

Dear Mr. Christianson:

It was a pleasure talking with you today. I look forward to meeting you and your managers on Monday, June 19, at 2 P.M. I will bring several samples of our products to examine and discuss.

In the meantime, I have enclosed the standard product specification and examples of the shrouded 1/16" hosebarb we currently manufacture, as we discussed.

Based on our preliminary discussion about your application, we would use a different valve than the one in these parts. This will allow us to maintain the overall package size you require, and it will also significantly reduce the inclusion (amount of ink that forms or drips after each connection) over the current design.

I have also talked to several of our design engineers. We have several product designs we have made for other ink-handling applications, some with little or no inclusion. Regardless of the results of this project, we would like the opportunity to review other connector needs you may have, whether current or conceptual.

We will provide pricing information for your purchasing agent at our meeting on Monday. Of course, if you have any questions about these or other parts that we manufacture, certainly give us a call.

Sincerely,

Forrest J. Henry

Forrest J. Henry
Application and Design Engineering Manager

Enclosure: Spec and samples

EXAMPLE 2:
Follow-up letter, modified block with standard format

Century Products, Incorporated
1800 Overlook Lane
Bismarck, ND 58501

June 20, 2000

Eric Christianson, President
Short Circuit Engineering
2110 Hotwire Drive
Minneapolis, MN 55455

Subject: Latex and our standard o-ring material
Reference: NASP

Dear Mr. Christianson:

This letter is in response to your voice message concerning our standard Buna-N o-ring and latex content. According to our supplier, the Buna-N o-ring used in our PLC220-04 does *not* contain latex.

I hope this confirms that the product is acceptable for your use. If you have any other questions or need further clarification, please feel free to give me a call at (701) 555-3131. Thank you for your continued interest in our products.

Sincerely,

Ken Byrnes

Ken Byrnes
Design Engineer

cc: Forrest Henry

EXAMPLE 3:
Abbreviated inquiry response, modified memo style

Century Products, Incorporated
1800 Overlook Lane
Bismarck, ND 58501

June 20, 2000

Eric Christianson, President
Short Circuit Engineering
2110 Hotwire Drive
Minneapolis, MN 55455

Dear Mr. Christianson:

Your purchasing agent, Alice Rostovich, asked me to provide some information on the differences between plastic parts that have been dyed and plastic parts with the colorant added to the material prior to molding.

It is common for dyed plastic parts to leach the colorant into liquid media it comes in contact with. While the amount that leaches out will decrease over time, it will never completely go away. We can eliminate this problem by adding the colorant directly to the base resin prior to molding. Because the colorant is integral to the resin itself, it will perform in the same manner as the natural base resin. Another advantage is a more consistent color without the mess associated with dying parts.

With only a few exceptions, the molded colors we use are all made from materials in compliance with requirements outlined by the FDA. We can specify a new colorant or select from colorants we have used in the past, depending on your lead-time, cost and color requirements.

I hope this information helps you in your product specification decisions. Thank you for you continued interest in Century Engineering products.

Sincerely,

Forrest N. Henry

Forrest N. Henry
Application and Design Engineering Manager

cc: Alice Rostovich, Purchasing, Short Circuit Associates
     Ken Byrnes, Design Engineer

Enclosure: Product specification

EXAMPLE 4:
Follow-up letter, modified block with indented paragraphs

Envelopes

Use envelopes that match the width of the stationery. Always include your name and return address in the upper left corner of the envelope. If your envelope has a preprinted company address, type or write your name just above or below the printed return address to ensure proper return delivery, if needed.

Verify addresses to avoid delivery delays. When writing to a specific person, type the person's name on the first line of the address. If you do not know the person's name, address the envelope to a department or title, such as "Human Resources Manager."

The USPS has increased its use of automated processing technology. It now uses machines to read addresses, similar to an optical character reader (OCR) used by a computer scanner. To take advantage of the increased efficiency, type envelopes to conform to standards developed by the USPS. (Envelopes not conforming to the standardization are sorted by hand, and so will take longer to be delivered). Some address-book software programs, such as TouchBase Pro for the Macintosh, print envelopes and bar codes that conform to the USPS codes, which noticeably increases mail-delivery time.

The following shows an example of a 3-line, standardized address:

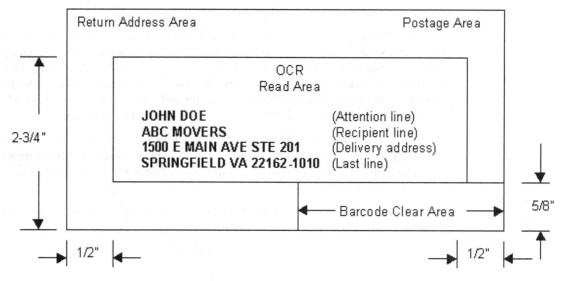

FIGURE 12–2
Standardized address

TIPS   The USPS recommends the following tips for addressing envelopes:

- Align all lines to a uniform left margin, allowing at least 1/2 inch on the right and left edge, with no more than 30 characters (spaces) per line.

- Type the name and address in the center, or OCR Read Area, of the envelope. Do not extend into the bar code area, beginning 5/8 inch from the bottom edge.

- Use upper-case letters. If lower-case letters are used, they must be readable by a standard OCR.

- Omit all punctuation marks, such as periods or commas, except for a hyphen in a ZIP + 4 address, or when a period or slash is an integral part of the address:
  101 1/2 MAIN ST
  or
  289-01 MONTGOMERY AVE

- Type directionals as a one- or two-letter abbreviation: **N, S, E, W, NE, NW, SE,** or **SW,** without punctuation.
- Add the alternate address, such as the floor or suite number, immediately after (on the same line as) the address.

MR M MURRAY
1401 MAIN ST APT C
FALLS CHURCH VA 22042-1441

- However, if the alternate address extends 30 spaces, add it in a line above the address:

MR M MURRAY
APT C
5800 SPRINGFIELD GARDENS CIR
SPRINGFIELD VA 22162-1058

- If adding an attention line, add it above the company name, as in the following:

ATTN JOHN DOE
ABC COMPANY
1401 MAIN ST
FALLS CHURCH VA 22042-1441
or
JOHN DOE
ABC COMPANY
1401 MAIN ST
FALLS CHURCH VA 22042-1441

- Fully spell out addresses or use the approved abbreviations for addresses, such as for street and avenue, and states (see Appendix 1: Common Symbols and Abbreviations).
- Do not include a "dual address" of a street address and a post office box, such as:

1201 BROAD ST E
PO BOX 1001
FALLS CHURCH VA 22062–1001

Instead, use one or the other, such as:

1201 BROAD ST E
FALLS CHURCH VA 22062-1001
or
PO BOX 1001
FALLS CHURCH VA 22062-1001

- Format the last line with at least one space between the city name, two-character state abbreviation, and ZIP + 4 code. Use the ZIP + 4 digits, when available, in the address.
- Spell out the full city name when spacing permits. Otherwise, use approved abbreviations for city names.

Preferred:	**WEST STOCKBRIDGE**	**NEWBERRY SPRINGS**
Acceptable:	**W STOCKBRIDGE**	**NEWBERRY SPGS**

Types of Letters

Letters are either informative, persuasive, or a combination of both. They can be positive or negative in tone, depending on your purpose. They can be formal or informal, depending on how well you know the reader.

Technicians may be expected to write several types of letters, such as a request for sales or technical information from a manufacturer, a price quotation to a customer, a letter of acknowledgment for good service, or a letter of complaint about poor service.

Effective letters must be written in simple, clear language and must include accurate facts. An incorrect part number or a poorly worded request may lead to confusion or errors.

Questions to Add Focus and Purpose

Some questions to ask yourself before beginning a letter are the following:

- What results do I want from this letter?
- Who is my reader?
- How familiar is my reader with the topic?
- What does my reader need to know?
- What do I need to know?
- What relationship do I want with the reader?
- How do I expect the reader to react to what I am saying and how I am saying it?
- What impression am I making on the reader?

The first question is the most basic, and without a clear answer, it is doubtful that the letter will accomplish any purpose. When you know what you want, you can communicate it. Knowing the reader will make some word choices easier. Knowing your purpose will help identify appropriate and important information to request or provide. Leaving out relevant information will delay results. The last three questions will affect the tone of the letter: how formal, how friendly, and how directly to write the letter.

Forbidden Words and Phrases

Many words and phrases are overused, evasive, redundant, or wordy. Avoid them in any kind of writing, particularly business letters.

Wordy phrases:

as I am sure you know	I would hope
as of this date	I would like to express
as you are aware	I share your concern
as you know	it is my intention
at this point in time	it is our (un)pleasant duty
at the present time	we regret to inform you
due to the fact that	we are cognizant of the fact

Overused words and phrases:

access	meanwhile
best wishes	more importantly
bottom line	mutually beneficial
delighted	needless to say

different from	ongoing
enclosed herewith	orient
finalize	parameter
glad	personally reviewed
great majority	prior to
happy	prioritize
herein	respectfully requested (submitted)
hereinafter	share
hopefully	specificity
however	subject matter
image	subject to your approval
inappropriate	therein
input	thrust
in the amount of	to impact
in the near future	to optimize
institutionalize	untimely death
interface	utilize
kindly favor us	very much
maximize	viable

Redundancies:

important essentials	serious crisis
final outcome	personally reviewed
end result	my own . . .
new initiatives	I, myself, . . .

Use clear, direct language, free of wordy and evasive phrases, to make letters effective and easy to understand.

Exercise 12.1   *Rewrite the following sentences in clear language that avoids redundant, overused, or wordy words and phrases.*

> *Example:*   The end result of the new initiatives will impact your department.
> *Rewritten:*   The results of these changes will affect your department.

1. From this point in time, this facility will utilize Ace Trucking for all ongoing shipments.
2. The great majority of personnel respectfully request your immediate attention to the following subject matter.
3. Kindly favor us with your input on this issue so that we can provide viable alternatives.
4. We regret to inform you of the untimely death of your colleague, Lee Shipman.
5. I would very much like to interface with you and finalize our plans.
6. Please submit payment in the amount of $150.

7. Needless to say, the end results will be mutually beneficial.

8. We would like to express best wishes for your achievements and hope that we can utilize your services in the near future.

9. The research document, enclosed herein, contains the important essentials of the product's ongoing development.

10. I have personally reviewed the plans and have decided to impact our image by maximizing our recent achievements.

Exercise 12.2   *Write a formal one-paragraph letter informing your employer of one of the following items:*

- A permanent address change
- A two-week vacation date
- A two-week termination notice

Exercise 12.3   *Write a formal letter including all ten parts and the envelope for one of the items in the exercise above. Type the letter in either the blocked or the semiblocked format.*

Exercise 12.4   *Write a formal letter to a company to request information on one of its products. Type the letter in either the blocked or the semiblocked format.*

Positive Letters

A positive letter expresses information in a neutral or positive tone. The writer could be making a request, providing information, or acknowledging good service. Some letters can be extremely brief—even only one or two sentences.

> Please send me your current catalog and price list of electronic components. Thank you.

Longer letters are made up of three distinct parts, usually presented in three or more paragraphs.

1. Introduction
2. Explanation
3. Resolution and closing

The introduction states the subject clearly and directly. It may also include the purpose for writing.

> You requested an itemized proposal for the installation of the IBM CAD/CAM system. The following list includes the specific modules, their costs, and the projected installation dates.

The middle part of a letter explains the request or provides the information. Usually, the information is organized and presented in an order that is easy to read and understand. Remember that readers sometimes skim letters, or read them quickly.

Complicated material must be explained as simply and accurately as possible. Some explanations contain specific points that need special attention. These points can be highlighted by individual paragraphs, numbering, or listing. Skipping lines between numbered or listed items adds visual emphasis.

The new location is desirable for the following reasons:

1. It has easy access to major traffic routes.
2. It is close to leading customers.
3. It includes reasonable land and construction costs.
4. Part-time employees are available from a nearby technical institute.

The resolution paragraph states the specific action that will accomplish your purpose for writing. It could be a request, date, or statement of commitment. State the specific response that you expect from your letter. Provide exact instructions or requests. Then close the letter with a statement of appreciation for the reader's attention, time, or interest.

I will inform you of our decision by January 31. Thank you for your interest in our plans.

Sometimes no resolution is required, and the writer can simply close the letter. Do not extend the letter with excessive, flowery language, but conclude the letter with a positive comment—either appreciation or acknowledgment.

Congratulations for your fine work.

Letters of "goodwill" are those expressing compassion—sympathy, appreciation, or congratulations—to business relations or associates (see Example 5 on the following page). These letters must be written carefully and simply.

People appreciate brief expressions of goodwill, but they may not appreciate a long, detailed account of your similar experiences and how you handled them. Try to keep messages of sympathy positive by not dwelling on the loss, but rather, by mentioning something positive (such as a pleasant memory) about the person. Letters of thanks can recount exactly the services you received from the reader, as well as an offer to return the favor. Letters of congratulations offer sincere praise for an achievement, without excessive praise that sounds phony. Avoid mechanical language in any of these letters.

Exercise 12.5   *Write a letter to an elected official (local, state, or national) informing the official of your opinion on a political issue. Begin by answering the "Questions to Add Focus and Purpose" on page 192. Gather the information you need to write an informed letter. Write the letter in three parts: the introduction, body, and conclusion. Include all the standard parts of a letter. Proofread and type the letter and envelope.*

Exercise 12.6   *Type a letter to a business associate for one of the following occasions:*

- The death of a family member
- Her promotion to vice-president of a company
- Thanks for referring a customer to you

600 Cornwallis Court
Roanoke, Virginia 24005

May 19, 2000

Ms. Jody Longley
3605 Winning Way
Seattle, Washington 98105

Dear Ms. Longley:

Congratulations on your announced promotion. After all your hard work and determination, you deserve the recognition. For a change, we have an effective computerized production system in the Seattle plant. I cannot help but think that the successful implementation of the three Puma robots was also due to your commitment and expertise.

I am very happy for you and proud to be working for a company that rewards and promotes employees of your caliber.

Regards,

Calvin M. Shargon

Calvin M. Shargon

EXAMPLE 5:
Letter of goodwill, blocked style.

Negative Letters

Letters are sometimes written to refuse service or employment, to turn down claims or credit, or to deny information. Since the reader will find the news disappointing, it is important not to confuse or insult the reader and make further association with the reader more uncomfortable than necessary. It is possible to continue a business relationship despite a negative situation.

These letters usually have three parts:

1. Introduction
2. Reasons for refusal
3. Refusal and closing

The introduction states the purpose for writing. It recounts the incident that occurred to initiate the letter. Write the introduction objectively, stating facts that both you and the reader accept as facts. Avoid accusatory language such as "you claim" or "you said that." Avoid any wording that is judgmental or belittling, such as "I should have expected this of you" or "How could you?" Accurately state the facts, as in the following example. In this example the writer is refusing the reader's request to redirect a service bill to the factory. The introduction recounts the facts:

I serviced an Apple IIc on March 19 at your Riverbend office complex. I replaced the motherboard and cleaned the drive heads. I tested the computer with your software and did not see any further problems. I left a bill for these services totaling $200.

Universal Communications Company   **1 Communications Plaza**
                                               Fort Walton Beach, FL 32549
                                               (404) 555-2345 www.unicom.com

March 4, 2000

Mr. Carlos Torres
Graduate Placement Office
South Florida Institute of Technology
Miami, FL 33157

Dear Mr. Torres:

[Introduction]   Thank you for your interest in the Universal Communications Company. I reviewed your résumé and application for a position as technician at Universal. Your education appears to cover all the areas of fundamental electronics and electricity.

[Reasons for refusal]   Presently, however, we are interviewing only technicians with experience in maintaining digital devices, an area that does not appear on your résumé.

[Refusal and closing]   For this reason, Mr. Torres, I cannot consider you for our present positions. If you are able to complete the basic and intermediate courses in digital devices before graduation, please consider resubmitting your résumé to Universal.

Best wishes for the future.

Sincerely,

Tracy R. Canfield
Human Resources Manager

TRC: na

EXAMPLE 6:
Negative (bad news) letter, blocked style

The reasons for refusal are stated to clarify your position and reassure the reader that you considered the request fairly and can justify your refusal. Often, company policies or procedures can be cited, which then provide guidance for the reader's actions in the future. Reasons for refusal should be consistent and legally sound. Some companies provide legal advice for these types of letters. Avoid sounding evasive or as though you are blaming the refusal on someone else, such as "this was the manager's decision, not mine." If you are unclear of the reasons, find out the reasons before you begin writing, and write with confidence and conviction.

> The Apple IIc that I worked on was out of warranty. The problems in the CPU and disk drive were caused by normal wear after heavy regular use. I found no evidence of equipment defects or malfunctions in either the CPU or the disk drive.

The next paragraph should clearly state what is being refused, and what, if any, further action can be taken. Acknowledging the reader's point of view, without changing your position, will make you sound more understanding and, thereby, make your decision more tolerable. Close with a positive, sincere statement that will keep the door open for further business or association with the reader.

> Your claim for factory reimbursement has been denied and payment for my services is expected from you by the 15th of this month. Be assured that our company would not bill owners for factory defects.
> Our company has always strived to give prompt, dependable service at reasonable rates. We value you as a customer and hope that you will continue to use our company to maintain your electronic equipment.

Remember that a firm "no" is easier to handle and tolerate than a runaround or foggy response that leaves the issue unresolved. Also, remember that letters written in an emotional state usually reflect that emotion. Writing letters in a biting, angry tone (no matter how justified) does not make good business sense since they produce negative results. Letters that are objective and rational leave readers their dignity and allow communication and business to continue. Example 6 on page 197 shows a negative letter that respectfully rejects an employment application.

Exercise 12.7   *Write a formal, negative letter for the following situation. Supply any missing information.*

You work for a large company that manufactures a highly competitive product. A competitor has written a letter to you requesting some design specifications about the product. You have met this competitor on several occasions, but you are not personal friends. You know that it is against company policy to discuss the product design specifications.

SPELLING: Suffixes *ible* and *able*

We add -*ible* or -*able* to turn words into adjectives. The choice between spelling the ending *ible* or *able* at the end of a word is usually determined by the Latin root word, as we found for the endings -*ance* and -*ence*. Few people these days are familiar with Latin, but we have some general guidelines to help make this spelling problem easier. Remember that a dictionary is always the best source because even the following guidelines have many exceptions.

Use -*able* with the following groups of words.

1. Most full English words (exceptions listed under -*ible*):

   available   correctable

   dependable   predictable

2. Full English words that had a final dropped *e:*

   excitable   presumable

   desirable   usable

   flammable

3. Roots that end in a final *i:*

   appreciable   satisfiable

   justifiable   sociable

4. Words that use the letter *a* in other endings:

   demonstrable   (demonstrate)

   hospitable   (hospitality)

   inseparable   (separate)

5. Roots that end with a hard *c* (pronounced "k" as in *cat*) or a hard *g* (pronounced "g" as in *get*):

   educable   navigable

   practicable   indefatigable

6. Roots that end with a soft *c* (pronounced "s" as in *cent*) or a soft *g* (pronounced "j" as in *gentle*) that require keeping the final *e:*

   replaceable   manageable

   traceable   salvageable

   serviceable   changeable

   noticeable   chargeable

Use *-ible* with the following groups of words.

1. Roots that are not full words.

audible	possible
invisible	terrible
indelible	feasible

2. Some full English words to which *-ion* could be added:

collectible	(collection)
connectible	(connection)
reversible	(reversion)
suggestible	(suggestion)
combustible	(combustion)
diffusible	(diffusion)

3. Roots that end in *-ns, -ss,* or *-miss:*

defensible	dismissible
sensible	permissible
accessible	admissible
depressible	

4. Roots that end in a soft *c* (pronounced "s" as in *cent*) or a soft *g* (pronounced "j" as in *gentle*):

forcible	tangible
deductible	illegible
reproducible	intelligible

5. Exceptions to *able* spellings:

flexible	resistible
collapsible	discernible
fusible	

Exercise 12.8 *Practice using* able/ible *correctly by adding the ending to each root, and then by spelling the entire word.*

able			*ible*		
avail	_____	_____	aud	_____	_____
depend	_____	_____	tang	_____	_____
excit	_____	_____	invis	_____	_____
us	_____	_____	feas	_____	_____
appreci	_____	_____	collect	_____	_____
soci	_____	_____	revers	_____	_____
demonstr	_____	_____	suggest	_____	_____
practic	_____	_____	flex	_____	_____
service	_____	_____	collaps	_____	_____
notice	_____	_____	resist	_____	_____
charge	_____	_____	reproduc	_____	_____

VOCABULARY: Roots *grad* and *gress*

The roots *grad* and *gress* both come from the Latin words meaning *gradus,* "a step." We generally use the *grad-* spelling when the root is at the beginning of the word, and *-gress* when the root is at the end of the word.

Gradual	going step by step, slowly
Gradient	the rate of change
Graduate	a person who completed a course of study
Aggressive	taking quick, sometimes violent action
Congress	a gathering
Digress	to go off in another direction
Progress	to go forward
Regress	to go backward
Transgress	to overstep or break a law

The familiar word *grade* also comes from the Latin word *gradus.* It literally means "a step, stage, or level."

Grade	a level or slope
Degrade	to lower in status or demote
Centigrade	divided into 100 degrees
Retrograde	going backward

Exercise 12.9   *Match the letter of the definition for each word.*

_d_	**1.** digress	**a.**	to overstep the law
_i_	**2.** congress	**b.**	slope showing the rate of change
_e_	**3.** gradual	**c.**	going backward
_h_	**4.** degrade	**d.**	going off in another direction
_b_	**5.** gradient	**e.**	moving ahead slowly
_A_	**6.** transgress	**f.**	person who finished a program of study
_j_	**7.** progress	**g.**	taking quick or violent action
_c_	**8.** retrograde	**h.**	to demote
_G_	**9.** aggressive	**i.**	to gather together
_F_	**10.** graduate	**j.**	to go forward

Exercise 12.10   *The following words are related to the words above. Write a definition (without using the dictionary) that relates each word to its root.*

Hint: The *-ion* suffix turns a word into a noun.

1. gradation _____

2. congregation _____

3. transgression _____

4. regression _____

5. graduation _____

6. degradation _____

7. progression _____

8. aggression _____

Hint: The *-ive* suffix turns a word into an adjective.

9. regressive _____

10. progressive _____

WORD WATCH: *stationary* and *stationery* *compliment* and *complement*

Stationary means "not moving" or "fixed." Think of "stAnding still" to remind you that this word ends with *ary*.

> Hold the probe *stationary* when reading the voltage.

Stationery means "writing materials." Think of "lettERs" to remind you that this word ends in *ery*.

> The letter was written on company *stationery*.

Compliment is a verb meaning "expressing courtesy, praise, or flattery," or it is a noun meaning "an expression of courtesy, praise, or flattery."

> I *compliment* you on your remarkable achievement. (*verb*)
> Please accept our *compliments* for your achievement. (*noun*)

The adjective form is *complimentary*.

> Include a *complimentary* closing in all letters. (*adjective*)

Complement is usually a noun meaning "something that fills or completes," or "something added to complete a whole." Use the word *complete* to remind you of the middle *e*.

> The *complement* of a 45-degree angle is another 45-degree angle—together they total 90 degrees. (*noun*)

The adjective form is *complementary*.

> *Complementary* colors, when they are combined in the right intensities, will produce white light. (*adjective*)

Exercise 12.11   *Use the correct word to complete each sentence.*

Use *stationery* or *stationary*.

1. The guard stood _____ while the crowd passed.
2. The _____ was white with black lettering.
3. The _____ machine collected dust.
4. The company's logo was printed on the _____.
5. How many envelopes were included with your _____?

Use *compliment* or *complement*.

6. Some people do not know how to react to a _____.
7. Carol _____ed her subordinate on his positive attitude.
8. A red tie will _____ a white shirt and dark suit.
9. A full _____ of personnel work the night shift.
10. His annual evaluation was _____ary, and he was promoted.

CHAPTER 13

Presentations

- Prepare an oral presentation.
- Prepare for a job interview.
- Spell words with a final *e* correctly.
- Use *sub* and *super* correctly.
- Use *used* and *supposed* correctly.

READING: John F. Kennedy's Address on the Space Effort

President John F. Kennedy gave his address on the space effort in the Rice University Stadium on September 12, 1962, at 10 A.M. In his opening words he referred to Dr. K. S. Pitzer, President of the University; Vice President Lyndon B. Johnson; Governor Price Daniel of Texas; Representative Albert Thomas of Texas; Senator Alexander Wiley of Wisconsin; Representative George P. Miller of California; James E. Webb, Administrator, National Aeronautics and Space Administration; and David E. Bell, Director of the Bureau of the Budget.

President Pitzer, Mr. Vice President, Governor, Congressman Thomas, Senator Wiley, and Congressman Miller, Mr. Webb, Mr. Bell, scientists, distinguished guests, and ladies and gentlemen:

I appreciate your president having made me an honorary visiting professor, and I will assure you that my first lecture will be very brief. I am delighted to be here and I'm particularly delighted to be here on this occasion.

We meet at a college noted for knowledge, in a city noted for progress, in a State noted for strength, and we stand in need of all three, for we meet in an hour of change and challenge, in a decade of hope and fear, in an age of both knowledge and ignorance. The greater our knowledge increases, the greater our ignorance unfolds.

Despite the striking fact that most of the scientists that the world has ever known are alive and working today, despite the fact that this Nation's own scientific manpower is doubling every 12 years in a rate of growth more than three times that of our population as a whole, despite that, the vast stretches of the unknown and the unanswered and the unfinished still far out-strip our collective comprehension.

No man can fully grasp how far and how fast we have come. But condense, if you will, the 50,000 years of man's recorded history in a time span of but a half century. Stated in these terms, we know very little about the first 40 years, except at the end of them, advanced man had learned to use the skins of animals to cover him. Then about 10 years ago, under this standard, man emerged from his caves to construct other kinds of shelter. Only 5 years ago man learned to write and use a cart with wheels. Christianity

FIGURE 13.1
President Kennedy at Rice University.

began less than 2 years ago. The printing press came this year, and then less than 2 months ago, during this whole 50-year span of human history, the steam engine provided a new source of power.

Newton explored the meaning of gravity. Last month electric lights and telephones and automobiles and airplanes became available. Only last week did we develop penicillin and television and nuclear power, and now if America's new spacecraft succeeds in reaching Venus, we will have literally reached the stars before midnight tonight.

This is a breathtaking pace, and such a pace cannot help but create new ills as it dispels old: new ignorance, new problems, new dangers. Surely the opening vistas of space promise high costs and hardships, as well as high reward.

So it is not surprising that some would have us stay where we are a little longer to rest, to wait. But this city of Houston, this State of Texas, this country of the United States was not built by those who waited and rested and wished to look behind them. This country was conquered by those who moved forward—and so will space.

William Bradford, speaking in 1630 of the founding of the Plymouth Bay Colony, said that all great and honorable actions are accompanied with great difficulties, and both must be enterprised and overcome with answerable courage.

If this capsule history of our progress teaches us anything, it is that man, in his quest for knowledge and progress, is determined and cannot be deterred. The exploration of space will go ahead, whether we join in it or not, and it is one of the great adventures of all time, and no nation which expects to be the leader of other nations can expect to stay behind in this race for space.

Those who came before us made certain that this country rode the first waves of the industrial revolutions, the first waves of modern invention, and the first wave of nuclear power, and this generation does not intend to founder in the backwash of the coming age of space. We mean to be a part of it—we mean to lead it. For the eyes of the world now look into space, to the moon and to the planets beyond, and we have vowed that we shall not see it governed by a hostile flag of conquest, but by a banner of freedom and peace. We have vowed that we shall not see space filled with weapons of mass destruction, but with instruments of knowledge and understanding.

Yet the vows of this Nation can only be fulfilled if we in this Nation are first, and, therefore, we intend to be first. In short, our leadership in science and in industry, our hopes for peace and security, our obligations to ourselves as well as others, all require us to make this effort, to solve these mysteries, to solve them for the good of all men, and to become the world's leading space-faring nation.

We set sail on this new sea because there is new knowledge to be gained, and new rights to be won, and they must be won and used for the progress of all people. For space science, like nuclear science and all technology, has no conscience of its own. Whether it will become a force for good or ill depends on man, and only if the United States occupies a position of preeminence can we help decide whether this new ocean will be a sea of peace or a new terrifying theater of war. I do not say that we should or will go unprotected against the hostile misuse of space any more than we go unprotected against the hostile use of land or sea, but I do say that space can be explored and mastered without feeding the fires of war, without repeating the mistakes that man has made in extending his writ around this globe of ours.

There is no strife, no prejudice, no national conflict in outer space as yet. Its hazards are hostile to us all. Its conquest deserves the best of all mankind, and its opportunity for peaceful cooperation may never come again. But why, some say, the moon? Why choose this as our goal? And they may well ask why climb the highest mountain? Why, 35 years ago, fly the Atlantic? Why does Rice play Texas?

We choose to go to the moon. We choose to go to the moon in this decade and do the other things, *not because they are easy, but because they are hard,* because that goal will serve to organize and measure the best of our energies and skills, because that challenge is one that we are willing to accept, one we are unwilling to postpone, and one which we intend to win, and the others, too.

It is for these reasons that I regard the decision last year to shift our efforts in space from low to high gear as among the most important decisions that will be made during my incumbency in the Office of the Presidency.

In the last 24 hours we have seen facilities now being created for the greatest and most complex exploration in man's history. We have felt the ground shake and the air shattered by the testing of a *Saturn C-1* booster rocket, many times as powerful as the *Atlas* which launched John Glenn, generating power equivalent to 10,000 automobiles with their accelerators on the floor.

We have seen the site where five F-1 rocket engines, each one as powerful as all eight engines of the *Saturn* combined, will be clustered together to make the advanced *Saturn* missile, assembled in a new building to be built at Cape Canaveral as tall as a 48-story structure, as wide as a city block, and as long as two lengths of this field.

Within these last 19 months at least 45 satellites have circled the earth. Some 40 of them were "made in the United States of America," and they were far more sophisticated and supplied far more knowledge to the people of the world than those of the Soviet Union.

The *Mariner* spacecraft, now on its way to Venus, is the most intricate instrument in the history of space science. The accuracy of that shot is comparable to firing a missile from Cape Canaveral and dropping it in this stadium between the 40-yard lines.

Transit satellites are helping our ships at sea to steer a safer course. Tiros satellites have given us unprecedented warnings of hurricanes and storms, and will do the same for forest fires and icebergs.

We have had our failures, but so have others, even if they do not admit them. And they may be less public.

To be sure, we are behind, and will be behind for some time in manned flight. But we do not intend to stay behind, and in this decade we shall make up and move ahead.

The growth of our science and education will be enriched by new knowledge of our universe and environment, by new techniques of learning and mapping and observation, by new tools and computers for industry, medicine, the home, as well as the school. Technical institutions, such as Rice, will reap the harvest of these gains.

And finally, the space effort itself, while still in its infancy, has already created a great number of new companies, and tens of thousands of new jobs. Space and related industries are generating new demands in investment and skilled personnel, and this city and this State, and this region, will share greatly in this growth. What was once the furthest outpost on the old frontier of the West will be the furthest outpost on the new frontier of science and space. Houston, your City of Houston, with its Manned Spacecraft Center, will become the heart of a large scientific and engineering community. During the next 5 years the National Aeronautics and Space Administration expects to double the number of scientists and engineers in this area, to increase its outlays for salaries and expenses to $60 million a year; to invest some $200 million in plant and laboratory facilities; and to direct or contract for new space efforts over $1 billion from this Center in this City.

To be sure, all this costs us all a good deal of money. This year's space budget is three times what it was in January 1961, and it is greater than the space budget of the previous 8 years combined. That budget now stands at $5,400 million a year—a staggering sum, though somewhat less than we pay for cigarettes and cigars every year. Space expenditures will soon rise some more, from 40 cents per person per week to more than 50 cents a week for every man, woman, and child in the United States, for we have given this program a high national priority even though I realize that this is in some measure an act of faith and vision, for we do not now know what benefits await us. But if I were to say, my fellow citizens, that we shall send to the moon, 240,000 miles away from the control station in Houston, a giant rocket more than 300 feet tall, the length of this football field, made of new metal alloys, some of which have not yet been invented, capable of standing heat and stresses several times more than have ever been experienced, fitted together with a precision better than the finest watch, carrying all the equipment needed for propulsion, guidance, control, communications, food and survival, on an untried mission, to an unknown celestial body, and then return it safely to earth, reentering the atmosphere at speeds of over 25,000 miles per hour, causing heat about half that of the temperature of the sun—almost as hot as it is here today—and do all this, and do it right, and do it first before this decade is out, then we must be bold.

I'm the one who is doing all the work, so we just want you to stay cool for a minute. [Laughter]

However, I think we're going to do it, and I think that we must pay what needs to be paid. I don't think we ought to waste any money, but I think we ought to do the job. And this will be done in the decade of the sixties. It may be done while some of you are still here at school at this college and university. It will be done during the terms of office of some of the people who sit here on this platform. But it will be done. And it will be done before the end of this decade.

I am delighted that this university is playing a part in putting a man on the moon as part of a great national effort of the United States of America.

Many years ago the great British explorer George Mallory, who was to die on Mount Everest, was asked why did he want to climb it. He said, "Because it is there."

Well, space is there, and we're going to climb it, and the moon and the planets are there, and new hopes for knowledge and peace are there. And, therefore, as we set sail we ask God's blessing on the most hazardous and dangerous and greatest adventure on which man has ever embarked.

Thank you.

Reading Comprehension Questions

1. Who is the audience of the address? _____

_____

2. What analogy did President Kennedy use to demonstrate the "breathtaking pace" of space exploration?_____

_____

3. Why did President Kennedy think the United States should be the first to explore space? _____

_____

4. Why did he choose to explore the moon? _____

_____

5. What were some of the immediate benefits to be gained from increased spending on the space program? _____

_____

6. What did he propose to do "before the decade is out?" _____

_____

7. What do you think President Kennedy's purpose was in giving the speech? Why at a college? Why in the state of Texas? _____

_____

_____

WRITING: Presentations

The reading article is a speech delivered by President John F. Kennedy in 1962, just after the former Soviet Union launched the first satellite into space, which stunned many Americans and propelled the nation into an era that is sometimes called the "race to the moon." He delivered the speech to announce his increased space budget and to challenge a newly emerging space industry to land a man safely on the moon by the end of the decade, "not because it is easy, but because it is hard." Although President Kennedy did not live to see it, on July 20, 1969, U.S. astronauts Neil Armstrong and Edwin "Buzz" Aldrin Jr., became the first men to walk on the moon.

Review the speech by President Kennedy again to find some common elements in effective public speaking:

- Introduction (greeting guests and the audience)
- Statements to generate rapport (relating to the audience)
- Lead-in to his purpose
- Humor to relax the audience, or lighten a serious topic
- Metaphors that make data come to life
- Conclusion with a restatement of the purpose

President Kennedy uses traditional techniques to engage his audience, inform them, and persuade them. Underline the words he used to establish rapport with Rice University and Houston, Texas. Find the places where he added humor to lighten a serious subject and give his audience a mental breather. Find an example of how he made large numbers, inconceivably large numbers to most people, take on meaning. Find the statement that explicitly states his purpose. Find examples of repeated phrasing that add verbal symmetry and a pleasing sound. Find the quote he uses from a renowned adventurer to strengthen his conclusion.

Many people openly admit that their biggest fear is public speaking. Whether it's in front of a class, coworkers, managers, or total strangers, they experience physical symptoms of sweating, trembling, rapid heartbeats, and worse. Their minds go blank, their mouths go dry, and their voices quiver—all classic signs of nervousness. For a few people, the fear is debilitating—we hear now and then of talented singers and musicians who are so paralyzed by stage fright that they cannot perform in public.

Most people experience some degree of fear when facing an audience, but they learn techniques to control their voice tone and body language to project confidence. Successful public speakers often attribute their apparent relaxation to rehearsal—practicing their speech or presentation enough times, and in a similar setting to the real one, that the words flow easily, despite a jittery stomach. Rehearsal appears to be the key solution for beginning speakers.

If a public-speaking class is available in your college, you will find it a good opportunity to practice different types of speeches and polish your delivery—your case of nerves might not go away entirely, but it can become manageable.

As an alternative, you can find a local chapter of Toastmasters International. This is a worldwide network of clubs in which people develop their public-speaking skills by practicing in a comfortable environment with other learners. With mentoring, structured activities, and international competitions, members gain experience and self-confidence they can carry back to the workplace.

Formal Presentations

In business, people in sales and marketing make formal presentations at customers' sites, annual meetings, and training seminars. When money is at stake or the audience is large, the presentation becomes more formal. For these situations, people take extra steps to guarantee a successful speech. Speakers sometimes videotape themselves giving the speech to observe facial expressions, body language, and delivery. They prepare visual aids that look polished and professional. They create handouts with detailed information and allow extra white space for notes. Many speakers enlist technical experts as backups to help answer detailed technical questions or operate the presentation equipment.

Informal Presentations

Most presentations are less formal. For example, some companies hold "brown-bag lunches," where a speaker discusses a topic while the audience eats lunch (the presentation might be recorded or telecast), or an employee presents an idea to a manager. While these situations are less formal than a sales presentation to a customer, the speakers usually prepare in similar ways, with research, an outline and notes, and handouts.

Even more informally, a manager might ask an employee to discuss a technical topic at a staff meeting. Sometimes these speeches are impromptu, meaning a speaker begins speaking without preparation or planning.

Guidelines

Generally, the better prepared you are, the better your presentation will be.

TIPS   Although you might never have the advantage of a professional speechwriter, as President Kennedy undoubtedly did, you can follow these guidelines to reduce nervousness and increase the effectiveness of your presentation:

1. **Research your topic.** Choose a topic and collect the data to back up your message or convince your audience. You might not use all the data, but keep them with you for questions from the audience.

2. **Analyze your audience.** Determine their level of technical background. Anticipate what your audience wants to know about the topic and what the audience intends to do with the information. Anticipate possible questions (general and technical) and prepare for them.

3. **Outline your message.** State the main point you want to make in one sentence. Then write the supporting points you want to make in a bulleted list. Sequence them in the order that makes sense for your purpose.

4. **Use a multimedia approach.** Most people are poor listeners, so provide visual aids for them to see and read. Handouts or transparencies not only visually organize and reinforce your message, but add interest, as well.

5. **Rehearse the speech out loud.** When you recite your speech, you can work out exact wording and possibly discover areas within the topic that need more research. Time yourself. Not only will you know, but you can inform others, if asked, of the time allotment needed. You can rehearse in front of the mirror (good for practicing facial expressions, as well), while driving the car, or at the front of an empty room. If possible, ask someone to listen and give you feedback. Ask the person to note any signs of nervousness, such as tight facial expressions or fiddling with hair or glasses.

6. **Arrive early.** Get comfortable in the room, take some deep breaths, and arrange your notes and visual aids. If you plan to use any electronic equipment, make sure it works and you know how to turn it on. To reduce stress, many professional trainers carry vital supplies in their briefcases, including their own markers, masking tape, and even an extra bulb for a projector. Greet people as they arrive, making eye contact and starting to establish rapport. For smaller audiences, this might be an opportunity to learn some of the names and backgrounds of your audience. Write down a few names in your notes, especially key people, so you can address people by name, if needed.

7. **Start with an introduction.** Instead of launching into the body of your speech, take a few minutes to introduce yourself and your subject and orient your audience to the scope of your speech. Experienced speakers include a "springboard motivator," such as an anecdote, question, or activity that captures the interest and attention of the audience and gets them involved with the subject.

8. **Use note cards or your visual aids** to keep on track and prevent your missing an important point. (But do not read your speech from a script.)

9. **Don't let questions digress from your main topic.** If someone in the audience asks a question that is unrelated to your topic, or strays too far from the scope of your presentation, tactfully ask the person to "hold that question" until the end of your presentation. If you have time later, respond to the question. Also, if you do not know the answer to a question, admit it, and establish how you will follow up with the person. For example, ask for the person's phone number or e-mail address. Or ask the person to send you an e-mail with more details about the question.

10. **End with a summary of your main points.** Your closing is an opportunity to reestablish your key points and show how they logically lead to your conclusion. Do not throw in new points or re-argue your prior points during your conclusion—just restate them and close.

Interviews

An interview is a special type of presentation that requires steps similar to those in other presentations. Although there are many types of interviews, this section (and the next chapter) focuses on the job interview. Job candidates research the prospective company, prepare by anticipating and rehearsing answers, create a résumé to leave with the interviewer (or send it in advance), establish rapport with the interviewer, and set the pace of the interview by answering questions clearly, without rambling or stumbling.

Interviewers also have a responsibility for the progress of the interview, but they don't have as much at stake. And they probably are not suffering from a case of nerves. Some interviews are conducted one-to-one, but more frequently, interviews are conducted with panels of two or more key staff, each with prepared questions for the applicant. All are taking notes when the applicant answers. While this is more efficient for the company, it puts more stress on the applicant, who must concentrate on one interviewer's questions and shift gears rapidly to answer the next interviewer's questions. Applicants must pay attention to introductions, remember names (at least a couple of them) and titles, and keep eye contact with the questioning interviewer. Job candidates find the interviewing process less stressful if they rehearse answers to typical interview questions in advance. Interview techniques are included in Chapter 14: The Job Search; however, you can begin by answering the top five questions:

- **Tell me about yourself.**

  This request is intentionally open-ended, and without a prepared answer, you could turn it into a rambling history of your life—not a good response. Unless something in your past is especially worth noting, keep your answer related to your professional interests, experience, and skills.

- **Why are you interested in this position?**

  Use this response to establish your long-lasting commitment to the field. Also, include any relevant research on the company—relate your skills to the direction in which the company is moving (such as your interest in an electronic device produced by the company).

- **Why should we hire you?**

  This question is difficult if your classmates are waiting outside the door, vying for the same job. Stress your technical skills, and also stress any technical experience, skills, or interests that might distinguish you from your classmates and make you the best candidate for the job, such as any independent projects you completed or former work experience that makes you a better candidate for this job.

- **What are your strengths?**

  Some people find it difficult to discuss their strengths—it seems like bragging. But you must search your soul for at least three defining characteristics that will make you a valuable employee. Consider how an admiring parent, instructor, or other adult might describe you. Try to back up your strengths with concrete evidence (such as leadership demonstrated in a former job or on an athletic team).

- **What are your weaknesses?**

  Be especially prepared for this one. The interviewer wants you to disclose something that could potentially eliminate you from the search. Use this formula: of the time you allot to this answer, spend 10% of it describing the weakness (say, procrastination), and the remaining 90% describing all the techniques you use to overcome this weakness and turn it into a strength (such as keeping lists and daily goals).

Above all else, when answering interview questions, be honest and professional. Your interviewer is taking notes, and any discrepancies can result in disqualification, or if you are hired and they are discovered later, termination.

Body Language

Body language consists of all the nonverbal messages we deliver to our audience. Nonverbal signals can be deliberate actions to support a message. For example, public speakers might pound a podium to emphasize a point, or walk into an audience to increase audience participation. Tapping fingers signal impatience. Clenched fists signal

FIGURE 13.2
Body language sends nonverbal messages.

anger. Open, uplifted palms signal a need for understanding or help. Waving arms signal intense emotion.

Other nonverbal messages can be physiological reactions to situations that we cannot easily control. For example, when someone is angry, lips get thinner, brows furrow, and faces get red and warm (hence the expression "hothead"). When someone is afraid, eyebrows go up, causing eyes to get wide ("wide-eyed with fear").

When the message delivered by body language contradicts the spoken message, listeners remember the body language. This means we must pay attention not only to what we say, but how we say it. Our entire appearance adds to our message, including our posture, where our eyes focus, how we move our hands, and how close we get to the audience.

Observe how professionals (actors or public speakers whom you consider convincing) use facial expressions and hands gestures to augment their words.

When you have written the content of your speech, practice speaking in front of a mirror, using expressions and gestures that support your message. Record yourself using a video camera, or ask someone you trust, to identify any distracting habits or mannerisms, such as words that you might overuse (saying "OK" frequently), wringing your hands, or fidgeting. Many times, we can break these habits just by becoming aware of them.

Follow three simple guidelines to get started:

- Smile occasionally, especially during introductions and conclusions. Usually a genuine smile can lighten the intensity of any information or news. It makes the speaker appear relaxed and confident, and that relaxes listeners, as well.

- Rest or fold your hands comfortably on the table or podium, or hold an appropriate object, such as a pointer. This reduces the chance that hand gestures will become distracting to listeners. With experience, speakers learn to use natural hand gestures that amplify the spoken message.

- Maintain eye contact with your audience. Move your eyes slowly from person to person. Watch out for staring at one person (which is bound to make that person uncomfortable) or staring at only a part of the room (the rest of the room will feel left out and possibly lose interest).

Visual Aids

Select the visual aids that are practical for you and appropriate for your audience, including transparencies, slide shows, videos, flip charts, eraser boards, demonstration models, and handouts. Consider the following factors when making your choice:

- Size of audience. As a general guideline: The larger the audience, the larger the visual aids. People in the back of the room want to see your visual aids. If you can't find a projection system to do that, consider handouts. For smaller audiences, your choices are broader.

- Location and logistics. Consider the size of the room, placement of chairs in the room, and equipment available. For example, auditoriums usually have projectors and screens available for far-away viewing. Other types of visual aids, such as flip charts or demonstration models, might not be visible by people in the back of a large room. Conference or seminar rooms, on the other hand, are usually smaller and have flip charts, eraser boards, and projectors readily available. All types of visual aids will be viewed easily.

- Subject matter. If your speech includes numerical data (such as statistical results or budgets) or detailed drawings (such as engineering drafts), provide the data on handouts for easier viewing—projections of detailed items are difficult to read. Bulleted lists of key points, however, are easily viewed on projections, flip charts, or eraser boards.

- Your resources, including software, hardware, time, and materials. Make the best of what you have to create a professional visual aid. If you have to learn a program or software application to create visual aids, allow enough time for experimentation.

Keep the audience, room, and subject matter in mind when creating visual aids. For example, use a large enough font for projections and transparencies that the people in the back of the room can read them. If the audience can read your message as well as hear it, you increase the chances that they'll remember it.

Multimedia presentation programs, such as Microsoft PowerPoint and Lotus Freelance Graphics, can incorporate photographs, slides, bulleted lists, and other text in exciting colors, fonts, and formats. Depending on the hardware available, you can project them on a screen from a computer or print them as transparencies and handouts. These programs might require a little training to use, although each contains ready-made templates, from which you can quickly choose the style and format for your presentation. If you are unfamiliar with the program, ask for assistance from friends, classmates, or instructors to get started, and allow a little practice time.

If no hardware will be available in the room (no transparency projector or computer), create posters or flip charts to take with you. You can hand-letter your lists and charts. Or you can purchase templates for letters, or even paste computer-created words and graphics for a more professional look.

Limit the scope of each visual aid to one point. If you pack too much onto a projection or page, it will be unreadable. Stick to one bulleted list, one chart, or one graphic per page. Include key words or phrases, not entire concepts.

If you want to interact with your audience, such as brainstorming for ideas, use equipment that you can write on and that will be visible to the audience, such as clear transparencies, flip charts, or eraser boards. Be sure you have the correct markers for each type. If you prefer, ask someone in the audience to write on the board while you lead the discussion.

Demonstrate with actual objects, when possible. For example, when discussing a software program, bring in a laptop computer and show a predeveloped and well-rehearsed demonstration of the program. Or when discussing how to take blood pressure (BP), bring in a BP cuff and demonstrate on a member of the audience.

Remember that audiences stay more attentive if they participate somehow. If time allows (sometimes it won't), ask for personal experiences, questions, or demonstrations—the audience will feel more involved.

In the following exercises, you will prepare and deliver short speeches. Notice the physiological responses you feel before, during, and after speaking, especially when you are in front of a room. Learn from your successes and mistakes—public speaking is a skill that can immediately transfer into your personal and professional life.

Exercise 13.1 *Prepare a brief speech on a technical topic. Use one of the following topics:*

- The technical device you described in Exercise 5.5
- The process you described in Exercise 5.7
- The procedure you wrote instructions for in Exercise 8.5
- The comparison/contrast you described in Exercise 9.2
- The report you completed in Chapter 10.

Exercise 13.2 *Prepare written answers to five typical interview questions:*

1. Tell me about yourself.
2. Why are you interested in this position?
3. Why should we hire you?
4. What are three of your strengths?
5. What are three of your weaknesses?

Exercise 13.3 *Conduct a mock (practice) job interview. Using the above questions and your prepared answers, in pairs, take turns interviewing each other in a setting similar to an actual interview, such as a conference room or office. Practice delivery techniques that demonstrate confidence. If possible, videotape the interviews. Then (respectfully) critique each other's interview, pointing out the good responses by the person and suggesting possible ways to improve.*

SPELLING: Dropping the Final *e*

In previous chapters you reviewed several rules for adding endings to words, including making words plural, changing the final *y* to *i*, and doubling the final consonant. One situation that has not been mentioned is what to do when a word ends in a final, silent *e*.

use   separate

practice   compute

Rule 1: When changing the word to its *s* form, just add the *s*.

uses   separates

practices computes

Rule 2: When adding a suffix beginning with a vowel, drop the *e*.

us e + ing   = using
practic e + al   = practical
separat e + ion   = separation
comput e + er   = computer

Some common exceptions are words that end with a soft *ge* (pronounced "j") or *ce* (pronounced "s").

change + able	= changeable
courage + ous	= courageous
notice + able	= noticeable
peace + ably	= peaceably

Rule 3: When adding a suffix beginning with a consonant, in most cases, leave the final *e*.

hope + ful	= hopeful
love + less	= loveless
move + ment	= movement
separate + ly	= separately

Some common exceptions are:

true + ly	= truly
nine + th	= ninth
wise + dom	= wisdom
awe + ful	= awful
argue + ment	= argument
judge + ment	= judgment
acknoweldge + ment	= acknowledgement

Exercise 13.4   *Combine the root word and the suffix correctly.*

1. He _____ reciting the formulas every chance he gets.
   (practice + s)

2. The students have become more _____ about research.
                              (knowledge + able)

3. The entire office has made _____ progress.
                  (notice + able)

4. The graduating students were interviewed _____.
                                      (separate + ly)

5. The resistor has become a _____ invention.
                  (use + ful)

6. We were _____ able to contain our _____.
         (bare + ly)   (excite + ment)

7. Her _____ has improved over the term.
   (write + ing)

8. The invention was his own _____.
                  (create + ion)

9. The situation seemed _____ until Dr. King took a _____
                        (hope + less)   (courage + ous)
   stand.

10. Thomas Edison claimed that genius was "1 percent _____ and 99 percent
                                            (inspire + ation)
    _____."
    (perspire + ation)

VOCABULARY: Prefixes *sub* and *super*

Super is a prefix that can mean "more than" or "over."

> superstar = a star more famous than most others
> supervisor = a person responsible for others

Sub is a prefix that means "less than" or "under."

> substandard = less than the standards
> submarine = underwater craft

Exercise 13.5   *Complete the words using* super *or* sub.

1. After experiencing a setback in business, some people become _____cautious.
2. A number printed slightly above the line is called a _____script.
3. When a liquid is at a temperature below its saturation point, it is _____cooled.
4. An image placed over another image is _____imposed.
5. A set of objects that is only part of a larger set is called a _____set.
6. The arc is below a blanket of granular flux in _____merged arc welding.
7. A first, quick look at an object or situation is called a _____ficial glance.
8. Saying something twice, repeatedly, or more than enough is _____fluous.
9. A routine used only for certain purposes, less important than the main program, is called a _____routine.
10. A _____structure is the part of a building above the foundation.

WORD WATCH: *used* and *supposed*

Used and *supposed* are the past-tense and past-particle forms of the verbs *use* and *suppose*. Because the final *d* is sometimes unvoiced (especially when followed by *to* as in *used to, supposed to),* many writers forget to add the *d* to the past-tense forms.

Warning: Remember to add the final *d* if the verb is followed by *to.* A helping verb, such as *is* or *was,* means that the verb is in its past-participle form.

> He's supposed to be at work.
> The supervisor is used to promptness.

The present tense of these verbs will not have a final *d*.

> I suppose you are right.
> I use many components every day.

Used is also an adjective meaning "worn" or "secondhand."

> Lynn bought a *used* car.

Use is also a noun meaning "the act of using," "an opportunity," or "a purpose."

> Mark could not find a *use* for the tool.

Other forms of *use* are *user, useful, usable, useless,* and *usage.*

Suppose means "imagine" or "assume." In the present tense, *suppose* makes the writer sound uncommitted or indecisive. Avoid indecision in technical writing.

> Weak: I suppose I'll start a career in sales.
> Strong: I'll start a career in sales.

Other forms of *suppose* are *supposedly* and *supposition* (another word for a hypothesis).

Exercise 13.6   *Fill in the correct form of the word.*

Use *use* or *used.*

1. The oscilloscope was _____ to troubleshoot the design project.
2. The team could not find a _____ for the design changes.
3. The _____ equipment needed repairs.
4. Mr. Jacobson was not _____ to delays in the schedule.
5. We are _____ to meeting our deadlines.

Use *suppose* or *supposed.*

6. He wasn't _____ to start replacing parts until he finished the troubleshooting procedures.
7. How do you _____ he arrived at the correct diagnosis?
8. The manager is _____ to notify the technicians of all design defects.
9. They are _____ to record the oscilloscope measurements.
10. If you have any questions, you're _____ to ask the project director.

CHAPTER 14

The Job Search: Résumés and Letters

- Résumés
- Job Applications
- Cover Letters
- Follow-up Letters
- Spell technical terms in résumés correctly.
- Use *meta* and *mega* correctly.
- Use *infer* and *imply* correctly.

READING: Web Spurs Change in Style of Résumé

by David Leonhardt

It is a tenet of the job search, passed down to college graduates for decades: Keep your résumé short, ideally no more than one page. A single sheet is more likely to hold an employer's attention and make an applicant look organized, not arrogant.

Now, however, the Internet is doing to one-page résumés what it has done to personal letters and travel agents. It is making them less relevant and perhaps even endangering their survival.

White space, brevity and verbs are out. Nouns and comprehensive descriptions— including obscure proper nouns, like the names of computer programs—are in. If the résumé continues page after page, or screen after screen, so be it. Even Headhunter.net, which has stricter rules than other Web sites, allows job seekers up to 6,000 characters, about three times the number in a typical one-page résumé.

Computer databases that screen job seekers are the culprit. With millions of applications sent by e-mail and the Web, employers lack the time to even glance at many. Instead, they sort through job-search Web sites like CareerMosaic or through their own virtual piles by asking computers to search for phrases in the résumés like "product manager" and "Microsoft Excel."

"It is a major change," said John A. Challenger, chief executive of Challenger, Gray & Christmas Inc., the Chicago firm that advises about 2,500 job searchers a year. "The more you can provide a company, in a way that makes sense, the better chance you have of being sorted out."

This movement of job hunters to the Internet is remaking one of the workplace's most enduring fixtures. "Résumés have been the same for 100 years," said

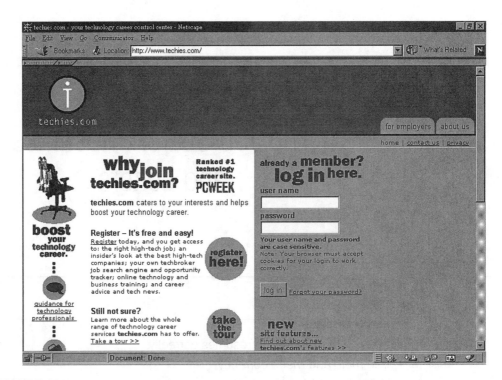

FIGURE 14.1
Internet job boards, such as techies.com, have individual résumé formats.

Jeff Taylor, the chief executive and founder of Monster.com, an Internet job site with 2 million résumés. Taylor recently retrieved résumés from decades past at universities around Boston and hung 30 of them on a wall in his Maynard, Mass., headquarters. "From five feet away, it was very difficult to delineate which was a 1922 résumé and which was a 1998 résumé," he said.

A printout of the typical online résumé, however, would be instantly recognizable from five feet because it might go on for four or five printed pages. Examined up close, it would show other contrasts. It would include long lists of accomplishments and skills, not necessarily written in full sentences. It would feature links to Web sites where employers could find more information. It might even have a section called "keywords" that unabashedly plays to the computer, seeking to pop up in as many employer searches as possible.

Some companies have tried to eliminate the traditional résumé altogether. Cisco Systems Inc., the computer equipment maker based in San Jose, Calif.,

says it must choose from among about 25,000 résumés that pour in each month to fill 750 jobs—except that many of the documents are not true résumés. They are "profiles," entered in data fields the company has set up on its Web site under the heading, "Because the best résumé is no résumé." The standardized format eliminates many worries, like what typeface or kind of paper to use for a résumé, the company says.

The one-page paper résumé is hardly on the brink of extinction, of course. Many job seekers now keep two versions on file: One for the Web, the other for actual interviews. Still, online résumés seem to be the wave of the future. And writing them does require linguistic skills, if of a new variety.

In his résumé, Enrique Ramirez, a 1998 graduate of Oberlin College, changed "ran experiments" to "researched" with the computer in mind.

"You have to word it just right," said Ramirez, 24, who landed a job with a health-care company near Chicago.

Similarly, Richard J. Carnoske, a software consultant in Tulsa, Okla., searches the Web to see how employers write job listings. When appropriate, he then uses their terms—like "e-commerce," which he added six months ago—in his résumé.

The new science of wording means that verbs, for years seen as the hallmark of a good résumé, can be less useful than nouns because recruiters more often search for nouns. Some job searchers even ponder the form of the nouns they are using, said Ben Bellimson, a senior vice president at CareerPath.com. "Management," for example, can be dangerous because it would be missed by a recruiter who typed in "manager." Job seekers who know about databases put down both words.

Internet job hunting can bring on its own headaches. Bullet points and boldface, often lost in the translation between software programs, can produce garbled text. So plain text is becoming more popular.

Still, Ramirez said, "You really don't know how it looks to the person receiving it."

There also is no guarantee that all that extra information will help. Many companies still take written applications more seriously than those submitted over the Internet because the written ones tend to indicate a higher level of interest, job search specialists said. And the size of some applicant pools has been vastly increased because the Internet has made the résumé-writing process much less time-consuming.

Anderson Consulting, for example, now receives an astonishing 4 million résumés each year. It uses computer searches to help determine which 350,000 candidates should be interviewed and which 15,000 should be hired. In other words, the statistical chances of landing a job at Anderson these days are about 1 in 250.

Reading Comprehension Questions

1. Who is the intended audience of this article? _____

   _____

2. What will the audience do with this information? _____

   _____

3. Because of the growing use of the Internet for job hunting, what are some changes happening to traditional résumés? _____

   _____

4. What are keywords? What is their function in résumés? _____

   _____

5. Why would job hunters keep two versions of their résumés? _____

   _____

6. What are the problems faced by résumé writers when using the Internet? _____

   _____

WRITING: The Job Search

The reading article discusses the job-hunting strategy of posting résumés on Internet "job boards," the Web sites that publicize résumés to potential employers and recruiters. Many employment recruiters use both traditional and electronic sources to find qualified job applicants who can fill their companies' employment needs. Good search strategies can prevent wasted time for both the employer and the job applicant.

This chapter focuses on preparing a traditional and online résumé and two sample letters (cover and follow-up letters) to use as models for future use. You have already practiced many of the skills needed for the job search, such as writing business letters

(Chapter 12) and interviewing (Chapter 13). You will also start gathering the work and education history needed to complete a job application.

As you approach graduation, you will begin the job search in earnest. People conducting a serious job search treat it like a job: they set daily objectives, keep records of all contacts and actions, and network with friends and classmates. They use discipline to stay focused and positive—disappointments are bound to occur. You can find many books on the topics of résumés and job search techniques. They contain current trends as well as traditional information, written by specialists in the employment field.

The Basic Résumé

Think of a résumé as a technical specification that describes you, your education, related skills, and experience in a format that is attractive and professional and distinguishes you from your classmates and other competitors in the job market.

A résumé has four basic parts:

- **Job objective**—state your desired job title in one abbreviated statement.

Objective: Technical support and computer networking
   or
Objective: Position in technical support and computer networking
   or
Objective: A challenging position in technical support and computer networking

---

- **Skills related to your job objective**—List the specific skills you have that make you a candidate for the job stated in your objective. Be specific and clear about your skills. For example, you might have one list for hardware, with devices or equipment you can use, and another list for software, with computer applications and programs you can use.

HARDWARE

Intel desktop	Server and workstation motherboards
Intel processors	IBM-PC Compatibles

SOFTWARE

Windows NT	Access	Corel Word Perfect
Windows 98	Lotus 1-2-3	Adobe Photo Shop
Windows 3.1	Pascal	Microsoft Word
MS-DOS, VB	Fortran	Microsoft Visual C++

- **Work experience**—List, in chronological order (starting with the most recent), the jobs you have held and a brief description, such as a bulleted list, of accomplishments, each beginning with a past tense verb.

The list can be selective, including only those jobs (or possibly volunteer work) that relate to the job objective. (The complete list of work experience goes on the job application form.) If you have no related experience, include past jobs in which you exhibited responsibility or other transferrable skills, such as time management, dependability, or leadership. Include measurable results, if possible. "Trained new coworkers to open and close the cash registers" is more specific than "Demonstrated leadership with co-workers."

Work Experience

Candia, Inc., Miami, FL – February 1996 to present
Technical Support Engineer, Team Leader

Provide technical troubleshooting and telephone support on the Intel account for all the integrators in Latin American region.

Resolve technical problems of hardware and software on Intel's desktops, workstations, servers, motherboards, and processors.

- **Education and training**—List, in chronological order (starting with the most recent), the schools or training programs you have attended and/or completed. Include your high school only if that is the last school you attended before this college. Include the year that degrees or certifications were awarded. Use the term "attended" if you began a course of study, but did not graduate or complete it. Include all technical certifications with the year and granting agency.

EDUCATION AND TRAINING

BS, Computer Information Systems, December 2000, University of Miami, Miami, FL

Microsoft Certified Professional, June 1999, Career Institute, Tampa, FL

Brokerage License Course, attended August-November 1995

Edit your drafts. Especially check your spelling—recruiters might interpret typos as a sign of professional carelessness and screen out the résumé.

Omit from Résumés

Federal laws prohibit companies from discussing or requesting information about the following, whether in interviews or on job applications (although you can volunteer the information):

- Age
- Religion
- Race or ethnic group
- Family or marital status
- Health, physical appearance, or disabilities (including photographs)

You do not have to reveal your living situation or whether you own a car (unless the job requires a car). The interviewer can ask only whether you have a way to get to work.

If you have a disability, the interviewer can ask only whether you are able to perform the requirements of the job. Companies must make "reasonable accommodations," but not all accommodations for people with disabilities.

To save space on the résumé, omit names of references. Include reference information on the job application form.

Résumé Formats

The scope of this chapter is limited to the standard guidelines for résumés, with one example. Most libraries and bookstores have résumé-writing books with multiple examples and styles of résumés and cover and follow-up letters. In addition, some graduate-placement offices and career counselors keep examples of résumés written by other students and can provide guidance for your specific job market.

General Layout

Generally, résumés have a similar overall layout: your name and address at the top in bold letters, the four types of information separated with bold headings and white space between them, and bulleted lists rather than paragraphs. One page is best, but two is acceptable with extended experience or other factors. Experiment with fonts, organization, and layouts that are easy to read. Use only one or two fonts, and limit the use of italics or underlining—too much variety is distracting and unprofessional. Simplicity is better. Your audience is reading, sometimes only scanning, for facts, such as skills or other job requirements.

Include the words you anticipate recruiters will scan for: a job title, certifications or degrees you possess, software and hardware you can use. If you are not a U.S. citizen, clearly state whether you are authorized to work in the United States.

Paper Version

Print résumés on good quality paper (over 25% bond), usually white, ivory, or pale gray, found at office-supply stores or printing shops. Do not use regular copier or printer paper (usually 20% bond). Also purchase matching envelopes and paper for cover letters and follow-up letters.

Many laser or ink-jet printers produce letter-quality résumés, and you have the added advantage of making ongoing edits and revisions to your résumé as your search progresses. See Figure 14.2 for an example.

Some people prefer to take their final draft to a professional printer. If you use a printing company, keep in mind that the staff will not edit your résumé—the clerk simply prints what you supply on disk or paper. Ask for a proof copy, and examine it carefully before signing the go-ahead to print the entire batch. Once it is printed, if you discover a typo or error, you will have to print it again, at your expense.

Some applicants include the "paper" version as an attachment to an e-mail. This is not advisable unless a company requests it. Because of the danger of viruses on attachments, many companies have a policy to delete attachments before opening an e-mail. Instead, "embed" (paste) the résumé in the e-mail, and then reformat it to fit the e-mail message window. For this, it would be better to follow the conventions of the electronic version, covered next.

Exercise 14.1   *Complete a sample résumé that includes the four basic parts.*

Electronic Version

You might also decide to post your résumé on one or more job boards on the Internet or use other services that scan résumés electronically. In most cases, you will have to make a second version of your résumé for electronic sources (see Figure 14.3). Each job board has a template and procedure, and it will include specific instructions and tips for submitting résumés.

Investigate several job boards before deciding which to use (although there is no limit to how many you can use). Some job boards are comprehensive, serving job searchers across many fields (for example, Monster.com or CareerPath.com). They might circulate your résumé to a wider audience of recruiters; however, your résumé might end up being sent to companies based on keyword matches unrelated to your job interests. Other Internet sites specialize in certain industries (for example, ConsultLink.com and techies.com specialize in jobs requiring technical training). The audience of these sites is smaller, but possibly closer to your intended career, making it

JANE MARIE DOE

555 Lakeside Lane
Miami, FL 33110

(H) (305) 555-1222
(W) (305) 555-4141
(E) jmdoe84@aol.com

OBJECTIVE

A challenging position in technical support and computer networking.

SKILLS SUMMARY

- Precise, self-motivated, and strong quantitative and interpersonal skills.

- Experienced in dealing with a diversity of professionals, clients, and staff members.
  Fluent in Spanish.

HARDWARE

Intel desktop	Server and workstation motherboards,
Intel processors	IBM-PC Compatibles.
	Macintosh

SOFTWARE

Windows NT 4.0	Access	Corel Word Perfect
Windows 98	Lotus 1-2-3	Adobe Photo Shop
Windows 3.1	Pascal	Microsoft Word
MS-DOS, VB	Fortron	Microsoft Visual C++

EDUCATION AND TRAINING

BS in Computer Information Systems, December 2000
University of Miami, Miami, FL

Microsoft Certified Professional, June 1999

Career Institute, Tampa, FL
Brokerage License Course, attended August-November 1995

WORK HISTORY

Candia, Inc., Miami, FL – February 1996 to present
Technical Support Engineer, Team Leader

- Provide technical troubleshooting and telephone support on the Intel account for all the integrators in Latin American region.

- Resolve technical problems of hardware and software on Intel's desktops, workstations, servers, motherboards, and processors.

FIGURE 14.2
Traditional résumé.

JANE MARIE DOE
555 LAKESIDE LANE
MIAMI, FL 33110

(H) (305) 555-1222
(W) (305) 555-4141
(E) jmdoe84@aol.com

OBJECTIVE: A challenging position in technical support and computer networking.

SKILLS SUMMARY: Precise, self-motivated, strong quantitative and interpersonal skills. Experienced in dealing with a diversity of professionals, clients, and staff members. Fluent in Spanish.

HARDWARE: Intel desktop, server and workstation motherboards. Intel processors. IBM-PC compatibles. Macintosh.

SOFTWARE: Windows NT 4.0. Windows 98. Windows 3.1. MS-DOS, VB. Fortran. Pascal. Corel Word Perfect. Adobe Photo Shop. Microsoft Office, Visual C++. Lotus 1-2-3.

EDUCATION AND TRAINING
BS in Computer Information Systems, December 2000
University of Miami, Miami, FL

Microsoft Certified Professional, June 1999

Career Institute, Tampa, FL
Brokerage License Course, attended August-November 1995

WORK HISTORY
Candia, Inc., Miami, FL – February 1996 to present
Technical Support Engineer, Team Leader

Provide technical troubleshooting and telephone support on the Intel account for all the integrators in Latin American region.

Resolve technical problems of hardware and software on Intel's desktops, workstations, servers, motherboards, and processors.

FIGURE 14.3
Online résumé.

a more targeted search. Résumés are usually visible only to subscribing companies or recruiters.

If you are using an Internet job-listing service, you might have to register and complete your résumé online. To do this, you make selections and type (or paste) information into boxes, and the service formats your résumé in a uniform manner.

Other services might request a résumé in ASCII format, free of all text attributes. Follow the online instructions carefully. For example, you might have to remove bullets, which are sometimes scanned as question marks. Margins might have restrictions. Generally, all headings and text should align with the left margin—no centering or double columns—to ensure correct scanning.

Because job boards rely primarily on keyword searches, think about the keywords related to the job you are seeking, and work them into your résumé, possibly using alternative terms or varied word endings. Experienced recruiters know to search for variations of job titles, but why risk it? Also, make sure that other keywords are available and prominent on the résumé, such as the type of degree or certification you possess.

Exercise 14.2   *Reformat your résumé for electronic transmission.*

Web Pages

If you have a Web page, you can make it available to recruiters and take advantage of the Internet tools. For example, you can provide a hyperlink in your electronic résumé to open your Web page. Or you can put your résumé on your Web page, and provide the Web address to prospective employers.

On the Web page, you can add hyperlinks to your former company or manager's name that will open the company's home page or the manager's e-mail for quick reference checks. Or add hyperlinks that open your portfolio, work samples, research papers, or projects demonstrating your skills. This is especially important if the job requires proficiency with computer programs, such as graphic arts or technical writing. Provide your résumé in multiple formats, such as .DOC, .RTF, and .HTM, and make sure the font is common for most computers, such as Arial or Times New Roman.

A note of caution about Web pages: If you plan to invite people to look at your Web page as part of your job search, take an objective look at your Web-page design before you do so. Make sure it projects the professional and businesslike image you want to convey.

If the Web page contains unrelated text or graphics, such as cartoons, jokes, or photos of family or pets, consider removing them until your job search is over. If the page looks silly to a recruiter, you might not be considered for a job.

Exercise 14.3   *Critique your own or a friend's Web page and recommend revisions to prepare it for use in a job search.*

References

Most companies expect a minimum of three references (none related to you). Generally, you must include at least one person as a professional reference—someone who can vouch for your technical ability or job performance, such as a former manager, teamleader, or instructor. Some companies specifically request the name and phone number of your current or most recent manager. If this is not possible or not in your best interests (perhaps your manager doesn't know you are in the job market), be ready to explain why and suggest someone else. The other two references can be character references, such as long-term friends, coworkers, instructors, ministers, or coaches.

Before using a person's name as a reference, ask the person's permission. If you see the person regularly, simply ask if you can use his or her name as a reference. Otherwise, call, e-mail, or write a letter to the person (see Figure 14.4). You might have to remind the person of the time during which you worked together.

A casual "OK" from the person is fine—written permission is not necessary—but use the opportunity to confirm the person's current address, phone number, and e-mail address.

If the person declines your request (perhaps the person will be unavailable during your job search), find another reference. Re-establish permission each time you enter the job market—do not assume permission given this year will apply next year.

Exercise 14.4   *Write a letter to a professor or former manager asking permission to use the person's name as a reference.*

555 Lakeside Lane
Miami, FL 33110
September 4, 2000

Professor D. B. Carr
University of Miami
Miami, FL 33186

Dear Professor Carr,

I would like to use your name as a reference during my job search this winter. I will graduate in December and plan to interview with software companies in the local area and in
northern Florida.

I was in your software design class for both CIS 330 and 332. My final project was the Token
Ring proposal that you implemented last summer. I am including this proposal and a summary of the completed project in my portfolio to show prospective employers.

You can respond to me either by e-mail (jmdoe@aol.com) or by telephone (305–555–1221).

If you are willing to be a reference for me, please confirm that the following telephone number and e-mail address are correct:

     (305) 555-3000, X336   **or**   dbcarr@mindspring.com

I look forward to hearing from you.

Sincerely,

Jane M. Doe

Jane M. Doe

FIGURE 14.4
Letter requesting permission to use a person as a reference.

Sensitive Situations

The job search can present unique problems and dilemmas for some people, but fore-
thought and preparation can help prevent negative consequences later. The most important guideline is to be honest—do not lie, mislead, or avoid the truth.

First, examine your résumé carefully, making sure each statement is honest. For example, if you claim to be "proficient" in a skill, imagine yourself getting hired to perform exactly that skill, on your own and under pressure. Some new-hires find out that
exaggerations can lead to embarrassment and even termination from a job.

Most companies check all or some work history, references, and schools. Sometimes
a new-hire can start working before the background check is complete. It is completed
eventually, and discrepancies can result in termination.

When you are invited for an interview, ask the human resources representative of
the company whether it requires any tests or screenings. In addition to employment
tests, some companies routinely administer drug screenings, polygraph (lie detector)
tests, and FBI fingerprinting and background checks. A few companies and government
agencies require a full security check to obtain a top-secret clearance rating. If you prefer not to take the tests, decline the interview.

If there are periods in your past that require explanations, such as a gap in work history, being fired from a job, or a legal situation, prepare an explanation that is as posi-

tive as possible. Rehearse the worst-case scenario: a direct question about it. Talk to an instructor or someone you trust for ways of handling difficult personal information. Then respond to questions directly and promptly in the interview.

Finally, once you sign a work agreement or contract with a company, stop your job search. If you get another job offer or become a final candidate at a second company, ask the second company to remove your name from further consideration. This is common courtesy.

Exercise 14.5 *Provide answers to the following.*

1. Were you ever fired from a job? If so, why?

2. Explain any gaps in your work history, starting from your first job.

3. Using the Internet and other resources, describe the purpose of and procedure for obtaining an FBI fingerprint and background check.

4. Using the Internet and other resources, discuss the Employee Polygraph Protection Act of 1988.

Job Applications

Most companies expect candidates to complete a job application and show a Social Security card and proof of citizenship. On the application form, you must complete all information (providing it does not relate to age, race, religion, health, or marital status). When completing the work history, include every job, and be prepared to account for gaps in your work history.

Include all education, usually starting with high school, and the years attended and degrees awarded. Include the names and phone numbers of three references. Begin now to start a work-history record of former managers' names, addresses, job titles and responsibilities, and starting and ending dates and salary—use additional paper if necessary. Take the information with you to interviews. Otherwise, you might have to take the application home with you, delaying the completion of your application package.

Exercise 14.6 *Complete the job application in Appendix 6, or obtain and complete a job application from a company.*

Cover Letters

Cover letters are formal business letters that accompany résumés. They provide an opportunity to introduce yourself and set the stage for the résumé. Include a cover letter with each résumé that you mail or fax, even when someone requests your résumé. (Cover letters are not submitted to Internet job boards.)

As with every letter, check the spelling and grammar carefully. Typos and carelessness in the letter can cause the recruiter to stop reading and discard your résumé.

Write a personalized letter to each company. A generic letter will not get the attention of the reader.

- State the purpose of the letter in the first paragraph. For example, state the job title you seek, the company's name, and how you heard of the job opening.

- In the next paragraphs, highlight the accomplishments in your work history or education that make you especially suited for the job. For example, describe a specific project that you completed that relates to the job opening. Keep the information in measurable terms.

- Finally, end the letter with a statement of your next action, such as when you will call to arrange an interview. This establishes your interest and takes the responsibility for moving the job search forward to the next step, the interview.

Figure 14.5 shows the main parts of a cover letter.

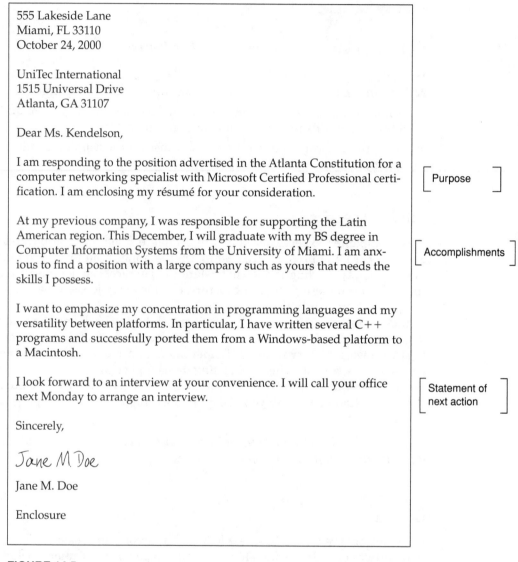

555 Lakeside Lane
Miami, FL 33110
October 24, 2000

UniTec International
1515 Universal Drive
Atlanta, GA 31107

Dear Ms. Kendelson,

I am responding to the position advertised in the Atlanta Constitution for a computer networking specialist with Microsoft Certified Professional certification. I am enclosing my résumé for your consideration.   [Purpose]

At my previous company, I was responsible for supporting the Latin American region. This December, I will graduate with my BS degree in Computer Information Systems from the University of Miami. I am anxious to find a position with a large company such as yours that needs the skills I possess.   [Accomplishments]

I want to emphasize my concentration in programming languages and my versatility between platforms. In particular, I have written several C++ programs and successfully ported them from a Windows-based platform to a Macintosh.

I look forward to an interview at your convenience. I will call your office next Monday to arrange an interview.   [Statement of next action]

Sincerely,

Jane M Doe

Jane M. Doe

Enclosure

FIGURE 14.5
Cover letter.

Exercise 14.7   *Write a sample cover letter to accompany your résumé for a position in your field to the following person:*

Mr. Richard Williams
UniTec International
1515 Universal Way
Atlanta, GA 30307

Job Interviews

In Chapter 13 you practiced answering typical questions asked during employment interviews. Review those questions and answers.

Before you begin to interview for jobs, use the following suggestions to pinpoint your career goals and help to narrow your job search:

- Consider your short-term and long-term career goals. Do you see yourself working to the top of a large organization? Or do you see yourself running your own business? Interviewers ask about career goals mainly to verify your sustained interest in the field. Of course, keep your response related to a professional career. If you state that your goal is to retire to a Caribbean island, you might not be considered as a serious candidate.

- Keep up with the industry. Read trade journals, both local and national. Attend trade shows, if possible. And start networking with professionals in your area. This will keep you informed of trends, events, and current developments in the field. It can provide employment leads, as well as useful information to discuss during an interview. Companies value employees who are willing to investigate current topics on their own.

- Research each top company in the industry. Find out about its history, origin, and company structure. Identify its main business, such as software, hardware, service, manufacturing, and so on. Make sure that your qualifications are in line with the mission of the company and that you are motivated to join the company. You might be asked why you chose to apply to the company, and an informed, thoughtful answer will demonstrate your interest.

- Decide whether you are willing to relocate, travel as part of the job, or work at multiple work sites or even at home. While this might not be a subject during the initial interview, relocation or travel constraints could become a deciding factor.

Just prior to the interview:

- Ask for directions to the interview location ahead of time, so you won't have the added stress of getting lost.

- Wear clothes that are appropriate for the company. For a large, conservative company, wear a traditional two- or three-piece suit. For a casual company, you might wear "business casual" clothing, such as a sports jacket for men, or a dress or pants suit for women. Consult your graduate placement office for the best advice. For unknown companies, go to the site and observe what employees wear to work. In most cases, however, do not wear denim or sports clothes to interviews.

- Arrive for the interview early. Give yourself some time to collect your thoughts and freshen up—take a glance in a mirror. You want to appear poised and confident, even if you are nervous and stressed. You might even have time to complete the job application form before the interview starts.

During the interview:

- Whether the interview is with one person or a panel of people, try to write down the name and job title of each person—you might not remember them otherwise. Then respond to people using their names, demonstrating your poise under pressure. In addition, you'll have names for your follow-up letters.

- Shake hands with each interviewer. Make sure your handshake is firm, but not a bone-crusher.

- Keep your answers focused. You might have only 30 minutes in which to sell yourself. Although the interviewer might ask a casual question as an ice-breaker, do not spend much time on the response. Try to move the interview into the real questions, and use every minute productively.

- Look people in the eye normally as you listen and respond. Evading eye contact can make you appear suspicious or even dishonest.
- Smile! Nervousness can sometimes cause faces to look pinched. A smile now and then relaxes the face and possibly will help you relax as well.
- If a work sample or performance test is required, ask about time limitations for the test. Watch the clock and pace yourself to complete it within the time limits.
- Typically, at the end of an interview, the interviewer will ask if you have any questions. This is an opportunity to ask about the company and its future plans.
- Resist the urge to ask about salary or benefits, even though these things might be your foremost questions. Interviewers do not normally discuss salary and benefits in the first interview.
- Ask what the next step in the process will be and when the interviewer will make a hiring decision. Most interviewers are sympathetic to the applicant's need for feedback and closure. But keep in mind that companies sometimes move slowly in the hiring process.
- Thank each person individually as you leave. Another handshake is appropriate. If the person has a business card, ask to take one with you. Again, it will be valuable when you write a follow-up letter.

Exercise 14.8   *Write answers to the following questions. Be specific.*

1. What are your short-term career goals?
2. What are your long-term career goals?
3. What motivates you?
4. What type of company are you most interested in working for? Discuss why.
5. Where would you most like to work (city, state, region, or country)?
6. What are your constraints in business travel?
7. What resources are available in your school or city library for researching companies? List up to five resources.

Follow-Up Letters

Job seeking is a formal process. Send a follow-up letter immediately after the first interview thanking the person for the interview. Use the letter to restate why you are the right candidate for the job (if it's true). If you do not know how to spell the person's name, call the company's receptionist. In the first paragraph, remind the person of the interview, including when it was. Then answer any open questions or provide follow-up information. Finally, restate your interest in the position and how you can be reached. Figure 14.6 shows an example.

Exercise 14.9   *Prepare a follow-up letter for an interview held with the following person:*

Mr. Richard Williams
UniTec International
1515 Universal Way
Atlanta, GA 30307

555 Lakeside Lane
Miami, FL 33110
November 1, 2000

UniTec International
1515 Universal Drive
Atlanta, GA 31107

Dear Ms. Kendelson,

I want to thank you for the opportunity yesterday to interview at UniTec International. It was a pleasure meeting you and Mr. Santoya to discuss the computer networking position available in January. | Reminder of interview |

When I arrived home, I located the networking proposal that I completed at my last position that I described to you, and I am enclosing it for your review. The transition to the new system at my former company was successful, thanks largely to the preparation and research provided by the proposal. | Answer open question |

I was impressed by the direction your company is taking, and I hope I can become part of it. I look forward to hearing from you about the position at jmdoe | Restate your interest in the job |

Sincerely,

Jane M. Doe

Jane M. Doe

Enclosure: Proposal

FIGURE 14.6
Follow-Up letter.

SPELLING: Technical Terms in Résumés

Technical terms and jargon are appropriate only for technical audiences. In most cases, the people reviewing résumés will be technical and will scan or search for specific jargon in résumés. Without the use of technical terms, and more importantly, the correct spelling and punctuation of technical terms, Internet searches might miss your résumé.

Frequently, technical jargon is based on abbreviations, acronyms, or registered or trademarked words. It is important to spell jargon in the standard way, using upper- and lowercase letters and spacing correctly. Before writing technical terms, observe the spelling carefully in textbooks, manuals, and industry magazines. Or conduct an Internet search to find the correct product or company spelling.

Examples of technical terms in computer programming include:

Fortran   C, C++, C-language program   Java, J++, Jscript, JavaScript

Exercise 14.10   *List 10 technical terms (including abbreviations and acronyms) used in your career field that you could include on your résumé. Confirm the spelling and punctuation.*

1. _____

2. _____

3. _____

4. _____

5. _____

6. _____

7. _____

8. _____

9. _____

10. _____

VOCABULARY: Prefixes *meta* and *mega*

Mega is a prefix meaning "one million" (in science) or "large" (in general language).

Megabyte – One million bytes
Megamall – Large shopping mall

Meta is a prefix meaning "changed" (in science) or "after, beyond or higher" (general language).

Metabolism – The changing of food into living tissue
Metaphor –   A figure of speech in which one thing is spoken of as
            if it were another, such as "all the world's a stage"

Exercise 14.11   *Use the prefixes* mega *and* meta *to complete the following words.*

1. The candidate used a _____search engine to conduct a large search.
2. The _____ton bomb had the force of a million tons of TNT.
3. A _____phone can increase the volume of the voice.
4. The cancer _____stasized to one part of the body after another.
5. Her experiences in college caused a complete _____morphosis in her attitude about science.
6. The researcher collected data from many studies and analyzed the results at a higher level, called a _____nalysis.
7. After a decade of phenomenal growth, the Atlanta has become a _____lopolis.
8. He took several capsules, a _____dose, of vitamin C to ward off a cold.
9. The program requires over 110 _____bytes of free disk space.
10. The part of the foot over the flat sole (between the ankle and toes) is called the _____tarsus.

WORD WATCH: *infer* and *imply*

Infer and *imply* are both verbs that refer to making assumptions based on hints or indirect information. The difference in usage is subtle: writers imply; readers infer. Or speakers imply; listeners infer.

Use *imply* for suggesting without stating (sending out the information).

> What does the brochure imply about the company?
> The brochure implies that the company uses a Linux operating system.

Use *infer* for drawing a conclusion (taking in the information).

> What can you infer from the brochure?
> From the brochure, I inferred that the company uses a Linux operating system.

Exercise 14.12   *Complete the sentences with* infer/infers *or* imply/implies.

1. The job board _____ that there are many positions available for technicians.
2. I corrected my résumé so no one would _____ that I had management experience.
3. The department manager didn't _____ that the position included a signing bonus.
4. The recruiter can only _____ that the company's opening is immediate.
5. Some job descriptions _____ that salaries are negotiable.
6. My cover letter _____ that I am willing to relocate.
7. The name of the job board _____ that it specializes in high-tech careers.
8. The term "passive job hunters" _____ that the candidates don't want their employers to know they are looking for another job.
9. As the name _____, "active job hunters" are willing to put their names on their posted résumés.
10. What did you _____ from the research about the company?

PART IV

Grammar Units

Grammar Unit 1: Subjects and Verbs

Grammar Unit 2: Subject and Verb Agreement

Grammar Unit 3: Prepositional Phrases

Grammar Unit 4: Pronouns

Grammar Unit 5: Pronoun References

Grammar Unit 6: Avoiding Shifts

Grammar Unit 7: Avoiding Sexism

Grammar Unit 8: Modifiers

Grammar Unit 9: Clauses and Simple Sentences

Grammar Unit 10: Compound Sentences

Grammar Unit 11: Complex Sentences

Grammar Unit 12: Fragments, Run-Ons, and Comma Splices

Grammar Unit 13: Transition Words

Grammar Unit 14: Parallelism

Subjects and Verbs

The most fundamental elements of a sentence are the **subject** and **verb.** Without both, a written idea is incomplete and cannot be called a sentence. A subject can be either a noun or a pronoun, and a complete verb may consist of up to three verbs, either action or linking. We begin this review with a careful look at nouns.

Nouns

Most people can remember the informal definition of a **noun:** a person, place, thing, or idea. The first three types (persons, places, and things) are the easy nouns to spot. They can be singular or plural. The plural is usually formed by adding *s* or *es.*

The final type of noun is the idea. In fact, the word *idea* is a noun. Other examples of idea nouns are found in the following sentence.

> *Example:* Your *education* and hands-on *training* will be useful in a technical *career.*

You can see that the italicized nouns in the example above are ideas, not things that we can see or touch. One test is to say the word with *a, an,* or *the* in front of it.

> an education
> the training
> a career

If doing so makes sense, it is a noun. If not, it is probably some other kind of word.

You will probably remember that **pronouns** are similar to nouns. They take the place of nouns. In these exercises, we do not consider pronouns as nouns. We deal with pronouns in a later unit.

Exercise 1   *Underline the nouns in the following sentences. The number of nouns in each sentence is given in parentheses.*

1. Pliers are used for many jobs. (2)
2. The sharp teeth and the strength of their grip can damage fine objects. (4)
3. The power must be appropriate for the type of wood or metal. (4)
4. Many companies have produced imitations of the original Vise-Grip built by Peterson Manufacturing. (4)
5. The factory in DeWitt, Nebraska, still uses the original logo. (4)
6. The first patent was issued in 1921. (2)

7. The first wrench was made of forged steel and had an adjusting bolt, but the lever that locks the jaw was missing. (5)

8. The hand-locking principle is now being used all over the world in hand tools. (3)

9. The hardest job for the Peterson family was marketing the new tool. (3)

10. The sons loaded their trunks and began traveling around the countryside, showing the pliers at county fairs. (5)

Exercise 2   *Write your own nouns in the following categories.*

People	Places	Things	Ideas
Name people famous for scientific principles.	Name cities where you would like to live.	Name tools or devices you use in your field.	Name qualities that companies look for in graduates.
_____	_____	_____	_____
_____	_____	_____	_____
_____	_____	_____	_____
_____	_____	_____	_____
_____	_____	_____	_____

Subjects

Subjects are almost always nouns or pronouns (words that take the place of nouns). They are usually found near the beginning of a sentence. They tell who or what the sentence is about. In the following example, the subject is italicized.

> *Pliers* have many applications.

The sentence is about *pliers* and thus we call that word the *subject* of the sentence. There are other nouns in the sentence, but only one subject.

Some sentences have more than one subject, and we call these compound subjects. The compound subjects are italicized in the following example.

> The locking *jaws* and gripping *teeth* make the tool useful.

Subjects are usually written at the beginning of a sentence, although they can appear at other places in the sentence, too. The delayed subject is italicized in the following example.

> With the Vise-Grip wrench, *William Peterson* hit upon a universal tool.

The preceding sentence is not about the wrench (the first noun in the sentence), but about William Peterson.

Exercise 3   *Draw a line under the subject(s) of each sentence.*

1. The first pliers sold for $1.95 in the 1930s.

2. At that time, 200-pound hogs sold for $2.50.

3. Nickel and silver have been used as coatings.

4. The uses are as assorted as the users.

5. The product is exported all over the world.

6. The bar and C-clamps are especially good for woodworking.

7. Motorcyclists use locking pliers for emergency foot pegs and kick-start levers.

8. The pliers have been used as temporary faucet handles.

9. Other locking tools available today are imported.

10. The quality can vary, and some tools don't hold up at all.

Verbs

A **verb** is usually defined as a word showing action—what is, was, or will be happening. All the following words are action verbs—things you can do.

    learn   connect   follow   wire   refer   display

Verbs have the special job of indicating the time of the action by **verb tense.** There are three main tenses: past, present, and future. The tense is changed when the writer changes the time:

Years ago, technicians *recorded* on notepads.
Today, technicians *record* on service report forms.
In the future, technicians *will record* on computers.

Helping Verbs and Linking Verbs

Another group of verbs, called **helping verbs,** are commonly used in combinations with each other or action verbs. Some of them, called **linking verbs,** can be used alone, while others (such as *be* and *been*) cannot be used alone.

am	is	are	was	were	be	been
have	has	had	do	does	did	will
shall	can	must	would	could	should	might

Whenever you see these verbs, they are either the verb or part of the verb in a sentence. It is the only job these words do, unlike most other English words. Sometimes they are followed by action words, and other times they are used alone.

Pliers *can damage* wood and metals.
(*helping verb + action verb*)

The grandson of William Peterson *is* the current president.
(*linking verb used alone*)

Participles

There are two other common types of verbs, both called participles. One is the **present participle,** or *-ing* form, and the other is the **past participle,** or *-ed* form. Both of these verb forms require a helping verb. They add a continuous time frame, an extended present or past time period, to sentences.

Present participle: Technicians *are using* integrated circuits.
(*extended present use*)
Past participle: Technicians *have used* integrated circuits.
(*extended past use*)

Regular verbs have predictable patterns.
To change from present to past tense, add *-ed* to the end of regular verbs.

Today we *tune* the car. (*present*)
Yesterday, we *tuned* the car. (*past*)

To change to the future tense, add *will* in front of the present-tense verb.

We *will tune* the car tomorrow. (*future*)

To use the past-participle form, add a helping verb to the *-ed* (past-tense) verb.

We *have tuned* the car several times. (*past participle*)

To use the *ing* form, use a helping verb and add *-ing* to the present-tense verb.

We *are tuning* the car. (*ing form*)

Note: Remember to write the final *ed* on past- or participle-tense verbs even if you do not pronounce them when you speak. Leaving them off in writing makes you sound un-educated and careless.

Wrong: I am *suppose* to be in class.
Correct: I am *supposed* to be in class.
Wrong: We *use* to have lab on Wednesday.
Correct: We *used* to have lab on Wednesday.

Irregular verbs have unusual forms in the past- and past-participle tenses and have to be remembered. If you have ever heard a young child say, "We runned down the street," you recognized that the child overgeneralized the *ed* rule and will eventually learn about irregular verbs. Common irregular verbs are listed in Appendix 5. The past tense is formed in some other way than by adding *-ed*. Notice that the last column is the participle form, which is used following a helping verb. *Bring* is an example of an irregular verb.

Today, I *bring* suggestions. (*present*)
Yesterday, I *brought* nothing. (*past*)
I *will bring* all the connectors. (*future*)
I *have brought* everything for the circuit. (*past participle*)
I *am bringing* the power supply. (ing *form*)

Exercise 4   *Underline the complete verb twice in each sentence. Underline the subject once.*

1. Seams can spoil the look of countertops.

2. A new, special joining system is now being used.

3. The system consists of a special router bit and router baseplate.

4. The worker adjusts the step in the baseplate.
5. A straightedge guides your cuts.
6. Each rout cuts a smooth wavelike profile along the edges to be joined.
7. The additional surface area gives you a stronger glue joint.
8. The two surfaces will be dead flush.
9. The result is a smooth, nearly invisible seam.
10. For professionals, the system is unbeatable.

Inferred *You*

In the following sentence, the subject is not written as a word.

> Connect the two lines.

The action, or verb, is *connect*. The person who will do the *connecting* is the reader. We call this type of subject the **inferred "you,"** which is known as the imperative mood. The verb is always a command in the present tense. We use this type of sentence only for instructions.

Exercise 5   *Underline the subjects once and the verbs twice in the following sentences. If the subject is the inferred "you," underline the present-tense verb.*

1. Unlock the toggle joint.
2. The spring pulls open the movable jaw.
3. Close the pliers to grip the object mechanically.
4. Push down on the release lever.
5. The spring pulls the lower jaw open again.

Exercise 6   *Supply the correct form of the given verb to fit the sentence. Refer to Appendix 5 for irregular verb forms.*

1. *(see)*   I have _____ a great change in woodworking tools.
2. *(swear)*   I _____ I would always use safety glasses.
3. *(write)*   I have not _____ anything longer than a three-page report.
4. *(cut)*   I _____ part of the report after I read it.
5. *(take)*   It has _____ me several hours to read the operating manual.
6. *(break)*   I _____ my concentration when a friend came.
7. *(do)*   The report was _____ before midnight.
8. *(give)*   I have _____ up my habit of putting things off.
9. *(prove)*   The argument could not be _____.
10. *(throw)*   After the report was returned, I _____ my backup copy away.

Exercise 7   *In the following article, underline subjects once and verbs twice. Then, rewrite the article by changing the present-tense verbs to past tense.*

Planning Something Big?

Architects make a model of a project pretty early in the planning stage. Builders and homeowners do this, too. No special skills are required. The designers place a floor-plan grid over the existing floor plans. They build a three-dimensional model. The kit contains scale building materials. An architectural designer creates and sells the kit. Remodeling costs less and goes more smoothly. The owners feel more satisfied with the final result.

_____

_____

_____

_____

_____

_____

_____

Exercise 8   *Use each present tense verb in a sentence. Then rewrite the sentence, changing the verb to past, future, past participle, and* ing *forms.*

Example: Come

a. (*present*)	We *come* to the site.
b. (*past*)	We *came* to the site.
c. (*future*)	We *will come* to the site.
d. (*past participle*)	We *have come* to the site.
e. (*ing* form)	We *are coming* to the site.

1. Break

a. (present) _____

b. (past) _____

c. (future) _____

d. (past participle) _____

e. (*ing* form) _____

2. Go

a. (present) _____

b. (past) _____

c. (future) _____

d. (past participle) _____

e. (*ing* form) _____

3. Write

a. (present) _____

b. (past) _____

c. (future) _____

d. (past participle _____

e. (*ing* form) _____

4. Do

    a. (present) _____

    b. (past) _____

    c. (future) _____

    d. (past participle) _____

    e. (*ing* form) _____

5. Bring

    a. (present) _____

    b. (past) _____

    c. (future) _____

    d. (past participle) _____

    e. (*ing* form) _____

Sentence Combining

Sentences can be as short as one word. In the sentence "Stop!" there is only a verb, and the subject is the inferred "you," the person or people being addressed. Most sentences are longer. Sentences can be much longer as a result of combining ideas. The following is an example of a long sentence.

> Many businesses are exploring the multitude of applications that a computer system can offer to make the businesses more productive, efficient, and profitable.

This sentence could be divided into the following smaller sentences, more like the writing style in grade-school science books.

> Many businesses are exploring applications.
> There are a multitude of applications.
> The applications are for a computer system.
> The computer system can make the businesses more productive.
> The computer system can make the businesses more efficient.
> The computer system can make the businesses more profitable.

The ideas in the short sentences deliver the same message as the one longer sentence. Notice that the group of short sentences has more words, and some ideas have been repeated. Although simplicity is often effective, technical writers use as few words as possible and avoid unnecessary repetitions.

One method of practicing efficient, clear wording is to combine short sentences into complex, powerful sentences. The purpose is to reduce the number of words. In the next exercise, you will be given a group of short, related sentences, and you will reword them into one sentence. Remember that there is no one correct way of combining. In fact, you will perhaps see several possible ways of combining each group. The best combination is the one that delivers the message clearly and correctly. Consider the following short sentences.

> *Example:*   a. A computer system consists of areas.
>   b. There are three main areas.
> *Combinations:* A computer system consists of three main areas.
>   There are three main areas in a computer system.

Both of these sentences are grammatically correct; however, the first sentence is written in the active style, which is preferred in technical writing. It introduces the subject (computer system) first, has an active verb, and has one fewer word. Avoid using "There are" at the beginning of sentences.

When combining sentences, *use the first sentence as the core.* You may add or eliminate some words, but do not change or add to the intended meaning. You will frequently be reordering and reducing the number of words. For instance, "the man is young" could easily be reordered "the young man . . ." to allow more ideas to be added.

Caution: It is possible to link each sentence with *and,* but this would not reduce the number of words.

Exercise 9   *Combine each group of sentences into one complex sentence. Do not link them only by using* and, *and do not change or add to the meaning. You may add or change words if necessary, and you may reorder the words and ideas.*

1. a. A computer system consists of areas.
    b. There are three main areas.
    c. One area is input.
    d. One area is processing.
    e. One area is output.
    Combination: _____

    _____

2. a. There is a fourth area.
    b. This area is storage.
    c. It backs up the computer system.
    Combination: _____

    _____

3. a. The system is made up of hardware.
    b. The hardware is machines.
    c. The system is made up of software.
    d. The software is programs.
    Combination: _____

    _____

4. a. The programs turn data into information.
    b. The data are unprocessed.
    c. The information is usable.
    Combination: _____

    _____

5. a. Data may be collected.
    b. Data may be processed.
    c. It may happen immediately.
    d. It may happen later in groups.
    Combination: _____

    _____

Subject/Verb Agreement

Every complete sentence has a subject and a verb. The subjects can be plural or singular, and verbs have tenses. Another consideration in writing standard, complete sentences is subject/verb agreement.

Agreement is a concept similar to that used when matching the right battery to an object. A car battery does not "agree with" a wristwatch. We decide which type of battery we need for a certain object, such as a car or wristwatch, and use only the correct type. In writing, we use the type of verb that agrees with its subject.

Subject/verb agreement is a troublesome problem for some people. We might expect that subjects and verbs follow the same singular and plural rules. They do not—but the rules are easy to follow once they are analyzed.

One comforting fact is that subject/verb agreement is necessary only when writing in the present tense. Also, agreement does not seem to be a problem when the subject is *I* or *you*.

This problem centers on the letter *s*. Read the following sentences.

> The *laser excites* electrons.
> The *lasers excite* electrons.

We usually write the subject first and then make the verb agree with it. For this situation, let's call the plural form of the noun the **s form** (which usually has an *s* at the end). Verbs also have *s* forms; however, the rule for when to add an *s* to a verb is the opposite of the rule for nouns.

Follow these general rules:

Rule: Add an *s* to a verb when the subject is singular. Do not add an s to a verb when the subject is plural. Or stated another way: *Either the subject or the verb will have an s.*

> The laser excites electrons.
> (s *form*)
> The lasers excite electrons.
> (s *form*)

Rule: In most present-tense sentences, either the subject or the verb will have an *s*, but not both.

As you know, some plural nouns do not end in *s* (*men, women,* and *children*), but we consider them to be *s* forms anyway.

> The men *supervise* the test.
> (s *form*)
> The man *supervises* the test.
> (s *form*)

Similarly, some nouns always end in *s* (*communications, physics, news,* and *scissors*). In these cases, determine whether you would refer to the noun as *it* or *they*. *Communications* is considered a singular noun (*it* is changing rapidly), and *scissors* is considered a plural noun (*they* are in my desk). Then use the correct verb.

Warning: Use the standard spelling rules for adding an *s* to a verb:
> carry—carries   display—displays

Changing the verb to the past tense eliminates the agreement problem.

> The engineer *supervised* the test.
> The engineers *supervised* the test.

Changing into the present (*-ing*) or past (*-ed*) participle form does not eliminate the problem, since helping verbs also have *s* forms.

> The men *were* supervising the test.
> (s *form*)
> The man *was* supervised during the test.
> (s *form*)

Notice the *s* forms of the following verbs.

s Form	Non-s Form
is	are
was	were
has	have
does	do

Exercise 1   *Rewrite the following sentences changing the subject from singular to plural, or from plural to singular. Then change the verb to agree with the subject. Keep the verbs in the present tense.*

> ***Example:***   *Circuits are* paths for current.
>   (s *form*)
> ***Changed:***   A *circuit is* a path for current.
>   (s *form*)

1. Tiny semiconductor lasers are the heart of grocery store scanners.   _____

_____

2. A scientist is working to develop a computer memory that uses a laser.   _____

_____

3. The technique increases computer memory and provides faster access to it. ____

_____

4. Each card is etched on a thermoplastic square. _____

_____

5. The laser has worked into many facets of society. _____

_____

6. They have transformed the science of eye surgery. _____

_____

7. The laser has been used to keep tunnels true. _____

_____

8. Laser disks store and play back up to 100,000 images. _____

_____

9. An experimental computer memory uses laser disks to store information. _____

_____

10. A laser processes the images for robots and missile guidance systems.

_____

_____

Special Case 1: Sometimes a sentence begins with an indefinite word such as *there, here,* or *where.* In sentences like this, the subject will come after the verb, and the verb must agree with what comes after it.

> *Example:* There is an illustration.
> There are two illustrations.

Note: The contraction for *there is* is *there's. There are* cannot be contracted.

Exercise 2 *Using* is *or* are, *select the correct form for each sentence.*

1. There _____a similar rule.
2. Where _____ the executives?
3. Here _____ the reason.
4. There _____ three ways of determining power.
5. What _____ the source of power?

Special Case 2: Compound subjects joined by *and* are considered plural.

> The scientist and the researcher agree.

In sentences with subjects joined by *or* or *nor*, the verb will agree with the closer subject.

> Neither the scientist nor the *researchers agree.*
> Neither the researchers nor the *scientist agrees.*

Exercise 3   *Underline the correct verb.*

1. The laser and the vacuum tube (has/have) been significant developments.
2. Neither prior tests nor this test (is/are) surprising.
3. Either a laser disk or several hard disks (is/are) used in automated teaching systems.
4. Fiber optics and laser communications (has/have) progressed incredibly quickly.
5. Neither the sophisticated equipment nor increasing development costs (has/have) slowed down the research.

Special Case 3:   When the verb is part of a negative contraction, the verb must still follow the *s* form rule.

s Form	Non-*s* Form
(he) doesn't	*(they)* don't
hasn't	haven't
isn't	aren't

Example: He *doesn't* like commuting to work.
(s *form*)
They *don't* like commuting to work.
(s *form*)

Exercise 4   *Underline the correct verb contraction.*

1. Lasers still (hasn't/haven't) worked their way into every facet of society.
2. They (doesn't/don't) know of any concept that has had a bigger impact on society.
3. It (doesn't/don't) have many limitations.
4. Many applications (hasn't/haven't) become popular yet.
5. There (isn't/aren't) many technologies developing as quickly as the laser.

Note: There is no correct way of contracting *am not.*

Sentence Combining

This exercise will give you more practice in forming efficient sentences.

Exercise 5   *Using the example as a model, combine each group of sentences into a single sentence.*

Example

a. The laser beam is light.
b. The light is a special kind.
c. The light is intense.
d. It differs from ordinary light.

Combination: The laser beam is a special kind of intense light that differs from ordinary light.

1. a. Ordinary light has many colors.

    b. Its waves move.

    c. The movement is random.

    d. The movement is in many directions.

    Combination: _____

    _____

2. a. Laser light is monochromatic.

    b. Its waves move.

    c. The movement is in a single direction.

    Combination: _____

    _____

3. a. Photons in ordinary light spread out.

    b. They diffuse energy.

    c. Laser beam photons are concentrated.

    d. They focus their energy.

    Combination: _____

    _____

4. a. The beam is generated by a device.

    b. The beam is amplified by a device.

    c. The device is called a laser.

    d. The device emits light waves.

    e. The light waves are spaced.

    f. The light waves are parallel.

    Combination: _____

    _____

5. a. The beam has characteristics.

    b. The characteristics are unique.

    c. The characteristics can lead to applications.

    d. The applications can be peaceful.

    e. The applications can be destructive.

    Combination: _____

    _____

6. a. A laser can be focused.

    b. It can be transmitted.

    c. It can heat a TV dinner.

    d. The dinner can be a thousand miles away.

    Combination: _____

    _____

7. a. But a laser could be focused on a tank.

b. It could be focused on a ship.

c. It could be focused on an aircraft.

d. It could burn a hole.

e. The burning would be almost instantaneous.

f. The burning would be through armor.

g. The armor would be metal.

Combination: _____

_____

8. a. A laser beam can be used to weld retinas.

b. The retinas have loosened from the eye.

c. The welding does not destroy the tissue.

Combination: _____

_____

9. a. But a laser could be used to destroy.

b. Human life would be destroyed.

c. The destruction would be in a flash.

Combination: _____

_____

10. a. A laser beam can be used to carry signals.

b. The signals are for radio.

c. The signals are for television.

d. The signals are for communication.

e. The communications are by telephone.

Combination: _____

_____

11. a. But a laser could also be used from space.

b. The use would be as a ray.

c. The ray would cause death.

d. The death would be to enemies.

Combination: _____

_____

12. a. Thus, laser light is not ordinary.

b. Its uses are not ordinary.

Combination: _____

_____

Prepositional Phrases

Prepositions are words that show relationships between two or more words. They are different from other types of words, and sometimes they are difficult to find.

There are about 30 to 40 common prepositions, some of which are combinations of words. To get an idea of how to identify them, consider the photo of a circuit board shown below.

Think of all the different words you can use to tell where the components are in relation to each other. The words that express these relationships (*on, above, next to*) are prepositions.

Common Prepositions

in	into	beyond	in front of
on	onto	among	in back of
at	from	between	in between
to	with	around	beside
by	within	after	next to
of	amid	before	out of
for	through	above	about
		below	

Photo of circuit board courtesy of Analog Devices, Inc.

A phrase is more than one word. A **prepositional phrase** begins with a preposition and ends with the next noun (or pronoun) or series of nouns (or pronouns). There could be any number of modifiers inside the phrase.

in the board
for each board

Note: The noun at the end of a prepositional phrase is never the subject of a sentence.

Exercise 1   *Put parentheses around each prepositional phrase. The number at the end of each sentence tells you how many prepositional phrases are in that sentence.*

1. Noise has recently become a topic of great interest. (1)
2. Legislative bodies have responded to public awareness of the detrimental effects of noise. (3)
3. All communication, with the possible exception of divine revelation, takes place in the presence of noise. (4)
4. For this talk, we will take the usual physics definition of noise. (2)
5. In written communication, noise can occur at each of three elements. (3)
6. The writer can contribute linguistic noise at his semantic encoder. (1)
7. The author must make a selection of data and arguments, but the selection must correspond to the mass of evidence. (3)
8. The most common source of noise is not the willful distortion of intent, but innocent lapses of attention. (3)
9. Our daily work abounds with good examples of bad habits, all sources of noise. (3)
10. An author who writes with no misspelled words has already dismissed the first grounds of suspicion of the limits of his scholarship. (4)

Prepositional Phrases and Subject/Verb Agreement

Sometimes a prepositional phrase will occur between the subject and the verb. The subject will not be inside a prepositional phrase.

Wrong: The leaders of our *company wants* to expand.
Correct: The *leaders* (of our company) *want* to expand.

Remember to make the verb agree with the real subject, not a noun in a prepositional phrase. Placing parentheses around the prepositional phrase, or drawing a line through it, will help you see the real subject of the sentence.

Example: The *values* (of power) *are measured* in watts.
             The *value* (of power) *is measured* in watts.

Exercise 2   *Underline the subject of each sentence once. Choose the correct verb by underlining it twice. Put parentheses around phrases that are between the subject and verb.*

1. An insurance policy for all types of accidents (is/are) available.

2. Misfortunes in the home (happens/happen) every day.
3. A buyer of a policy (has/have) to read it carefully.
4. The coverage of different policies (was/were) different.
5. Valuable items in the home (has/have) to be appraised.
6. The replacement value of some structures (is/are) less than the actual cost.
7. Unique older homes under special policies (is/are) an exception.
8. Renters without many possessions (buys/buy) renter's insurance.
9. An umbrella policy for extended personal liability (increases/increase) coverage.
10. Premiums in many states (varies/vary) among companies.

Exercise 3 *Rewrite the sentences in Exercise 2, changing the subject from plural to singular or from singular to plural. Use the correct verb.*

1. _____
   _____
2. _____
   _____
3. _____
   _____
4. _____
   _____
5. _____
   _____
6. _____
   _____
7. _____
   _____
8. _____
   _____
9. _____
   _____
10. _____

Sentence Combining

Now practice putting sentences together again, using all the ideas in each group to make one, powerful sentence.

Exercise 4 *Using the following example as a model, combine each group of sentences into a single sentence.*

Example

 a. A friend is a management consultant.
 b. He is successful.
 c. The success is exceptional.
 d. He is my close friend.

Combination: A close friend of mine is an exceptionally successful management consultant.

1. a. You can walk into his office.

 b. You have a feeling.

 c. The feeling is of being "uptown."

Combination: _____

_____

2. a. The furniture is fine.

 b. The furniture tells you something.

 c. The carpeting tells you something.

 d. The people are busy.

 e. The people tell you something.

 f. They tell you his company is prosperous.

Combination: _____

_____

3. a. A critic would say something.

 b. A critic would call him a man.

 c. The man is a "con."

 d. The critic would be wrong.

Combination: _____

_____

4. a. The man was not brilliant.

 b. The man was not wealthy.

 c. The man was not lucky.

 d. The man was persistent.

 e. The man never thought he was defeated.

Combination: _____

_____

5. a. Behind the company is a story.

 b. The company is prosperous.

 c. The company is respected.

 d. The story is of a man.

 e. The man is fighting.

 f. The man is battling his way upward.

Combination: _____

_____

6. a. In his first six months things happened.

    b. He lost his savings.

    c. The savings were from 10 years.

    d. He lived in his office.

    e. He lacked money to pay rent on an apartment.

    f. He turned down numerous job offers.

Combination: _____

_____

7. a. He wanted something more.

    b. He wanted to stay with his idea.

    c. He wanted to make it work.

    d. It meant hearing prospects say no.

    e. It happened 100 times as often as they said yes.

Combination: _____

_____

8. a. It took seven years.

    b. The years were unbelievable.

    c. The years were hard.

    d. During that time, I never heard something.

    e. I never heard my friend complain.

Combination: _____

_____

9. a. He would explain something.

    b. He explained it to me.

    c. He was learning.

    d. The learning was how to sell.

    e. The selling was in a business.

    f. The business was competitive.

Combination: _____

_____

Pronouns

Pronouns are words that take the place of nouns. Used carefully, pronouns provide a graceful substitution for repeated nouns. Even though the dictionary contains thousands of nouns, we have only a small number of pronouns, and each pronoun has a specific function in a sentence. There are four main groups of pronouns: subject, object, possessive, and reflexive.

Subject Pronouns

A subject pronoun is used to replace the subject of a sentence. The subject pronoun must agree with the "person" and "number" of the subject. The subject pronouns are listed below.

Person	Singular	Plural
First	I	we
Second	you	you
Third	he, she it, one	they
Question form	who	who

Exercise 1   *Underline the subject(s) in each sentence. Write the pronoun that would replace the subject.*

1. _____ Science covers the broad field of human knowledge.
2. _____ Scientists discover and test facts and principles by using the scientific method.
3. _____ Nikola Tesla pioneered the development of radio and high-temperature electricity.
4. _____ Electronics is a branch of physical science.
5. _____ Which people are on the research team?

Hazard: Pronouns should not be used as the subject in the first sentence of a paragraph. A subject pronoun should be used only after the noun has been introduced in each paragraph.

Object Pronouns

Another group of pronouns is used to replace nouns that are objects, such as those in prepositional phrases or after verbs.

Person	Singular	Plural
First	(to) me	(to) us
Second	you	you
Third	him, her	them
		it, one
Question form	whom	whom

Exercise 2   *Replace each italicized object noun with a pronoun.*

Example: Mr. Reeder gave the broken component to *our team.* (*us*)

1. _____ First we tried to isolate the cause of the problem, and then we tried to fix *the problem.*

2. _____ We were not sure to *which people* we could turn for advice.

3. _____ After locating the special tools, we used *the tools* to find the faulty unit.

4. _____ The technical manuals that we borrowed from *Mr. Reeder* contained schematic diagrams.

5. _____ The results were sent back to the *developers* for further evaluation.

Compound Pronouns

If more than one pronoun is used as a subject or object, be sure that each pronoun is the correct form for the situation. When deciding which form to use, try each pronoun individually. Always refer to yourself last.

> *Example:* He and I arrived at the same conclusion.
> *He* is the subject form—*He* arrived.
> *I* is the subject form—*I* arrived.
> The *I* pronoun comes last in the combination.

> *Example:* The prize money was shared by them and us.
> *Them* is the object form—by *them.*
> *Us* is the object form—by *us.*
> The *us* pronoun comes last in the combination.

If a pronoun is used with nouns, be careful to use the correct form of the pronoun for the situation.

> Give the results to Chris and *me.*
> Chris and *I* will evaluate the data.

Exercise 3   *Circle the correct pronoun in each sentence.*

1. Tim, Angela, and (I/me) are lab partners.
2. Tim didn't do as much work as Angela and (I/me) did.
3. We showed our data to Bill, Nick, and (he/him).
4. Our instructor gave a warning to (he/him) and (we/us).
5. He said that Angela and (I/me) could no longer share results with (they/them) and Tim.

Possessive Pronouns

This group of pronouns shows ownership. The choice of a **possessive pronoun** varies depending on whether it is followed by a noun or whether the possessive pronoun is used by itself.

Example: That is *my* hypothesis.
The hypothesis is *mine*.

Person	Singular	Plural
First	my, mine	our, ours
Second	your, yours	your, yours
Third	histheir, theirs	
	her, hers	
	its, one's	
Question form	whose	whose

Exercise 4   *Circle the correct possessive pronoun in each sentence.*

1. (You/your) experience in troubleshooting is limited, but (their/theirs) goes back many years.
2. Each scientific field regulates (it/its) own professional standards.
3. The choice is (your/yours).
4. I had to make up (my/mine) own mind.
5. The team leader claimed that the responsibility was (her/hers).

Hazard: Do not put apostrophes in possessive pronouns except in the impersonal pronoun *one.*

Example: A career choice is one's own decision.

In other cases, adding an apostrophe changes the function of the word.

It's stands for *it is.*
You're stands for *you are.*
Who's stands for *who is.*

Reflexive Pronouns

Reflexive pronouns refer back to someone or something. They end in *self* or *selves*. They are used sometimes to add emphasis to the person or people already named in the sentence.

Example: Check the results yourself.
(Referring to the understood "you")

Person	Singular	Plural
First	myself	ourselves
Second	yourself	yourselves
Third	himself	themselves
	herself	
	itself	
	oneself	

Do not use reflexive pronouns if the noun has not been used in the same sentence.

Wrong: The results were checked by myself.
           Problem—myself does not refer to any person in the sentence.
Correct: The results were checked by *me.* (*using the object pronoun*)
Correct: *I* checked the results. (*using the subject pronoun*)

Hazard: In standard English, there are no such words as *hisself, ourself, theirself, themself, selfs,* or *theirselves.* Be careful to spell pronouns correctly.

Exercise 5   *Fill in the correct reflexive pronoun.*

1. He cheated _____ out of a valuable learning experience by not completing the experiment.
2. You must discipline _____ to be systematic and logical.
3. It is one's responsibility to teach _____.
4. The determined technicians trained _____ to use the six-step troubleshooting method.
5. We could not bring _____ to admit that most of our tests were unnecessary.

Special Case 1:   Some nouns always end with a final *s*. Some of them are considered singular and are replaced by a singular pronoun. Others are considered plural. Use your ear to decide which pronoun is right.

Singular Nouns (Pronoun: *it*)	Plural Nouns (Pronoun: *they*)
communications	eyeglasses
fiber optics	pants
physics	scissors
electronics	pliers

Special Case 2:   Indefinite pronouns are always considered singular. Some indefinite pronouns (*everybody, everything*) may actually be referring to several people or things, but each person or thing is being referred to individually. The following list includes several of these singular pronouns.

anyone	anything	anybody
everyone	everything	everybody
no one	nothing	nobody
someone	something	somebody

Indefinite pronouns as subjects of sentences require *s*-form verbs (in the present tense) and *s*-form helping verbs.

> *Example:*   Everyone is invited to attend. (every *one* person)
> No one has completed the assignment. (not *one* person)
> Everything is finished. (every *one* thing)

Exercise 6   *Circle the correct verb in each sentence.*

1. Everyone (has/have) finished the experiment.
2. My pliers (was/were) left on the lab bench.
3. Someone (was/were) going to ask the instructor.
4. Until now, mathematics (has/have) been my easiest subject.
5. Everything about this experiment (was/were) confusing.
6. No one in my classes (has/have) heard the news.
7. The news (was/were) all about the discovery.
8. Everybody in my class (is/are) interested.
9. Nothing in my classes (was/were) this exciting.
10. (Has/Have) anyone researched this subject?

Exercise 7   *Write complete sentences that use the following pronouns correctly.*

1. *their* _____
2. *mine* _____
3. *themselves* _____
4. *he* and *his boss* _____
5. *who* _____
6. *hers* _____
7. *one's* _____
8. *itself* _____
9. *you* and *her* _____
10. *everybody* _____

Pronoun References

Pronouns are singular or plural, and they have subject, object, possessive, and reflexive forms.

Exercise 1   *As a review, fill in the table of pronouns.*

Person	Singular				Plural			
	Subject	Object	Poss.	Reflex.	Subject	Object	Poss.	Reflex.
1st	I	____	my/mine	myself	____	us	our/ours	____
2nd	____	you	____	yourself	you	____	your/yours	____
3rd	he	____	his	himself	they	____	____	themselves
	she	her	____	____	____	____	____	____
	it	it	____	____	____	____	____	____
	one	one	____	____	____	____	____	____

Every noun can be referred to by a pronoun.

Dr. Thompson = *she / he*
gentleman = *he*
dog = *it* (or *he, she*)
players = *they*
resistor = *it*
resistors = *they*

Each pronoun must refer to its noun in the same "number" (singular or plural) as the noun is last written. If the noun is plural the pronoun referring to it must also be in the plural form.

The *lady* asked where *she* could find the library.
Tom and Larry found that *they* needed more information.

Exercise 2   *Circle the correct pronoun. Draw an arrow to the noun that the pronoun refers to. Circle the noun also. The first one has been done for you.*

1. Put the (tools) back in (its/ (their)) compartments.
2. The research department met all (its/their) deadlines for the month.

3. The scientist applied for (his/their) first patent.
4. Ted asked the guests to make (theirselves/themselves) at home.
5. We tried to finish the experiment (ourself/ourselves).
6. The student grabbed (his/one's) notebooks as he ran out the door and put (it/them) in (his/their) briefcase.
7. The students complained that Dr. Johnson spoke too quickly to (him/them).
8. This scientific law can be useful to students because (it/they) will provide the starting point for many experiments.
9. The research, design, and engineering departments are experiencing the results of (its/their) effort.
10. The research and design effort has surprised (its/their) critics.

Collective nouns are words that have a singular form but may be singular or plural in meaning. Some examples of collective nouns are *team, group, class, audience,* and *series*. When we use a collective noun, we are referring to the "collection" as one unit, and we refer to it with a singular pronoun.

> The *band* played *its* theme song.
> The *research team* completed *its* final report.

However, if we are referring to individual members of the unit, we use a plural pronoun.

> The *members* of the band played *their* theme song.
> *Most* of the team turned in *their* results.

Exercise 3   *Circle the correct pronoun. Draw an arrow to the noun to which it refers.*

1. A team was selected in January, but (it/they) did not meet until the following June.
2. The technicians on the team received (its/their) instructions from the team leader.
3. The team presented (its/their) report at the end of the year.
4. This group of resistors is known for (its/their) durability.
5. A visiting group of scientists from Japan finished (its/their) tour of the factory before meeting with the president.
6. Two people in the group discussed (his/their) observations with the chief engineer.
7. The class of 1980 held (its/their) reunion at the college.
8. Four people from the class sent (his/their) regrets.
9. One class of technician trainees became well known for (its/their) final projects.
10. The first series of tests was completed before (its/their) deadline.

Avoiding Shifts

Keeping a consistent tone and voice is important in effective writing. Abrupt changes in tone or, even worse, illogical shifts in voice or number make readers uneasy and confused. The three most common shift errors in writing occur with verbs, pronouns, and voice.

Verb and Pronoun Shifts

Verb shifts can occur whenever two or more verbs are used in a sentence. When the verbs are used to express a single time frame, they must remain in the same tense.

> *Wrong:* Computer simulations are a new method of discovering truth, and they were being used in many laboratories.
> (*present to past tense shift*)
> *Correct:* Computer simulations are a new method of discovering truth, and they are being used in many laboratories.
> (*consistent present tense*)

There are some instances in which different verb tenses are logical and necessary.

> The computers allow simulations of what, until now, had been conducted only in the laboratory.

Do not use consistent tenses when clarity and logic demand other forms.

Pronoun shifts occur when pronouns do not agree in number (either singular or plural) with the nouns to which they refer. Remember that pronouns have singular and plural forms whether they are in the subject form (*I, we*), object form (*me, us*), possessive form (*mine, ours*), or reflexive form (*myself, ourselves*). For a review, refer to Grammar Unit 4.

> *Wrong:* A *model* can be seen on the screen after *they are* programmed to resemble the real thing.
> *Correct:* A *model* can be seen on the screen after *it is* programmed to resemble the real thing.

Exercise 1   *Correct the incorrect verb and pronoun shifts in the following sentences. Cross out each incorrect form and write the correct form above it.*

1. Some experiments would be so costly that it could not be performed in real life.
2. The computer is playing a larger role since they are capable of handling large amounts of data.
3. The Los Alamos laboratory conducted one of the first simulations when they tested the effects of nuclear fission during World War II.

4. Stanislaw Ulam, a Los Alamos mathematician, performed an experiment themselves.

5. Ulam and his colleagues used some of the earliest computers, the ENIAC and MANIAC, which are very slow.

6. Supernovas are simulated because it is rare and inaccessible.

7. In retrospect, the solution seems simple, but it was a challenge at the time.

8. A computer worked out the consequences of a physical law, but these are not new undertakings.

9. In the eighteenth century, an apparatus called an orrery simulates motions of planets and their satellites.

10. Clockwork gears turned balls in rhythmic patterns, and it predicted future configurations of the solar system.

Voice Shifts

Voice shifts refer both to the person (first, second, or third) and to active and passive voice. Remember that the active voice shows a subject performing the action of the verb.

> The technician calculated the capacitance.

The passive voice shows an inactive subject, and the verb includes a helping verb.

> The capacitance was calculated by the technician.

Determine your voice before beginning to write. Choose the appropriate voice for your audience and purpose. If your readers need directions or instructions, use the second person, active voice. If your readers need information, use the first or third person (first for informal communications, third for more formal communications), active voice. Then proofread for consistency after the rough draft is finished.

The easiest way to describe "person" is to look at the types of pronouns used in each style. The first person uses *I* and *me* because the writer is describing personal experiences.

> I programmed the model.

The second person uses *you* or the inferred "you" because the writer is directing the reader.

> Program the model.

The third person uses *he, she, it,* or *they* because the writer is informing the reader of someone or something else.

> They programmed the model.

The most common voice shift error is using *you* after starting with another pronoun.

Wrong: I found that a simulation works well when you study supernovas.
Correct: I found that a simulation works well when I study supernovas.
Correct: I found that a simulation works well in the study of supernovas.

Hint 1:   Use the first person (*I, me*), active voice for memos, letters, and lab reports.

> *Poor:*   After I corrected the problem, the report was written.
> *Correct:*   I wrote the report after I corrected the problem.

Hint 2:   Use the second person (*you* or inferred *you*), active voice for directions.

> *Poor:*   Time-cards will be turned in to supervisors on Friday.
> *Correct:*   Turn in your time-card to your supervisor on Friday.

Hint 3:   Use the third person (*it, they*), passive voice for formal types of writing, such as research papers or formal proposals, if it seems to be the conventional format.

> *Poor:*   The problem was determined to be the same one I had corrected earlier.
> *Correct:*   The problem was determined to be the same one that had been corrected earlier.

Exercise 2   *Edit the following paragraphs to eliminate voice shifts. Choose one voice, circle the shifts, and add the correct forms.*

1. When you begin your job search, it is important for one to avoid making several common mistakes. Any one of these mistakes can ruin his chance of getting the job you want.

2. The first error is not knowing what you want to do. Employers don't want to hear, "What's available? I'll do anything!" A qualified and motivated person is one whom the employer wants.

3. Job candidates must take the initiative to obtain your ideal job. They can start in small ways, such as making lists of steps or discussing job-hunting strategies with a professional. Consider the job hunt your immediate job.

4. Another mistake is going to too few prospects. Some new job hunters approach only a few leads at a time. I was discouraged when those leads didn't work out and felt like I had to start the search over. Job hunters who contact only a few leads seem to take rejections more personally.

5. You must be able to view employment from the employer's perspective. I have reasons for a job, but they are of no concern to the potential employer. One must be able to focus on the needs, objectives, and problems of the employer.

Avoiding Sexism

In traditional writing, authors often used masculine pronouns (*he, him*) in titles and examples that were actually referring to both men and women.

> When an employee works hard, he is rewarded.
> Each student must submit his homework on time.

Years ago, this was accepted and tolerated. Currently, however, with the rising number of women in the working world and increasing awareness of bias and sexism, it is no longer acceptable to use just the masculine forms.

Even though we may suspect that the people we refer to in writing are either predominantly male or predominantly female, we must make attempts to include both genders. There are several ways to avoid sexism in writing.

1. *Use plural (nongender) forms.* The best neutral references are plural pronouns. Plural pronouns do not indicate gender.

> When people work hard, they are rewarded.
> Students must submit their homework on time.

Remember that when subjects are changed from singular (*he* or *she*) to plural (*they*), sometimes the verb must be changed. "He or she *is* rewarded," or "They *are* rewarded."

2. *Use singular, nongender pronouns.* Some pronouns are impersonal, such as *one* or *that person.*

> When someone works hard, that person will be rewarded.

This method is useful only in short pieces of writing or in occasional references.

3. *Use pronouns of both genders.* The pronouns can be separated by a slash (he/she) or a conjunction (*or, nor*) or varied throughout longer documents.

> A student must submit his or her homework on time.

This method can become wordy and awkward with repeated uses.

4. *Rephrase to avoid pronoun references.* Another way to eliminate the problem of pronouns is to reword the sentence into the passive voice so that no pronouns are needed. Remember, however, that the passive voice weakens a statement since there is no "doer."

> Hard work is rewarded.
> Homework assignments must be submitted on time.

5. *Use inferred "you" statements.* Write directions, steps, or instructions using the inferred "you." Use only present-tense verbs.

Work hard to be rewarded.
Turn in homework on time.

6. *Use nonsexist terms, and titles.* The following sex-specific terms and titles are being revised due to increased sensitivity to nonsexist language.

Sexist	Nonsexist
chairman	chair, chairperson
committeeman	committee member
postman	postal carrier
policeman	police officer
Dear Sir	Dear Sir or Madam, Dear Personnel Manager
his	his or her, his/her (or change to plural)
he	he or she, (or change to plural)
man, mankind	person, humanity, people, the average person
manpower	workers, employees, staff, personnel

Exercise 1   *Edit the following passage for sexism. Do not change references to specific people.*

Thomas Alva Edison was probably the greatest inventor in the history of mankind. After a mere three months of public school education, during a period when a schoolteacher considered herself aloof from questioning boys, "Al's" mother removed him from school and began teaching him at home. He asked an endless number of "why's" and received his first chemistry book at the age of nine. Soon he began to learn so fast that his mother could no longer teach him. Eventually, he was to patent 1093 inventions. Edison always tried to develop devices that would be useful to man, such as the electric light and the phonograph. He also improved the inventions of other men. These included the telephone, the typewriter, the motion picture, the electric generator, and electric-powered trains. He is known for his "brute force" method of solving problems: he would try endless methods, materials, and experiments until he found the answer, often working 18 hours a day in his lab. Edison believed that if a scientist was a genius, he would display "1 percent inspiration and 99 percent perspiration."

Modifiers

Modifiers are words that limit and describe other words. We use modifiers in writing to add specific details to objects, people, feelings, and actions. Compare the following descriptions:

> The man wearing jeans, a shirt, and shoes walked into the lab.

> The bearded young man wearing faded Levis, a red Nike T-shirt, and high-top sneakers confidently walked into the fast-paced development lab.

The two sentences express the same basic idea. But the first sentence sounds flat and boring when compared to the second sentence, which is full of modifiers. Used wisely, modifiers add power and life to writing. The most common modifiers are called *adjectives* and *adverbs*.

Adjectives

Adjectives are modifiers that describe nouns or pronouns. Some sentences have no adjectives, whereas other sentences have many.

> The salesperson wore a suit.

This simple sentence has a subject and verb; however, the picture communicated by the writer is brief and incomplete. Two people reading this sentence may have completely different "pictures" of this salesperson. Neither the person nor the suit is described. If description makes a difference, such as in describing a fashionable suit or a certain salesperson, adjectives are needed.

> The *young* salesperson . . .
> The *tailored* suit . . .

Although adjectives are usually in front of nouns, another way of writing these same ideas uses linking verbs.

> The salesperson is *young*.
> The suit is *tailored*.

Supply adjectives of your own:

> The _____ salesperson . . .
> The _____ suit . . .

The new salesperson wore a (an) _____ suit.

The _____ salesperson wore a three-piece suit.

Note: The articles *a, an,* and *the* are also adjectives since they are written in front of nouns and pronouns.

Exercise 1 *Underline the adjectives in the following sentences. The number of adjectives in each sentence is indicated in parentheses. Draw an arrow to the word each adjective describes.*

1. The consultant advised men to wear a blue or gray suit to the first interview. (6)
2. A black suit is overpowering. (3)
3. Black accessories often complement a business suit. (3)
4. A woman has several more color choices for a business suit. (6)
5. Blue is the favorite color of most men and women. (3)
6. Brown is a good, basic color for women, but should only be used as a background or accessory color by men. (6)
7. Burgundy is one of the most flattering, authoritative colors. (4)
8. Women can wear burgundy suits or accessories. (1)
9. Men can wear burgundy ties, either as a background color or a print, but they should not wear burgundy suits or shoes. (5)
10. Gray and navy business suits are the most authoritative colors for men or women. (6)

Hyphenated Adjectives

Sometimes two words are joined by a hyphen to become one adjective, as in the following examples.

middle-class neighborhood   navy-blue three-piece suit

top-level management   well-polished image

It would sound ridiculous to talk about a "middle neighborhood" or a "class neighborhood." Instead, the two words have been joined with a hyphen to become the single adjective "middle-class."

Rule: If two words require each other to describe the noun, use a hyphen between them.

Exercise 2 *Add a hyphen between the adjectives that require each other to describe the noun.*

Example: Graduation is a decision-making time.

1. He wears a two tone suit to interviews.
2. A lemon colored suit is not appropriate for an interview.
3. Some businesses allow an open collared, button down shirt.
4. The interviewer was impressed by her well polished image.
5. He bought a long sleeved, white shirt.

Adjectives for Comparison

We often indicate comparisons with adjectives by using *more* or *most*.
    A camel suit is a *popular* choice.
    A blue suit is a *more popular* choice.
      (comparing blue to camel)
    Gray is the *most popular* choice of all.
      (comparing gray to all others)

Sometimes we add the suffix *-er* or *-est* to the end of the adjective.
    The blue tie is bright.
    The blue striped tie is brighter.
      (comparing blue to blue striped)
    The red tie is the brightest.
      (comparing red to all others)

General Rules:

Use *more* or *-er* to compare one thing to one other thing.

Use *most* or *-est* to compare one thing to all others.

If the adjective has one syllable, add *er* (a cleaner suit).

If the adjective has three or more syllables, use *more* (a more expensive suit).

If the adjective has two syllables, use either *er* or *more* (a more useful suit, but an easier choice).

Note: It is redundant (and incorrect) to use both *more and -er* or *most and -est*.
    Wrong: This suit is *more bluer* than that suit.
    Correct: This suit is *more* blue than that suit.
    Wrong: The navy-blue suit is a *more usefuler* choice than black.
    Correct: The navy-blue suit is a *more useful* choice than black.

Exercise 3   *Compare the following adjectives by choosing the correct comparative form.*

Example:

harmful	more harmful	most harmful
	More/-er	**Most/-est**
1. hard		
2. serious		
3. easy		
4. slow		
5. fine		
6. accurate		
7. precise		
8. sharp		
9. correct		
10. complete		

The three common exceptions are *good, less,* and *bad.* The comparative forms of these words use neither *-er* or *more* nor *-est* or *most.* They use different words. See if you can figure them out.

	er-form	*est*-form
good	better	
bad	_____	worst
less	lesser	_____

Check with your instructor or the dictionary if you're not sure.

Note: Some words cannot be used to compare, such as dead, unique, and only. Each of these simply states a condition.

Wrong: This color looks deader than that color.
Correct: This color looks dead compared to that color.

Wrong: This tie is more unique.
Correct: This tie is unique.

Also, technical writers must be careful about overusing comparative words such as *real, really, nice, pretty good, pretty well,* and *very.* These words do not have a clear meaning and should be replaced by more specific terms.

Poor: That tie looks really nice.
Correct: That tie looks professional.

Exercise 4   *Edit the adjectives in the following sentences to make them correct and clear.*

1. Blue looks more better on you than gray.
2. The reception was a really nice party.
3. My manager is a really nice person.
4. That is the baddest-looking outfit I have ever seen.
5. Meeting the president was a most unique experience for me.
6. We had the seriousest conversation of the evening.
7. I think she was more sharp than anyone else at the party.
8. I gave him my most firmest handshake.
9. We were all using our correctest manners.
10. This clock is much more slower than my watch.

Adverbs

An **adverb** is a word that describes a verb, an adjective, or another adverb. It is used to answer certain questions in a sentence:

How?   When?   Where?   To what degree?

Henry never wears a blazer to an important meeting.
  (When? Never!)
He did not fully accept the concept of a "power suit."
  (To what degree? Fully!)
The college graduate dressed professionally.
  (How? Professionally!)

Note: Many adverbs end in ly. We add *ly* to most adjectives to make them function as adverbs.

He wore *attractive* clothes.
  (*adjective* describing the clothes)
He dressed *attractively*.
  (*adverb* describing how he dressed)

Exercise 5   *Underline the adverb that answers the question after each sentence.*

1. He decided to wear the suit again. *(When?)*
2. She relied heavily on her navy-blue suit. *(To what degree?)*
3. When he examined his wardrobe, he realized that he had almost no blue. *(To what degree?)*
4. The manager told him not to wear a three-piece suit here. *(Where?)*
5. Some lab technicians always wear a white jacket. *(When?)*
6. Many professionals pride themselves on dressing well. *(How?)*
7. There is currently an emphasis on certain conservative styles. *(When?)*
8. The expense of a suit means students will shop for clothes carefully. *(How?)*
9. Green hasn't ever been a popular choice for men's suits. *(When?)*
10. Dressing professionally will increase your chances of being successful. *(How?)*

Double Negatives

In mathematics, two negatives make a positive. In language, two negatives also make a positive.

It is not unlikely means "it is likely."

Used carefully, a **double negative** can be used to add emphasis to a point. However, some writers use a double negative without intending the meaning to be positive. Most people are aware of the rule against using two negative words for the same idea, but some get careless or lazy about making negative statements correctly.

Caution: People who intend to project an educated image will never use *ain't*. If you use *ain't*, break the habit quickly and substitute *isn't, aren't, haven't*, or *am not*.

Wrong: He ain't dressed like a professional.
Correct: He isn't dressed like a professional.

Some words have a negative meaning and should not be used with another negative word:

hardly	no one	rarely
never	none	scarcely

Wrong: *No one scarcely* attends the early labs.
He *never* has *none* of his work finished on time.

Some words have a positive meaning and are often used correctly with negative words:

any	anyone	ever

Correct: *No one ever* attends the early labs.
Hardly anyone attends the early labs.
He *never* has *any* of his work finished on time.

The following examples correct a double negative, each adding a slightly different emphasis.

Wrong: I haven't hardly started thinking about clothes.
Correct: I have hardly started thinking about clothes.
I haven't started thinking about clothes.
I hardly think about clothes.
Wrong: We don't never shop there.
Correct: We never shop there.
We don't ever shop there.
We don't shop there.

Exercise 6 *Rewrite the following sentences to correct the double negatives.*

1. I haven't never seen a man wear a yellow suit. _____

_____

2. It ain't no proper color for a professional. _____

_____

3. I can't never find the time to shop. _____

_____

4. They don't hardly know where to begin. _____

_____

5. Don't let no one talk you into buying a suit you don't like. _____

Exercise 7 *Proofread the following paragraph for modifier and double-negative errors. You should find six errors. Write the correct forms above them.*

Our response to color is as much emotional as physical. Certain colors deliver messages that haven't nothing to do with the individual wearing them. More darker colors generally convey authority. Medium-range colors like blue or tan make us look more friendlier and approachablier. Large expanses of pastels make us look less seriouser, sometimes unprofessional, and are more suited for off-the-job looks. Our response to colors may not be hardly deliberate, but colors contribute to our general impression of the person wearing them.

Misplaced Modifiers

A modifier can be a phrase that describes a word in the sentence. Be careful to place the modifying phrase close to the word it describes. **Misplaced modifiers** can be silly, as in the following sentence:

Misplaced: I saw a quarter walking down the street.
        (Was the quarter walking?)
Correct: Walking down the street, I saw a quarter.

Some misplaced modifiers, such as *only, just,* and *almost,* can make the meaning unclear, as in the following sentence:

He just planned the circuit.

This sentence could be interpreted two ways—he planned the circuit but left the construction for someone else, or he planned the circuit recently. Be careful to add enough information for the reader.

Correct: He planned the circuit just now.
        His part of the project was planning the circuit.

Exercise 8   *Rewrite the following sentences to eliminate misplaced modifiers.*

Example: Despite their intensive training, robot failures are sometimes difficult for maintenance technicians. (The robots didn't have the training.)
Revised: Robot failures are sometimes difficult for maintenance technicians, despite their intensive training.

1. A major appliance manufacturer only purchased one industrial robot. _____

_____

2. After installation, the company had a series of problems with the robot. _____

_____

3. The robot was just designed to load and unload a stamping press. _____

_____

4. After breaking down, the maintenance crews pulled the robot off the line and replaced it with a human operator. _____

_____

5. Finally, the robot was pulled away by the supervisor from the stamping press permanently. _____

_____

6. When not properly designed and installed, the factory is a rough environment for a robot. _____

_____

7. Unlike other factory machinery, human beings can quickly replace problematic robots. _____

_____

8. Early failures tended to discourage further investments in robotics technology of time and money. _____

_____

9. Today, experienced consultants can be hired to ease the transitions that are experts at building robotics systems. _____

_____

10. Robotics systems operate far more successfully that are built by systems manufacturers than those built and installed in-house. _____

_____

Clauses and Simple Sentences

A **clause** is a group of words that contains at least one subject and complete verb. There are two kinds of clauses: the **independent clause** (a complete idea, a sentence) and the **dependent clause** (an incomplete idea related to the independent clause).

This unit focuses on the single independent clause, also called a **simple sentence.** (In the following units, you will practice combining independent and dependent clauses into compound, complex, or compound-complex sentences.)

> clause = group of words with a subject and verb
> independent clause = sentence

An independent clause, or simple sentence, states one idea. It can contain many other parts of speech, including prepositional phrases and modifiers, and might have the inferred "you" as the subject, a compound subject, or a compound verb, but it must have at least one complete subject and verb.

The following are examples of simple sentences of one or more subjects (underlined once) and verbs (underlined twice):

> <u>Open</u> the door to the universe. (The subject is the inferred "you.")
> The <u>telescope</u> <u>opens</u> the door to the universe.
> The <u>telescope</u> <u>scans</u> at low power and <u>adjusts</u> to higher power.
> <u>You</u> and your <u>friends</u> <u>can</u> <u>sit</u> in your patio chairs, <u>enjoy</u> the heavens, and <u>explore</u> the solar system.

Exercise 1   *In each sentence, underline the subject once and the verb twice. Optional: Put parentheses around the prepositional phrases (see Grammar Unit 3, if needed). The subject and verb are not part of a prepositional phrase.*

1. Computers are manufactured in many colors.
2. Designers have created dozens of designs and continue to develop more.
3. The latest models and concept designs show interest in consumer trends.
4. Makers of PC systems want to generate sales and get consumers excited by new colors and designs.
5. Users want new PCs with a modern look and state-of-the-art features.
6. Serial and parallel connectors for peripherals might disappear and be replaced by higher-speed USB connectors.
7. New designs and prototypes are shaped like bunnies, wheels, beans, and tablets.
8. Slimmer profiles, all-in-one designs, and smaller but more powerful motherboards characterize the new technology.
9. Consumers and businesses alike seem interested in modernizing their computers.
10. Watch for the new designs at retail stores.

Compound Sentences

If we wrote only simple sentences, our writing would appear choppy and dull. To make our writing smoother and more interesting, we vary our sentence structure. One way of doing this is joining two sentences together to make compound sentences.

A **compound sentence** has more than one independent clause joined with a connecting, or coordinating, word.

> Maiman's first laser generated only red light, but today lasers come in many colors.
>
> two or more independent clauses = *compound sentence*

Coordinating words show that the two complete sentences are of equal importance, and they are joined because the author wants to relate them.

Exercise 1   *Read each compound sentence. Underline the two independent clauses in each sentence, then circle the coordinator. Notice that each clause has a subject and verb.*

1. The laser beam is a tool in today's welding industry, and it is gaining in popularity.
2. Lasers can be used to weld hard-to-access areas, or they can be used to weld extremely small components.
3. The production of laser light is a complex process, yet we are learning more about lasers every day.
4. Lasers for welding or cutting are solid-state lasers, or they are gas lasers.
5. Gas lasers are capable of higher-wattage outputs, so they are applied to thicker sections.
6. Solid-state lasers use light energy to stimulate their electrons, but gas lasers use an electrical charge.
7. The laser can be directed over long distances, for lasers can be focused and directed by mirrors and lenses.
8. Today lasers are not difficult to generate, nor are they difficult to control.
9. Photon lasers are used to weld thousands of parts, and new applications are found daily.
10. The principles for generating a beam are similar for both cutting and welding, but with cutting we add an oxygen-assist gas.

The common coordinators, words used to join two sentences, are the following:

> and   but   for   now   or   yet   so

Do not confuse compound sentences with simple sentences, such as the following, that contain compound subjects or verbs.

Welders and surveyors use lasers for precision work.
(simple sentence with compound subject)
Scientists research and study new applications.
(simple sentence with compound verb)
Distances and angles can be measured and read electronically.
(simple sentence with compound subject and verb)

In the following exercise, practice recognizing the difference between compound sentences and simple sentences with compound subjects and verbs.

Exercise 2   *Read each sentence. If it contains two independent clauses, and thus is a compound sentence, write C in the blank. Circle the comma and coordinator. If it is a simple sentence, write S in the blank.*

Clue: Simple sentences may have compound subjects or verbs.
Compound sentences have two complete sentences.

_____ 1. The electronic distance measurement instrument (EDM) was introduced in the late 1950s, but it has undergone continual changes.

_____ 2. EDM instruments use either microwaves or light waves, including laser and infrared.

_____ 3. Distances are now being measured electronically, and positions are being determined by satellite receivers.

_____ 4. New EDMs have built-in or add-on calculator-processors.

_____ 5. Microwaves or light waves (lasers) are used by EDM instruments.

_____ 6. Laser instruments have a reduced capability in daylight, yet long-range land measurements can be taken.

_____ 7. Most short-range instruments are still infrared, but short-range laser EDMs are becoming competitive.

_____ 8. Low-powered lasers present little or no danger to the operator, nor do they present any danger to the public.

_____ 9. The visible light beam of laser instruments is an advantage over infrared and makes lasers more helpful.

_____10. Microwave systems are often used in hydrographic surveying, and they have a usual upper limit of 60 km in range.

Punctuating Compound Sentences

Punctuation within sentences gives readers the same kinds of clues as the periods and question marks do at the ends of sentences. To "fuse" a compound sentence correctly, we have two choices: (1) a comma and coordinator, or (2) a semicolon.

Commas and Coordinators   To complete a fusion, we can put a **comma and a coordinator** between the two independent clauses.

Example: The first laser was a pure red. Now lasers come in many colors.
Fusion: The first laser was a pure red, but now lasers come in many colors.

Warning: Remember that connecting compound sentences with only a comma or only a coordinator improperly fuses the two sentences.

Comma splice error:	Laser light is monochromatic, it has a single wavelength.
Run-on error:	Laser light is monochromatic and it has a single wavelength.
Correct:	Laser light is monochromatic, and it has a single wavelength.

Exercise 3   *Fuse the two sentences into a compound sentence by adding a comma and coordinator. Use a logical coordinator. Smooth the fused sentences when necessary above.*

1. Laser devices are now used in construction work.

   Laser devices are also used in welding and surveying.

2. Lasers are normally used in a fixed direction.

   Lasers can be used in a revolving, horizontal pattern.

3. Care must be taken of the laser beam.

   The beam can be deflected by dust or high humidity.

4. Lasers must not directly strike the eyes.

   Lasers should not be used in poor weather conditions.

5. We don't normally think of lasers being used underground.

   Lasers are commonly used in trenches.

6. To prevent underground humidity, blowers are used.

   The laser could also be placed above ground with a signal-sensing target rod toward the trench.

7. A pipeline laser is mounted in a storm-pipe manhole.

   Then the laser is used with a target for the laying of sewer pipe.

8. A laser sensor can be mounted on a backhoe.

   The sensor gives a proper trenching depth.

9. Working above ground eliminates the humidity factor.

   Working above ground allows for more accurate horizontal alignment.

10. Laser devices can be disturbed from their settings.

    Automatic shutoffs are installed to prevent injuries.

Semicolons in Compound Sentences   Another method of fusing compound sentences is using a semicolon (;) between the two clauses.

Maiman's first laser generated only red light; today lasers come in many different colors.

Notice that the second clause does not start with a capital letter and does not use a coordinator. The second clause is considered a continuation of the first clause. Semicolons should not be overused or their effectiveness will be diminished. If semicolons are used carefully and appropriately, they provide an abrupt, striking coordination; overused, they become weak and showy.

Remember that semicolons can be used if the following conditions are met.

1. The two clauses are not coordinated by *and, but, or, nor, for, so,* or *yet.*

> *Wrong:* Maiman's first laser generated only red light; but today lasers come in many different colors.

2. The ideas of the two clauses are closely related.

> *Wrong:* Maiman's first laser generated only red light; the most powerful laser fills a large building.

The second clause may begin with connective or transition words and phrases (which are followed by a comma) such as the following:

therefore	in addition	nevertheless
however	on the other hand	
in fact	consequently	

> *Example:* Lasers are being used in military weapons; in addition, they are being used in recreation, manufacturing, and medical industries.

However, these words or phrases may also be used to begin a new sentence. Sentence length and sentence variety are the determiners. If both clauses are long, use a period. Vary the use of coordinate conjunctions, semicolons, and simple sentences to make the message more interesting.

Exercise 4   *Rewrite the compound sentences in Exercise 3. Fuse them with semicolons.*

1. *Example:* Laser devices are now used in construction work; in addition, they are used in surveying and welding.

2. _____

_____

3. _____

_____

4. _____

_____

5. _____

_____

6. _____

_____

7. _____

_____

8. _____

_____

9. _____

_____

10. _____

_____

Complex and Compound-Complex Sentences

In this unit, you will practice combining clauses that are not both independent. One clause will be independent; it is called the **main clause.** The other will be dependent, and it is called a **subordinate clause** because the idea that is expresses is subordinate to or dependent on, the main idea. The prefix *sub* means "beneath" or "low". A subordinate clause will be less important than the main clause.

> *sub* = low
> *subordinate* = lower in importance
> *subordinate clause* = incomplete idea fragment
> independent + dependent clauses = *complex sentence*
> main clause + subordinate clause = *complex sentence*

We call sentences with unequal clauses **complex sentences.** One purpose for using subordination is to add interest and variety to sentence structure. Another purpose is to add unity to sentences by explaining the relationship of one clause to another.

The words we use to signal subordinate ideas are called **subordinate conjunctions,** but you might remember them more easily as **signal words,** because they give direction to the meaning of the sentence. Read the following sentence and notice how the signal word *because* relates the two clauses.

> We must be able to detect useful sounds because we are surrounded by environmental noise.

If you take this complex sentence apart, you will find two clauses:

> We must be able to detect useful sounds
> (*main clause*)
> because we are surrounded by environmental noise.
> (*subordinate clause*)

The signal word *because* makes the second clause dependent on the first. The main clause of the sentence is independent and could correctly be written as a simple sentence. A subordinate clause by itself is a fragment because it does not express a complete idea. It needs an independent clause to complete it.

A subordinate fragment can also be corrected simply by removing the signal word.

> Because we are surrounded by environmental noise.
> (*an incomplete fragment*)
> We are surrounded by environmental noise.
> (*a complete sentence*)

Remember that every complex sentence needs at least one independent clause. A subordinate clause written alone is a fragment.

When we use subordination to show a relationship, that relationship is determined by the signal word. The subordinate clause can introduce or follow the main clause.

> Because we are surrounded by environmental noise, we must be able to detect useful sounds.

Writers usually alternate between the two methods throughout a report to add variety to their sentences. As a rule of thumb, however, remember that long, complicated technical sentences are usually more understandable if they begin with the main clause.

The most common signal words are listed below. Notice that each one is used to show a relationship.

SUBORDINATE CONJUNCTIONS (SIGNAL WORDS)

Why	When	How	Where
because	after	although	where
if	as	as if	wherever
since	before	as though	
so that	once	how	
	since	unless	
	until	though	
	when	even though	
	whenever		
	while		

Special Signal Words

that	which	who	
	whichever	whoever	

There are many types of complex sentences. Some have more than one main clause or more than one subordinate clause. Often, these sentences can be rearranged in several different, correct ways. The writer must decide which way is most effective.

Exercise 1   *Underline the main clause in each complex sentence. Circle the subordinate conjunction. Draw a wavy line under the subordinate clause.*

> *Example:* (When) Lars Ulrich and James Hetfield got together in Los Angeles in 1981, Metallica was formed.

1. The band played speed-metal music, although it was also inspired by British heavy metal.
2. Kirk Hammett is the lead guitar player, although he was not part of the original group.
3. Bassist Cliff Burton was killed in Scandinavia when the band's bus crashed.
4. After Jason Newsted was added, the group toured Japan.
5. While the speed is extraordinary, the volume is ear-shattering.
6. Even though Metallica's albums are popular, few radio stations play their songs.

7. Unless you feel the impact, the lyrics sound bizarre.

8. Metallica made metal-music history when the group played "One" at the Grammy Awards ceremony in 1989.

9. Although the group did not win the first hard rock/heavy metal award, the nomination gave Metallica new respect.

10. Because Metallica keeps a consistently unconventional style, its fans remain loyal.

Commas in Complex Sentences

One of the clauses in a complex sentence is the main (independent) clause. Recognizing this main clause is the trick to knowing when and where to place commas in complex sentences.

The general rules for using commas are as follows:

General Rules:

- If the subordinate clause introduces the main clause, put a comma after the subordinate clause.

- Do not put a comma after the main clause when it is followed by a subordinate clause.

Exercise 2   *Complete the sentences by adding your own dependent clause beginning with the signal word. Be sure that your added clause has a subject and verb. Make each sentence different and true.*

1. Service technicians have interesting jobs because _____

   _____

2. Service technicians have interesting jobs after   _____

   _____

3. Service technicians have interesting jobs unless   _____

   _____

4. Service technicians have interesting jobs although   _____

   _____

5. Service technicians have interesting jobs when   _____

   _____

In the "picture" form below, the signal word represents a subordinate conjunction, and the lines represent clauses. The signal word makes the clause that follows it dependent (DC).

SIGNAL   DC   ,   IC   .
____   IC   SIGNAL   DC   .

Exercise 3   *Underline the main clause in each complex sentence. Draw a wavy line under the subordinate clause. Draw a box around the signal word. Add a comma, if necessary, to punctuate the sentence correctly.*

   Example: The music was too loud until I adjusted the dials.

1. Music is pleasant noise until harmonic distortion colors the sound.

2. When distortion occurs the speakers sound like they are "breaking up."

3. Distortion can be a desired result if the coloration becomes part of a musician's song.

4. "Exciters" are used when musicians want to "brighten" the sound.

5. If distortion is not controlled the speakers can be ruined.

6. Generally distortion is avoided although there are exceptions.

7. A filter is used when owners want to test their speakers.

8. Manufacturers print a spec sheet since some buyers want more information.

9. It is easy to detect distortion if the equipment is measured using a wave analyzer.

10. Intermodulation distortion occurs when two frequencies interact to form a new output frequency.

Exercise 4   *Rewrite the sentences in Exercise 3. Reverse the order of the clauses, and punctuate the new sentences correctly.*

Example: Until I adjusted the dials, the music was too loud.

1. _____

_____

2. _____

_____

3. _____

_____

4. _____

_____

5. _____

_____

6. _____

_____

7. _____

_____

8. _____

_____

9. _____

_____

10. _____

_____

Exercise 5   *Use the given signal word to combine the two simple sentences into a complex sentence. Add commas if needed. Make sure that the new sentence is smooth and accurate. Write some of the sentences so that the subordinate clause introduces the main clause.*

Example: Dizzie Gillespie is 83 years old.
He can do pretty much what he pleases. (Because)

Because Dizzie Gillespie is 83 years old, he can do pretty much what he pleases.

1. Dizzie became the trumpet prince of bebop in 1948.
   He was playing with Thelonious Monk and Charlie Parker. (When)

   _____

   _____

2. He wears eyeglasses for reading and his hair is turning gray.
   Gillespie hasn't changed much. (Although)

   _____

   _____

3. He is still actively playing in New York.
   He jams, performs, and records albums. (Where)

   _____

   _____

4. His own development has not stopped.
   His style is still familiar. (Even Though)

   _____

   _____

5. Dizzie promotes his new albums.
   He starts another project. (Until)

   _____

   _____

6. His eyes are glowing with delight.
   He uses all manners of expression. (While)

   _____

   _____

7. He calls his wife, Lorraine, often.
   They have been married for 50 years. (As)

   _____

   _____

8. They met.
   Lorraine was a dancer. (When)

   _____

   _____

9. Members of his various bands play well together.
   They learn to communicate with each other. (If)

   _____

   _____

10. Dizzie is a self-taught musician.
    He admires the education and training of the new, young musicians. (Although)

    _____

    _____

that, which, and *who* Clauses

Another type of subordinate clause begins with *that, which,* or *who.* These clauses are similar to adjectives because they describe or modify a noun or pronoun in the main clause. In subordinate clauses, these words are both the **signal word** and the **subject** of the subordinate clause.

> People *who compose music* must be sensitive to the listener.
> Composers write music *that is communicating feeling.*
> A piece of music may have technicalities, *which may have to be explained.*

Remember to use *who* and *that* to refer to people and *which* and *that* to refer to objects. They should be placed directly after the noun or pronoun they modify. Notice in the following two sentences how the meaning changes because of the placement of the dependent clause.

> Music *that is popular* is the first topic.
> Music is the first topic *that is popular.*

Use *that* when the modifier is restrictive, meaning it is required to complete the meaning of the sentence.

> The band *that won the grammy* has just put out a new CD.
> Give all the guitars *that require special tuning* to the guitar tech.

Use *which* when the modifier is nonrestrictive, meaning it is an added thought, like an "interrupter," and should be set off with commas. When we speak, we usually pause briefly before and after a nonrestrictive clause and use a slightly different pitch from that of the main clause.

> The band, *which won a grammy in* 1999, has just put out a new CD.
> He also plays a 12-string guitar, *which often surprises his audience.*

Remember, also, that the signal word is similar to a pronoun that refers to the noun in front of it. Even though the signal word is not exactly singular or plural, the *verb* following *that, which,* or *who* must agree with the noun that the clause modifies.

> A musician *who only mimics* is not a truly musical individual.
> Musicians *who only minic* are not truly musical individuals.

Exercise 6   *Combine the simple sentences into complex sentences using* that, which, *or* who. *Make sure that the new sentence is smooth.*

> *Example:*   Ken heard music.
>   He liked the music.
> *Combined:*   Ken heard music that he liked.
>   Ken liked the music that he heard.

1. An intelligent listener must be a person.

The person is able to recognize a melody.

_____

_____

2. The melody in the music is like the story.

The story expresses an emotion or idea.

_____

_____

3. The listening process has three planes.

The three planes are vital to understanding music.

_____

_____

4. The sensuous plane is listening for pleasure.

This plane is the easiest way.

_____

_____

5. Many people abuse that plane in listening.

These people claim to be qualified music lovers.

_____

_____

6. The second plane is the expressive one.

It is the meaning behind the notes.

_____

_____

7. Music expresses moods.

The moods are sometimes similar to emotional moods.

_____

_____

8. The third plane is the sheerly musical plane.

This plane consists of melodies, rhythms, and harmonies.

_____

_____

9. Melodies and rhythms catch our attention first.

Melodies and rhythms are exciting.

_____

_____

10. Intelligent listeners become aware of the musical plane.

Intelligent listeners want to become more alive to music.

_____

_____

Compound-Complex Sentences

Writers vary their sentence structure by using simple, compound, complex, and compound-complex sentences. A **compound-complex sentence** has at least one dependent clause and two or more independent clauses. Notice the combinations and punctuation in the following example:

> When it was introduced in 1939, the Gibson Electraharp was the first pedal-steel guitar,
>   *(dependent clause)*   *(independent clause)*
> and it had six pedals.
> *(independent clause)*

Exercise 7   *Combine each group of sentences to make one compound-complex sentence. Be sure to punctuate correctly.*

1. a. Electric guitars were a new innovation in 1938.
    b. Hawaiian-style guitar playing dates back to the 1800s.
    c. The playing had inherent problems in the metal bar.

    _____

    _____

2. a. Pedals have been around for a long time.
    b. Pedals were developed for harps.
    c. A true system for raising and lowering pitch had been perfected by 1810.

    _____

    _____

3. a. The Electraharp system was patented by John A. Moore.
    b. Moore brought the idea to Gibson.
    c. Moore refined the idea with Gibson engineers.
    d. Three were built in 1939.

    _____

    _____

4. a. Early Electraharp players had no teachers.
    b. They had to write their own sheet music.
    c. The sheet music corresponded to the eight strings and frets.

    _____

    _____

5. a. The war ended.
    b. Gibson stopped marketing the steel guitars.
    c. The public's demand didn't go away.
    d. Now the steel guitar has made a revival in blue-grass, folk, and country music.

    _____

    _____

Fragments, Run-Ons, and Comma Splices

This unit focuses on three types of sentence errors: the fragment, run-on, and comma splice. Fragments can result from missing subjects (verb fragment), verbs (subject fragment), or both (phrase fragment). To correct them, you must add the missing subject and/or verb.

The run-on and comma splice errors result from incorrectly joining two independent clauses. To correct them, you must add the needed punctuation or conjunction. These sentence errors are troublesome for many writers and require careful editing. The practice provided in the following exercises will help you focus on the problems, but you must then learn to edit your own writing.

Do not rely on the grammar check included with your word processor to highlight these sentence errors. Grammar checks (at the time of this publication) highlight only the most obvious fragments and overlook all run-ons and comma splices.

Fragments

A sentence must have a complete subject and a complete verb, and it must express a complete idea. A group of words without a complete subject, verb, or idea is called a **fragment.**

We can speak in fragments and still communicate effectively. This is due mainly to other cues besides our words, such as eye and hand movements and voice inflection. We can observe and question our listeners, and we have the luxury of repeating and rewording our message until we are satisfied it has been understood correctly.

In writing, however, we cannot observe our readers. In professional writing, we must make every attempt to follow standard English rules to ensure effective, efficient communication. Using complete sentences is an important rule.

A fragment may be caused by many types of errors: a missing subject, a missing or incomplete verb, or an incomplete thought. In this unit you will learn to identify and correct three types: missing subject, missing verb, and phrase fragments (see Grammar Unit 11 for a review of incomplete idea fragments).

Missing-Subject Fragments

The subject of a sentence may appear in various places in the sentence.

> Personal *robots* hold promise of creating greater excitement than the computer.
> So where are *all* these robots?
> Once found only in industry and the laboratory, over the past decade the *computer* has become a fixture in the majority of households.

The subject tells what the sentence is about. The subject may be a noun (person, place, thing, or idea) or a pronoun (*he, she, it, they*). Pronouns are used in paragraphs only after the noun has been stated. A verb ending in -*ing* can function as a noun and the subject of a sentence.

> *Mowing* the lawn, *washing* clothes, and *taking* out the garbage are some of the obvious uses.

Note: The inferred "you" subject is used only with present tense verbs.

> *Wrong:*   Took a chance.
> *Correct:*   (You) Take a chance.

Exercise 1   *Supply a subject for each sentence.*

1. _____ is the career field I am studying.

2. _____ taught my first technical class.

3. _____ is a company at which I intend to apply for a job.

4. _____ likes to study with a partner.

5. _____ is my favorite class this term.

Exercise 2   *Underline the subject of each sentence. If the subject is missing, rewrite the sentence to make it complete.*

1. Computers have many uses.

2. By the mid-1980s, bought hundreds of thousands of early computers.

3. After the novelty wore off, wound up back in the closet.

4. The reality was that only a few people had a real need.

5. For everyone else, was a device without an application.

6. With personal computers is very different.

7. Notice the development of the software market.

8. Is capable of performing many applications.

9. The personal-computer field continues to grow.

10. Learned more about my future employment opportunities.

Missing-Verb Fragments

The verb in a sentence states an action or condition (state of being) about the subject.

> Heath Company introduced a personal robot, the Hero Junior, in 1984. (*action*)
> The Hero Junior *was* basically a toy. (*state of being*)

Including the helping verb with a participle makes the sentence complete. In the following fragment, notice that the absence of a helping verb makes it incomplete.

> The Hero Junior *introduced* by Heath Company.

Often in speaking, we slide over helping verbs (or the final *ed* or *s*), which starts a careless habit of leaving them out in writing. A present (*ing*) or past-participle verb without a helping verb is a fragment. The helping verb can be added as a word or contraction.

Exercise 3 *Underline the complete verb. If the verb or part of the verb is missing, rewrite the sentence and make it complete.*

1. Of the companies currently producing robots, Heath done well.
2. The personal robots industry unstable.
3. New manufacturers appearing and disappearing all the time.
4. All the mechanical parts were designed to have multiple functions.
5. All the products available for general sales.
6. In November 1985, a division of Ideal responsible for a personal robot, Maxx Steel.
7. We followed the same steps.
8. A time line of the history of personal robots versus that of the PC.
9. Firms beginning to enter the market.
10. A Texas-based marketing-research firm seen the investigation.

Linking Fragments to Sentences

Another way of correcting missing-subject or missing-verb fragments is to combine the fragment with a related sentence. A sentence may have two subjects or two verbs, but there must be a logical connection in which the fragment adds to the meaning of the sentence.

Wrong:	The robots of today.	And workers perform side by side.
	(missing-verb fragment)	*(sentence)*
Correct:	The robots of today and workers perform side by side.	
	(sentence with two subjects)	
Wrong:	Robots are available.	And used for many purposes.
	(sentence)	*(missing-subject fragment)*
Correct:	Robots are available and are used for many purposes.	
	(sentence with two verbs)	

Warning: Do not add commas between a pair of subjects (*compound subjects*) or a pair of verbs (*compound verbs*).

Exercise 4 *If the following groups contain one or more fragments, combine the fragment(s) to make a complete sentence, and write the combination. Change the punctuation to make the combination correct. If there is no fragment, write* **correct**.

1. The installed base of robots in the United States. Is seen in many applications and industries. _____

   _____

2. The robot is a powerful tool in the manufacturing area. And requires specialized environments. _____

   _____

3. Robots are unforgiving. They reveal inefficient, ill-conceived, and outmoded processes. _____

_____

4. Robot manufacturers are aggressively looking to improve their products. And broaden their market. _____

_____

5. Enhancements such as vision, sensors, and communications systems. Are being applied almost routinely. _____

_____

6. Robot manufacturers are not sitting on their haunches. They want to assure the robot's position on the market. _____

_____

7. Too often the robot has been looked on as a human replacement. And subjected to environments designed for humans. _____

_____

8. The conditions of the workstation. And inherently human skills contribute to the human operator's low productivity. _____

_____

9. The robot, in a very real sense. Forces the manufacturer to look at how process operations are organized. _____

_____

10. Industry leaders hope for sustained growth of robot usage. And are committed to increased development. _____

_____

Phrase Fragments

A third type of fragment, the **phrase fragment,** suffers from the lack of a subject and a verb. It may be made up of one or more prepositional phrases. Remember that subjects are never inside prepositional phrases. These fragments can be corrected by combining the fragment with its parent sentence.

    Wrong: Robot sales have fallen short. Of earlier forecasts.
           (sentence)   *(phrase)*
    Correct: Robot sales have fallen short of earlier forecasts.
           (sentence with phrase)

Exercise 5   *Write the combination necessary to correct any phrase fragments. If there is no fragment, write* correct.

1. To understand the situation properly. One must keep the current state of robotics in perspective. _____

_____

2. First, the technology continues to advance. To the benefit of the user both in terms of reliability and capabilities. _____

_____

3. A sure sign of industry maturity is the serious work being done. Several committees protect the standards for robot design. _____

_____

4. Second, the robot becomes an integrator. Of a variety of advanced manufacturing technologies. _____

_____

5. Third, the growth of the robotics industry is tied to an attempt. To improve and modernize manufacturing processes. _____

_____

6. We are aware that the manufacturing process will be renewed. But we lack the commitment to get the job done. _____

_____

7. The robot is a tool. Much like CAD/CAM, machine vision, and a host of other technologies. _____

_____

8. They are interrelated. Through marketing and development, their benefits are known. _____

_____

9. They will proliferate only at the rate we choose. To integrate and apply them. __

_____

10. As we learn more about robotics. This technology can grow and prosper. _____

_____

Exercise 6   *There are six fragment errors in the following paragraph. Find the fragments. Then edit the paragraph to correct the fragments by combining them with sentences and adding or removing periods and capital letters.*

The first International Personal Robot Congress and Exhibition held in 1984 in Albuquerque, New Mexico. Isaac Asimov, author of hundreds of books. Opened the event. Spoke of robots as they have appeared in his science fiction and as he thinks they should be in reality. Asimov laid down the "three rules of robotics," which Asimov claims still stand: (1) A robot not injure a human being, or through inaction allow a human being to come to injury. (2) A robot must obey orders given to it. By a human being, except when those orders would violate the first law. (3) A robot must protect its own existence. Except when that would violate the first or second law.

Run-Ons

A run-on is a sentence error in which two independent clauses (grammatically complete sentences) are run together without any separating punctuation.

To correct the run-on, you can add the missing comma, or you can reconstruct the sentence and add a semicolon, semicolon and transitional word, period and capital letter, or subordinate conjunction. You can also revise it to be a simple or complex sentence.

Run-on error:	The Chandra is an X-ray telescope and it is orbiting Earth.
Revised:	The Chandra is an X-ray telescope**,** and it is orbiting Earth.
	The Chandra is an X-ray telescope**;** it is orbiting Earth.
	The Chandra is an X-ray telescope**; furthermore,** it is orbiting Earth.
	The Chandra is an X-ray telescope**. It** is orbiting Earth.
	The Chandra is an X-ray telescope **that** is orbiting Earth.
	The Chandra**,** an X-ray telescope**,** is orbiting Earth.

Comma Splices

A comma splice is a sentence error in which two independent clauses (grammatically complete sentences) are joined only by a comma and are missing the required coordinating conjunction, such as *and, but, for, or, nor, so,* or *yet.* Correct a comma splice by adding the missing conjunction or by revising the sentence the same way you would a run-on.

Comma splice error: The Chandra was launched in 1999, it will operate for 20 years.
Revised: The Chandra was launched in 1999, **and** it will operate for 20 years.
The Chandra was launched in 1999**;** it will operate for 20 years.
The Chandra was launched in 1999**; furthermore,** it will operate for 20 years.
The Chandra was launched in 1999. **It** will operate for 20 years.
The Chandra was launched in 1999 **and** will operate for 20 years.
The Chandra**,** launched in 1999**,** will operate for 20 years.

Exercise 7   *Revise the following sentences to correct the comma splice errors. Vary the way you revise the sentences.*

1. The Chandra is an X-ray telescope it has 10 times greater resolution than other telescopes.
2. It has 50 to 100 times more sensitivity than any other telescope and it is orbiting Earth.
3. The telescope orbits Earth every 64 hours it was launched by a space shuttle in July 1999.
4. People devoted many years of their lives to the Chandra they wanted the Chandra to function perfectly.
5. The telescope's mirrors must be perfect it will send back continuous data.
6. It is a dream come true for scientists it will advance the science of astronomy.
7. The Chandra is an X-ray observatory it produces images of high-energy X-rays.
8. The information will be sent back to astronomers it will answer questions about the formation of the cosmos.
9. Scientists hope this is only the beginning Chandra should have a long life.
10. Future space launches might visit the Chandra the Chandra could have mechanical difficulties.

Transition Words

Writing a unified report begins with careful outlining and continues with writing headings and subheadings in the report. Because these methods are so essential, no other method can compensate for their loss. However, other methods can supplement them, and the easiest one is using **transition words** as you move from one idea or example to the next. The following lists include words that are commonly used to identify sequences of ideas, examples, contrasts, and conclusions.

1. *Sequence words* indicate not only number order but also emphasis or priority.

first	after that	in addition	especially
first of all	next	furthermore	particularly
second	later	last	moreover
third	then	finally	also, too

Hint: When listing reasons or cases, writers usually save the best for last. Begin with the weakest or least significant idea, and work up to the strongest or most important.

The example below shows one use of sequence words.

> I follow the same steps when I write. First, I research my topic. Then, I write an outline. After that I start writing. The final and most important step is editing and proofreading. This last step, although it takes less time than the others, is the one I dread most of all.

2. *Examples* need to be clearly identified to prevent confusion and misunderstanding. When you are moving from an abstract theory or principle to a concrete application, provide immediate, obvious signals for your reader.

for instance	for example
that is	such as
let us say	in the case of

> Stress can have many side effects. For instance, a supervisor feeling pressure to complete a difficult project may become irritable, develop an ulcer, or possibly even succumb to illnesses such as a cold, flu, or more serious disease.

3. *Contrast* transitions show differences and similarities.

however	nevertheless	moreover	still
yet	instead	otherwise	even though
on the one/other hand			

> The pressures of work and school were affecting my health.
> Nevertheless, I continued to keep up the pace.

4. *Conclusions* will often identify the end of a section or report.

after all	in conclusion	consequently	finally
anyway	in fact	therefore	in summary
at least	in short	thus	
at any rate	in other words	hence	

The problems were not quickly resolved. Consequently, the project missed its deadline.

Exercise 1   *Arrange the following transition words into the four categories listed below.*

then	for instance	therefore	thus
nevertheless	however	finally	such as
consequently	instead	particularly	for example
let us say	in addition	in short	otherwise

Sequence	Examples	Contrast	Conclusion
then	let us say	nevertheless	consequently
in addition	for instance	however	therefore
particularly	such as	instead	finally
	for example	otherwise	in short
			thus

Exercise 2   *Choose logical transition words for the following passages. If more than one word could be used, think about the difference in meanings, choose one, and write the others in the margin.*

1. The vacation provided some much-needed and long-overdue relaxation. _____, after one week back at work, I felt like I had never been away.

2. The vacation provided some much-needed and long-overdue relaxation. _____, when I returned to work, I felt enthusiastic and energetic again.

3. I didn't drink as much coffee. _____, I listened better to what my colleagues were saying. And _____, I regained my sense of humor and found I could laugh at myself without getting defensive and hostile.

Exercise 3   *Use sequence words to describe, in one paragraph, how you perform one of the following activities.*

- Study for a test
- Accomplish your work
- Complete a lab

Parallelism

When you first studied geometry, you learned the definition of parallel lines: lines that extend in the same direction at the same distance apart. Parallel circuits, in electronics, are constructed so that an equal voltage is applied to all components. In writing, we use a similar concept of parallelism.

Parallelism in writing is a method of using similar words or sentence structures to identify equal ideas. A series of nouns, verbs, or phrases must be expressed in a parallel form. To keep ideas parallel, we state them in similar forms and word orders. Sentences that are not in parallel form sound awkward and confusing.

The problems in the following awkward example lie in the verb tense and word order.

Awkward: Management communication can promote effectiveness or be destroyed.

The following corrected sentence puts the verbs in a parallel form (present tense) and reorders the words to make the idea logical and clear.

Parallel: Management communication can promote or destroy effectiveness.

Putting a sentence into a clear, parallel form often gets rid of clumsy wording and "noise."

Awkward: Even the best communicators learn as much as possible about their listeners and tailor their remarks to their interests, attitudes, and what their values are.

Parallel: Even the best communicators learn as much as possible about their listeners and tailor their remarks to the listeners' interests, attitudes, and values.

Occasionally, repeating a similar form can add emphasis to the equality of the ideas. The following examples show different types of parallelism.

Parallel Words and Phrases

While all we want is to *sound good*
                         look good or
                         appear to be intelligent,
some people get carried away.
(*all present-tense verbs*)

Failing to close the feedback loop can take several forms:
         not listening to our own messages, and
         not listening to our receiver's feedback.
    (*repeated* not *and* -ing *forms*)

Parallel Clauses

> When an unhappy employee bellows, "I'm not mad!" it's pretty
> obvious that *what he or she is saying and*
> > *what he or she is communicating*
> are two different things.
> > (*repeated* what, is, *and* -ing *verb forms*)

Parallel Sentences

> *Tailor your language to your audience.*
> Or, when in doubt, *use simpler language.*
> > (*repeated understood* "you" *and present-tense verbs*)

Exercise 1   *Underline the parallel ideas in the following sentences.*

1. Remember that what you do speaks louder than what you say.
2. You should tell the reader what is coming up and make it easy for that person.
3. Much of the meaning we convey to other people, we convey through our tone of voice, appearance, timing, and many other nonverbal factors.
4. People will not communicate effectively if they fail to adjust language to their audience or fail to word ideas in an understandable way.
5. Your presentations will be more successful if you state your purpose, define unfamiliar terms, present information logically, and restate important ideas in a conclusion.
6. Transitions in presentations are more easily identified if you state your objectives clearly at the beginning and if you use sequence words, such as *first* and *second,* during the talk.
7. Messages that use simple language will be better understood by your co-workers, by your managers, and by your friends and family.
8. There are kernels of information that are important to understand and necessary to convey.
9. In written communication, *access* means using enumeration, white space, short paragraphs, highlighting, and bullets to identify bits of information.
10. No one wants to wade through a confused piece of writing, and no one wants to find a surprise twist at the end.

Suggestions for Writing Parallel Ideas

The following suggestions will help you to make parallel ideas more clear.

Hint 1: Repeat a preposition, an article, or the introductory word of a phrase or clause.

Examples

People ensure effective communication
    by adjusting language to their audience or
    by wording ideas in simple ways.

They fail to adjust language to their audience or
    to word ideas in ways that others can understand.

Two factors lie at the root of this error:
    a failure to recognize differing purposes and
    a desire to exert power.

A message is effective when it
    is reached by its intended audience,
    is understood by the receiver,
    is remembered for a reasonable time, and
    is used when appropriate.

Hint 2:   Use correlatives such as the following:

both . . . and	not only . . . but also
either . . . or	whether . . . or
neither . . . nor	

Examples

Communication includes both the delivery of a message
    and the understanding of a message.

Actions not only speak louder than words,
    but also can confuse the meaning of or contradict words.

Hint 3:   Be sure that items in a series are in a similar form.

Examples

The confused listener was filled with questions,
                        frustration, and
                        rage.
                              (*series of nouns*)

An effective communicator can present a message that is clear,
                           complete, and
                           concise.
                              (*series of adjectives*)

Exercise 2   *Supply a parallel word or phrase for each sentence.*

Examples

The father and son loved to play catch and <u>*go fishing*</u>.
(*play* and *go* are both present tense verbs)

Sue searched for a fitness center in the newspaper and <u>*in the phone book*</u>.
(*in the newspaper* and in *the phone book* are both prepositional phrases)

1. Some people think that exercise is boring and _____.
2. People use excuses such as being too busy or _____
   _____.
3. To make exercise fun, choose exercises that you enjoy and _____.
4. It is a question of what fitness means to you and _____.
5. See a doctor if you have dizzy spells or _____
   _____.

Exercise 3   *Make the main ideas of these thesis statements parallel.*

1. Earth is in danger.
   a. Industries dump raw sewage into rivers
   b. Toxic wastes
   c. Cutting down forests
2. My vacation was a disaster.
   a. Lost my return airline tickets
   b. Spending too much money
   c. The rental car broke down twice
3. A technical career is challenging.
   a. Machines and instruments are complex
   b. Learn new skills to keep up
   c. Companies try to stay ahead of the competition

Exercise 4   *Rewrite the awkward sentences in a parallel form*

Example

Awkward: Computer systems called "computer-aided design" (CAD) and
"manufacturing" (CAM) are useful to engineers.

Revised: Computer systems called "computer-aided design" (CAD) and
"computer-aided manufacturing" (CAM) are useful to engineers.

1. Engineers can design mechanical parts and analyzing them.
   _____

2. Companies can reduce the development time and cost-wise.
   _____

3. The advantages of a CAD system are its speed and being accurate.
   _____

4. The system can help conceive a product and having it redesigned.
   _____

5. Computers make calculations for optimum shapes and in the right size.
   _____

6. Designers are not only free from the repetitious task of drawing lines but also calculating workpiece sizes is difficult. _____
   _____

7. A CAM system communicates with the robot to handle, process, and to produce a product. _____
_____

8. CAD drawings can be stored in the computer or the hard copies can be stored, too.
_____

9. CAD/CAM can be done on stand-alone computers or the four or five workstations use a central computer. _____
_____

10. Computer programs are the form of communication between the programmer and when the computer is used. _____
_____

Exercise 5   *The following paragraphs need revision because of errors, due to awkwardness, fragments, and redundancy. Rewrite the paragraphs to make them clear and effective.*

1. The personal computer is the consummate tool. It can work for you and linked to other computers. When linked to other computers, it has access to a tremendous range of information services to retrieve, store, reorganization, and shipment of data.

_____
_____
_____
_____

2. People from many professions and interests use computers every day. These people report making more money, they had more fun, or being more powerful as a result of using personal computers. Their success stories started by calling up information utilities that maintain data banks, or they called up other users to ask questions and have gotten immediate answers._____

_____
_____
_____
_____
_____
_____

3. Computer communication is a complex topic. There are several things you need to know, or someone else who knows them. If you know someone who already understands computer communication. Just sit back and listen. But if you are on your own. You must learn something about computers on your own. _____

_____
_____
_____
_____
_____

4. Computers store information in their internal memory or CDs or hard disk, but no one can own all the information he could need. Now databanks can collect, maintain, will purge dated information, and can update all files. _____

_____

_____

_____

5. Databanks are either narrow, well-defined systems that serve a specific target group. Or more people use multipurpose systems that provide broad types of information. Multipurpose utilities usually perform at least three major functions: communications, allowing transacting, and information access. _____

_____

_____

_____

_____

PART V

Mechanics Units

Mechanics Unit 1: Capital Letters

Mechanics Unit 2: Abbreviations and Acronyms

Mechanics Unit 3: End Punctuation

Mechanics Unit 4: Commas

Mechanics Unit 5: Semicolons and Colons

Mechanics Unit 6: Parentheses, Dashes, Brackets, Ellipses,
Slashes, and Hyphens

Mechanics Unit 7: Apostrophes

Mechanics Unit 8: Quotations

Capital Letters

Several capitalization rules that are most often used in technical writing will be reviewed. Style guides are available in libraries and bookstores for specific or unusual situations.

A popular trend these days is using a company logo (emblem or symbol) of lowercase (small) letters. This appears most frequently in "artistic" advertisements for products. Most writers, when referring to these same items, use standard capitalization rules regardless of the company's stylized logo.

phase linear®
I bought Phase Linear speakers for my car.

Rule 1: Capitalize the first word of a sentence, the first word of a quote, and the pronoun *I*.

The elderly man and I didn't hear the last call.
We heard the conductor say, "Last stop coming up."

Rule 2: Capitalize specific names of people, including initials and nicknames (although nicknames are rarely used in formal writing).

Honest Abe Lincoln	John F. Kennedy
Robert "Skip" Simons	B. F. Skinner

Rule 3: Capitalize names of countries, states, cities, and other geographic regions. Do not capitalize directional words.

the Southwest	New York City
turn southwest	European vacation

Rule 4: Capitalize names of races, religions, nationalities, and any word derived from the name of a country.

Chinese	Puerto Rican
Episcopal	Indian

Note: Some words derived from country names are used so often that we sometimes forget to capitalize them.

French door	Italian dressing
Scotch eggs	English ivy

Many Americans use *white* and *black* to refer to races. These words do not have to be capitalized, though some writers prefer to capitalize them. Decide how you will treat them and be consistent.

Exercise 1   *Change any incorrect lowercase letters to capital letters.*

1. although robert grew up in florida, he now considers himself a georgian.
2. we learned that after the catholics were driven out of europe, they immigrated to northeastern regions in the united states such as new york, massachusetts, and connecticut.
3. my mother ordered italian pizza, french bread, and boston-lettuce salad.
4. martin luther king said, "i have a dream."
5. there are many muslims and jews in new york city and many baptists in atlanta.
6. her sales territory covers the entire northeast.
7. my family and i moved from the north to miami, florida.
8. jazz is a popular form of music in new orleans.
9. thomas a. edison, credited for saying "there is no substitute for hard work" invented the thermionic diode.
10. silicon valley is a nickname for an industrialized area in northern california surrounding san jose.

Rule 5:   Capitalize a person's title if a proper name follows it, but not when the title is used alone.

Professor Henry Higgins   former President Ford

the professor   a former president

Note: Abbreviated titles (Mr., Ms., Jr., Ph.D.) begin with a capital letter and end with a period.

Rule 6:   Capitalize names of days and months, but not seasons.

Thursday   spring

March   fall

Rule 7:   Capitalize names of languages and specific courses, but not general subject areas.

electronics   mathematics

Digital Electronics   Technical Mathematics

English   Pascal

Rule 8:   Capitalize a person's name that is part of a general theory or principle, but not general electronic or scientific terms, even if derived from a name.

Ohm's law   5 ohms

Einstein's theory of relativity

Exercise 2   *Change any incorrect lowercase letters to capital letters.*

1. freshman orientation was taught by a new professor.
2. the doctor spoke both english and spanish fluently.
3. all labs will meet on tuesdays and thursdays during the fall term.

4. after the class studied ohm's law, the problems used ohms to measure resistance.

5. our class will graduate with an associate of applied science degree in june.

6. the electron was discovered by french physicist jean baptiste perrin and british physicist sir joseph thompson.

7. i had no problem in math until i took physics II from professor levine.

8. former president bush attended the spring conference.

9. my cousin was supposed to appear before judge carlson on tuesday, but the judge rescheduled the hearing.

10. my professor earned a ph.d. last fall.

Rule 9:   Capitalize the first word and all major words in titles of books, articles, movies, and television shows, and names of businesses.

The Cosby Show	*Electronics Review*
In Search of Excellence	Scientific Atlanta

Note: Titles of books, magazines, movies and television series are either italicized, (if possible) or underlined. Titles of articles from newspapers and magazines are enclosed by quotation marks.

Rule 10:   Capitalize historic events, famous places, holidays, and organizations.

Tuskegee Institute	Labor Day
World Trade Center	Space Age
United States Senate	Mississippi River

Rule 11:   Capitalize brand names but not product names.

Panasonic turntable	Pontiac sportscar
Epson printer	IBM typewriter

Note: Many companies and products on the market have unusual spellings of common words. Be careful to spell them as they are used by companies.

Compaq Portable II	LANLink
B&K-Precision	Printonix

Exercise 3   *Change the incorrect lowercase letters to capital letters.*

1. jim watches *good morning america* every day.

_____

_____

2. lee bought a hewlett packard printer for her computer system.

_____

_____

3. our family celebrated independence day by visiting the smithsonian museum.

_____

_____

4. after reading several issues of *popular mechanics,* i ordered a subscription.

_____

_____

5. nancy and tom were inducted into alpha beta tau last term.

_____

_____

6. all the teenagers in london were wearing levi jeans.

_____

_____

7. my english teacher told us to read *the great gatsby* and any two plays by shakespeare.

_____

_____

8. i looked through the heathkit catalog to find a project for my digital electronics class.

_____

_____

9. the student government association sponsored a halloween party at northside children's hospital.

_____

_____

10. he gave his wife a siamese kitten, and she gave him an airedale puppy.

_____

_____

Abbreviations and Acronyms

Shortened forms of words and phrases are referred to as **abbreviations** or **acronyms.** For all but the most common shortened forms, technical writers provide the expanded meaning of an abbreviation or acronym the first time it is used. Either the expanded meaning or the acronym may be placed in parentheses.

> The circuit needed a *direct current (DC)* voltage source.
> The computer has *64 K (kilobytes)* of *ROM (read-only memory).*

Many technical symbols are abbreviations or acronyms. They do not follow the standard rules for capitalization, since sometimes a lowercase letter may stand for one word and the uppercase letter may stand for another, such as *M (mega)* and *m (milli).* Many common electronics symbols and abbreviations are listed in Appendix 1. Common abbreviations and acronyms are found in some dictionaries and textbooks. Be sure to use technical abbreviations, acronyms, and symbols properly.

Abbreviations are shortened forms of words or phrases. Each letter (as in CIA) or each unabbreviated word (as in titles) is pronounced. Abbreviations of titles such as *Mr.* and *Jr.* begin with a capital letter and end with a period. Pronounce the title as though the full word were written.

Other abbreviations, such as *TV, AC,* and *FBI,* may stand for one or several words. Each of the letters in such abbreviations is pronounced.

The abbreviations for states have undergone a recent change. Currently, we use the two-letter postal code for states, such as *GA (Georgia)* and *CA (California).* Notice that both letters are capitalized and no period is added. Before this code was established, people abbreviated states by using the first few letters, capitalizing only the first letter, and ending with a period, as in *Geo. (Georgia)* or *Calif. (California).* Some states were abbreviated several ways (*Cal., Calif., Ca.*). Although some people still use the old method, the new postal code provides a uniform system.

Exercise 1   *Find the postal abbreviations for the following states (see Appendix 1).*

Alabama _____   Alaska _____   Arizona _____   Arkansas _____

Write the postal abbreviation for your state. Write the postal abbreviations for three states in which you would like to live.

Your state _____   Others _____   _____   _____

Other abbreviations, such as days of the week and months, usually consist of the first syllable of the word, beginning with a capital letter, and ending with a period. If the word is short or has only one syllable, such as *May* or *June,* do not try to abbreviate it.

Exercise 2   *Write the standard abbreviations for the days of the week (see Appendix 1).*

Sunday_____   Monday_____   Tuesday_____   Wednesday_____
Thursday_____   Friday_____   Saturday_____

Exercise 3   *Write the expanded form of the following abbreviations.*

1. TV _____   **2.** mike _____

3. stereo _____   **4.** amp _____

5. CD player _____   **6.** scope _____

Acronyms are words created by combining the first letter of most or all of a combination of words to make a pronounceable word. Many times these word groups are arranged and contrived to make the pronunciation easier, such as *bit (binary digit)* and *laser.*

Exercise 4   *Write the expanded meaning of laser. Use a dictionary if necessary.*

l_____   a_____   by s_____
e_____   of   r_____

Even though acronyms are types of abbreviations, no periods are placed after the letters, and all the letters are usually capitalized. When acronyms, such as *laser,* become commonly known and used, they are written in lowercase letters. Watch your textbooks and technical magazines for accepted forms.

Exercise 5   *Write the expanded meaning of the following acronyms. Use a dictionary if necessary.*

1. CD-ROM _____

2. RAM _____

3. sonar _____

4. scuba _____

5. radar _____

Write five common acronyms in your area of study and their expanded meanings.

6. _____

7. _____

8. _____

9. _____

10. _____

End Punctuation

Sentences end with a **period, question mark,** or **exclamation mark.** The end punctuation provides a meaningful signal to the reader. For example, notice the difference in meaning in the following sentences expressed by the end punctuation:

> You have a second PC.
> You have a second PC?
> You have a second PC!

Periods for Ending Sentences

Use a period to end most sentences, including statements, commands, and indirect questions.

Statement:	I installed the program on my PC.
Command:	Install the program on your PC.
Indirect question:	He asked if I installed the program on my PC.

If entries in a bulleted list are complete sentences, end each entry with a period.

> To install the software:
> * Close all running applications.
> * Insert the CD in the CD-ROM drive.
> * Follow the on-screen instructions.

Note: Do not end entries with a period if the entries are phrases.

> Before installing the software, close all running applications, including:
> * Word processors
> * Graphics programs
> * Internet connections and e-mail carriers

Other Uses of the Period

Use a period to follow most abbreviations, including days, months, and titles, even if a comma also follows the abbreviation.

The class meets Tues. and Thurs. from 7–9 P.M., and the lab meets on Wed. from 7–10 P.M.
The semester runs from Sept. through Dec.
Either Dr. Manning or Prof. Gallagher will teach the class.
The recruiter from Scientific Creations, Inc., is on campus today.

Note: Do not use periods after abbreviations for company names or organizations, after most acronyms, or when spelling out a title.

IBM	NASA	DNA	RAM	GB
Professor	Doctor	Mister	Miss	

Follow conventions when using periods in some abbreviations, acronyms, and decimals (observe current styles and conventions in textbooks and industry magazines).

Ph.D.   .bmp   .exe   8.5 × 11

Question Marks

Use a question mark to end direct questions. Direct questions often begin with words such as *who, what, where, when,* and *how.*

Who purchased the software?
When did you register the software?

Some questions begin as a statement and end with a question.

The program runs correctly, doesn't it?

Use a question mark to signal a request.

Would you please complete the registration for me?

Use a question mark within parentheses to indicate uncertainty (in informal communication only).

I want to register the program (available online?) after finishing the installation.

Note: Indirect questions are statements that include a question but are not asked as a direct question. Do not use a question mark to end an indirect question:

He asked how to find the library.
She wants to know if the instructions worked.
I wonder if I have the most recent version.

Exclamation Points

Use an exclamation point to end sentences expressing genuine excitement or strong emotion.

Oh, no!   I passed!   Help!

Overuse of the exclamation point weakens its effectiveness and makes the writer sound immature or even hysterical:

> This summer, the giant airship will lift a 160-ton load! It can cruise from 50–60 miles per hour! It has a range of 6,000 miles! The load can remain stable even in heavy winds. This may revive the airship industry!

Many writers agree that the exclamation point is rarely appropriate in business communication, especially in formal correspondence, because it changes the tone from neutral to emotional. If an exclamation point is used at all, the writer should monitor the surrounding wording to keep the remaining tone businesslike and professional.

> We are pleased to announce a rebate to our established customers!

Never use multiple exclamation points in technical writing—it appears childish and overly dramatic.

Exercise 1   *Add the correct end punctuation.*

1. Fingerprints provide clues about a person's health .
2. Have you ever been fingerprinted ?
3. Many people wonder if fingerprints are reliable .
4. We won !
5. Children's fingerprints vanish more quickly .
6. Would you please search for further research in this area ?
7. You found the follow-up study, didn't you ?
8. Run !
9. He asked, what fingerprints can detect today ?, period
10. Fingerprints might someday be used to detect the presence of drugs, alcohol, and diseases .

Commas

Commas are punctuation marks that are used for many different purposes. The rules for using commas are concrete, but since there are so many, remembering them requires some practice. This unit will review the basic rules for commas and give you practice in correct placement of commas. Knowing the rules will help make comma decisions easier in writing. Commas are used with the following:

1. Series of elements

2. Adjective pairs

3. Interrupters

4. Introductory information

5. Quotes

6. Compound and complex sentences

7. Miscellaneous uses

Note: Commas are always placed directly after a word. Leave one space after a comma before beginning the next word.

Commas in a Series

Two similar types of words written together are called a **pair.** They are separated by a conjunction such as *and* or *or.* A pair can consist of words or phrases.

> gray *or* blue (*pair of adjectives*)
> into the building *and* through the door (*pair of prepositional phrases*)

No commas are placed between pairs.

A **series** is three or more like words or phrases. Items in a series are separated by commas, with a conjunction before the last item in the series.

> gray, blue, *or* brown

We could add conjunctions between the items, but this would be wordy.

> gray *or* blue *or* brown

Instead, we eliminate all but the last conjunction and add commas between each item. Sentences can contain series of words and phrases.

> He bought new shirts, ties, shoes, *and* socks.
> He walked to the building, through the door, *and* inside the room.
> He introduced himself, shook my hand, *and* sat down in the chair.

Note: Some people prefer not to put a comma before the conjunction in a series. Most technical writers put a comma before the conjunction because it provides a visual cue, like a green light turning to yellow, that the series is ending. Decide how you will handle it, and be consistent.

Exercise 1   *Place commas in any series below. In this exercise, place a comma before the conjunction in the series.*

1. On his first second and third interviews, he wore the same suit with different shirts and ties.
2. Margaret was interested in finding a navy-blue suit a blue blouse and black shoes.
3. The clerk asked Jack for his size and color preference.
4. To determine a company's dress code, you must read the employee manuals ask questions and watch the people around you.
5. The book recommended a gray suit white shirt black shoes and burgundy tie.
6. A "power suit" will make you feel confident attractive and proud of your appearance.
7. People looking for a new job a promotion or a salary raise must choose their wardrobes carefully.
8. Dave's weight and height limited his choice of styles.
9. Finding the right suit took time energy money and patience.
10. The personnel manager commented favorably on Roger's résumé interview and appearance.

Commas Between Adjective Pairs

When two or more adjectives are used together, sometimes a comma is placed between them and sometimes it is not. Read the following examples.

1. a dark, simple suit
2. a dark business suit

In the first phrase, there is a comma between the two adjectives. Either adjective could modify the noun separately (*a dark suit, a simple suit*). The phrase could be written, *a dark and simple suit.* In fact, the two adjectives could be reversed, *a simple, dark suit,* and still make sense. In this case, a comma is placed between the two modifiers.

In the second phrase, *dark* is modifying a *business suit. A dark and business suit* would sound ridiculous, and so would reversing the two adjectives: *a business dark suit.* A comma is not placed between these modifiers.

Rule: If the adjectives can be reversed and still make sense, put a comma between them. Notice the difference caused by the comma in the two sentences below.

> She wore a soft blue dress.
> She wore a soft, blue dress.

The comma in the second sentence tells us that the dress was soft (in texture) and blue. In the first sentence, the blue color is a soft (pastel) shade.

Exercise 2   *Place commas between modifiers if they are needed.*

1. A handshake is a common business gesture.
2. A firm straight handshake gets a meeting off to a good start.
3. If you receive a brief abrupt handshake in return, the meeting is already in trouble.
4. Originally, a mutual clasp of forearms meant there were no small hidden weapons under the sleeve.
5. At an important business meeting, if everyone gets a handshake but you, be ready for bad news.
6. Years ago, fathers taught only their sons to shake hands because it was considered a required masculine gesture.
7. Today, women in business sometimes extend their hands first to eliminate any hesitation an unsure inexperienced man might have.
8. A vertical robust handshake signals you are ready to do business.
9. A frail limp handshake indicates lack of confidence or enthusiasm.
10. An overly aggressive bonecrushing handshake is a sign of fear, resentment, or extreme competitiveness.

Commas Around Interrupters

Commas are placed around words that interrupt the flow of the main sentence, particularly if the interrupting words are between the subject and the verb of the main sentence. You can usually "hear" a pause, even if the words are read silently. Words, phrases, and clauses can interrupt the main sentence. Test to see if the words are essential or interrupting by removing them from the sentence. If the sentence still makes sense, then the words are interrupters, and they should be set off with commas.

Words may interrupt the meaning of a sentence. Often they define or rename a noun. Information that is added because it is helpful, but not essential, is set off with commas.

> NASA, the space agency, has promised free space shuttle launches.

In the preceding example, *NASA* is written to rename *the space agency*. Notice the difference made by commas in the following examples.

> My brother Richard attended the launch.
> My brother, Richard, attended the launch.

The first sentence means that Richard, not another brother, attended the launch—the name is essential. The second sentence means that there is only one brother, his name is Richard, and he attended the launch—the name is not essential, therefore it is interrupting.

Phrases may be written between the subject and verb of a sentence. Commas are placed before and after phrases that interrupt the meaning of the sentence.

> Multiwire boards, unlike other boards, have a flexible design. (*interrupting phrase*)
> The design flexibility of multiwire boards creates an added benefit. (*essential phrase*)

Clauses may be written within the main clause of a sentence.

> The technicians, who are about to graduate, are in the lab.

In the preceding sentence, the main clause is *The technicians are in the lab*. The dependent clause, *who are about to graduate,* is written within the main clause. Determine the difference between the two sentences that follow.

1. The technicians who are about to graduate are in the lab.
2. The technicians, who are about to graduate, are in the lab.

In sentence 1, the dependent clause tells where you can find the graduating technicians. The clause is necessary for the meaning of the sentence, so no commas are used.

In sentence 2, the dependent clause states that the students in the lab are also those who are about to graduate. As you read the sentence, you can hear a pause at the commas. The dependent clause is interrupting the main clause with extra (nonessential) information, so commas are placed before and after the dependent clause.

Another type of interrupter is the name of the person or group who is reading (or hearing) the message. We set off the person or group being spoken to with commas.

> Students, go to the lab.
> Start the experiment, Jack, and record the data.

Caution: Remember that commas around interrupters are used in pairs. If you use one comma at the beginning of a word, phrase, or clause, use a second comma at the end.

> *Wrong:*   The people who were talking about space are outside.
> *Wrong:*   The people, who were talking about space are outside.
> *Correct:* The people, who were talking about space, are outside.

Exercise 3   *Add commas around interrupting words, phrases, or clauses.*

1. The manager of the company Mr. Mendez issued the new policy.
2. The workers many of whom were new employees seemed to accept the change.
3. Soon more policies unrelated to the first occasion were changed.
4. The manager knowing that the workers weren't fixed in old ways took the opportunity to make changes.
5. That situation workers accepting change without resistance was unusual.
6. Dealing with resistance something new managers are unprepared for takes some experience.
7. I want you Mr. Riley to learn from this experience.

8. The president meaning no harm asked the manager to defend the changes.

9. The board of directors too were interested in the manager's methods.

10. Mr. Mendez however was eager to discuss the positive experience.

Exercise 4   *Each of the following complex sentences has a dependent clause between the subject and verb of the main clause. Some are essential; some are interrupting. If the sentence is correctly punctuated, write* correct. *If not, add commas where they belong. Be prepared to justify your answer.*

1. Cellular telephone service which allows phones in motion to communicate with each other via radio is highly successful.

2. The service which is called the mobile satellite system (MSS) reaches every part of the continental United States.

3. With MSS, even an Idaho lumberjack who uses a cellular terminal can receive and transmit messages.

4. The messages even though they come from the wilderness travel via satellite to the telephone network used by the rest of the country.

5. Industry and emergency services which include nationwide paging, data collection, and position locating have become possible as a benefit of MSS.

Commas after Introductory Information

Words, phrases, and clauses are sometimes written to introduce the main sentence. Put commas after introductory information, as in the following examples.

> Students, gather up your equipment.
> Before long, the room was empty.
> Although they weren't finished, the students left.

Note: If an introductory phrase is short, a comma does not have to be added. (In the following exercises, however, add the commas.)

Exercise 5   *Add commas after introductory information.*

1. Although it was only 45 years ago the discovery of DNA has had a tremendous impact on science.

2. After years of study James Watson and Francis Crick worked out the double helix.

3. By the early 1970s researchers had begun transferring genes from the DNA of one species to the DNA of another.

4. Until the late 1970s the different branches of biology were isolated.

5. Due to this isolation biochemists, microbiologists, and geneticists all had different vocabularies.

6. Once the barriers were swept away the pace of discovery picked up.

7. Although the processes they study may be different researchers are all ultimately governed by genes.

8. Surprisingly questions from different areas were similar.

9. If one area holds more promise due to molecular biology it would be the study of cancer.

10. In the future cancer will be studied through genes.

Commas in Direct Quotes

Commas are used to separate a direct quote from the rest of the words in a sentence. The final comma is placed inside the ending quotation mark. For more information on quotes, see Mechanics Unit 8.

The committee stated, "Our recommendation is to widen our market," in its report.

Do not use commas to set off quoted material that is an integral part of a sentence.

The committee suggested that we should "widen our market."

Exercise 6 *Add commas to the direct quotes.*

1. The critic stated "This foul-up is the worst I've seen."
2. "What's more" he said "new developments surface every day."
3. "But, despite numerous setbacks, we're making progress" he said.
4. "The latest protest" he added "comes from within our ranks."
5. "Many times, we are our own toughest critics" he mused.

Commas in Compound and Complex Sentences

These rules are covered more thoroughly in the grammar units covering compound sentences and complex sentences. As a brief review, remember the following rules.

Rule 1 Use a comma and a coordinator to fuse two complete sentences into a compound sentence.

There are no perfect insulators, for all insulators have leakage current.

The common coordinators are *and, but, for, yet, or, so,* and *nor.*

Rule 2 Use a comma after a dependent clause that introduces the main clause in a complex sentence.

Because all insulators have leakage current, there are no perfect insulators.

Do not use a comma if the main clause comes first.
There are no perfect insulators because all insulators have leakage current.

Some common signal words for dependent clauses are *if, when, because, since, although, who, which,* and *that.*

Exercise 7   *Add commas where necessary.*

1. There is a new business trend and it is called globalization.
2. Because of a radical shift in the balance of global economic power companies all over the world are playing.
3. Once the dominant world force the United States is now learning to work with other nations.
4. The United States, Europe, and Japan are the major players and they have a strong link between them.
5. As the level of trade among these countries increases so does overseas investment.

Miscellaneous Uses of Commas

Commas are used in other ways, such as in dates, addresses, openings and closings of letters, numbers, and contrasted material. Read the following examples.

Dates:	He was hired on August 15, 1985, in Seattle.
	I read the March 1987 issue on Friday, May 5, 1987.
Or:	I read the March 1987 issue on Friday, 5 May 1987.

Note: The commas are not required when the day of the month is missing, or when the date is written in international style (day, month, year).

Addresses:	Miami, Florida, is the home of the Dolphins.
	The address is 4217 Lincoln Drive, Minneapolis, Minnesota 55455.

Note: Commas are not used between a state and the zip code.

Letter openings:	Dear Mr. Samuels,	Dear Emily,

Note: In formal letters, a colon is used after the greeting.

Dear Mr. Samuels: Dear Ms. Matthys:

Letter closings: Yours truly,   Sincerely,
Numbers: We received over 1,000 orders.
Contrasted material: The meter read 100 mA, not 1,000 mA.

Exercise 8   *Add commas where necessary.*

1. I applied for the job on May 19 1999.
2. I received a note that my résumé was being forwarded to Dr. Sherrit Ph.D. College Recruiter San Jose California.
3. She wrote back in November 1999 to set an interview in the Palo Alto California plant for January 15 2000.
4. Then she changed the date to February 1 not January 15.
5. She explained that over 1500 college students had applied for the 50 available positions and signed the letter "Sincerely Ms. Sherrit."

A Final Word about Commas

Many writers, especially beginning writers, recall being told to use a comma wherever they pause. Sometimes this generalization is useful, but more often it adds to the confusion and uncertainty of when to use commas and how many commas to use. It can also lead people to believe that there are no firm rules, that only breathing patterns determine comma placement. The truth is that you can pause wherever you see a comma, and that is about as far as breathing is involved in the matter.

Although it is true that exceptional cases can break the rules, and that professional authors seem to honor some rules and ignore others, commas are normally placed according to standard rules established to help the reader. Using or not using a comma can change meaning, as examples have shown. Use commas carefully as tools to make your message clear.

Colons and Semicolons

Colons

---

1. A colon (:) is used to introduce a list of items following a complete sentence.

Four types of communication are the following: listening, reading, speaking, and writing.

Note: The colon is never used following a verb. After a verb, simply state the list as a series of items.

The four types of communication are listening, reading, speaking, and writing.

2. Colons are used to express ratios of numbers. We sometimes read these colons as the word *to*.

His odds were 2:1.

3. Colons are used in formal business letters in place of the comma following the greeting.

Dear Sir or Madam: Dear President Bush:

4. Colons are used according to conventions when indicating times, biblical passages, and subtitles.

9:15 P.M.   12:00 noon   Mark 2:1–15
Modern Electronics: Basics, Devices, and Applications

Semicolons

---

1. The semicolon (;) is most frequently used to combine two related, complete sentences into one compound sentence. Think of a semicolon as a link between the period and the comma—both are included in the symbol.

A semicolon is placed at the end of the first sentence, followed by one space. The first word of the second sentence is not capitalized (unless it needs to be capitalized for another reason).

His new job was satisfying; he felt productive and challenged.

2. Semicolons are sometimes used between independent clauses (sentences) linked by words such as *therefore, however,* and *moreover.*

He worked at that company for six years; however, he held a variety of positions during that time.

3. Semicolons are used instead of commas between a series of items that contain commas within the items.

The courses in the first three terms are challenging: Digital I, Electronics I, and Technical Math I; Digital II, Electronic Devices, and Technical Math II; and Electronic Communications, Intro to Microprocessors, and Industrial Control Systems.

Exercise 1 *Add colons or semicolons where needed. Do not add any other type of punctuation.*

1. The United States operates five satellite tracking stations Goldstone, CA Kauai, HI Merrit Island, FL Fairbanks, AK and Rosman, NC.

2. The *Challenger* was launched before 8 00 A.M. in April 1983 with it was launched the first Tracking and Data Relay Satellite (TDRS).

3. The orbiting satellites have two sets of antennas the S-band and the Kμ-band.

4. The S-band antennas are located on the outside surface of the shuttle Kμ-band communications use a steerable dish antenna mounted in the cargo bay.

5. Before TDRS, the ratio of time in and out of contact with a shuttle was 1 4.

6. With TDRS, nearly constant global coverage is possible therefore, this tracking system can replace more than half of those on the ground.

7. Mission Control at Houston is the central point for shuttle flights here the communications flexibility is greatest.

8. The interface systems at Mission Control use computers for two reasons to format, compress, and route outgoing information and to reformat, decode, and forward incoming data.

9. Two other communications systems are connected to NASA Cape Canaveral, FL, and Edwards AFB, CA.

10. The Defense Department provides logistics, special studies, and search-and-rescue support these are performed by the Army, Air Force, Navy, and Coast Guard personnel and equipment.

Parentheses, Dashes, Brackets, Ellipses, Slashes, and Hyphens

This unit describes common uses of several types of punctuation.

Parentheses

1. Parentheses (plural form of *parenthesis*) are used to set off interrupting information. Because parentheses are used less frequently than commas, parentheses provide more of a separation than commas. In fact, the expression "parenthetical remark," which refers to related but nonessential information, comes from the function of parentheses.

Parenthetical remarks interrupt the flow of the sentence, and they usually can be "heard" as a pause even when read silently. They enclose extra information, comments, or facts.

Some people use big words to impress others (or so they think) at the cost of being clear.

A mass announcement (high efficiency) dealing with a sensitive issue that could better be dealt with through an individual discussion (low efficiency) can result in low effectiveness.

2. Parentheses are also used when the added material is too long to be inserted with commas, such as complete sentences or long phrases, particularly those that include commas.

We increase efficiency by cutting the message's cost (for example, running off multiple copies rather than typing original letters) or reaching more people (broadcasting a message to all workers in an office rather than selected audiences).

3. Parentheses can be used to enclose numbers or letters that enumerate ideas in a sentence. They can be used singly or in pairs for this purpose.

Failing to close the feedback loop can take several forms: (1) not listening to our own messages, and (2) not listening to our receiver's feedback.

Rules: Remember to follow the punctuation rules when using a set of parentheses:

A. Place no punctuation before the first parenthesis.

B. Delay the punctuation that would have come before the first parenthesis until after the second parenthesis.

C. Follow standard punctuation and capitalization rules within the parenthetical remark.

Dashes

The **dash** is the most abrupt separator of interrupting information.

1. A dash is used in places where commas or parentheses could also be used, but it adds the most emphasis.

Ask for feedback to see how well you're doing—kids are great teachers.

Studies have shown that as much as 78 percent of meaning is transmitted nonverbally—that is, without words.

2. A dash is occasionally used in place of a colon to introduce a list.

We have a variety of communication media available for our use—telephones, memos, letters, interviews, group meetings, and so forth.

3. A dash can also be used as bullets to highlight key ideas or bits of information.

There are four essential functions of nonverbal messages:
—to accentuate information conveyed verbally
—to express like or dislike
—to convey intensity of feeling
—to contradict verbal messages

Note: On a keyboard, a dash is made with two unspaced hyphens. In handwriting, make the dash longer than a hyphen.

Brackets

Brackets are squared-off parentheses that are usually used within direct quotations. Brackets are used only in pairs.

1. Brackets could be used because you want to add extra or specific information or modify a quote, which is often necessary when the quoted sentence includes pronouns that were defined elsewhere.

Original: "We will cover three major changes in our compensation plan."
Modified: "[The board of directors] will cover three major changes in [Acme Incorporated's] compensation plan."

2. Use brackets around the Latin word *sic* (meaning "thus") after an author's error to show your reader that it was not your error.

Spelling error: "Failing to close the feedback look [*sic*] can take several forms."

The rules for punctuation around or within parentheses also apply to brackets.

Ellipses

Ellipsis points (three spaced dots) ARE used to mark words left out of direct quotes. A fourth dot is a period.

Original: "The shuttle voice communications with the ground use the S-band 10-watt (2205.0 and 2250.0 MHz) transmitter."

Modified: "The shuttle voice communications with the ground use the S-band 10-watt . . . transmitter."

Slashes

The **slash** is used to indicate choices or options.

1. The slash is generally used to replace the word *or*. Occasionally, it is used in ambiguous situations to allow for the possibility of singular or plural events (*result/s*), or male or female individuals (*she/he*). The slash should not be used too often for these purposes.

Some courses are not taken for a grade, but just for a pass/fail credit.

A manager will cause the death of his/her effectiveness by committing the seven deadly "sins" of management communication.

2. The slash is frequently used in dates consisting of numbers.

The order was dated 11/20/99.

Note: An enumerated date should be written and read as month/day/year.

3. In technical notation, the slash symbolizes division or *per*.

$R = VII$ $10/hour

Hyphens

1. The **hyphen** (-) is usually used to join two words. Some compound words always are hyphenated. Dictionaries include required hyphens.

ampere-turn   high-pass
hard-core   full-scale

Rules

Dictionaries sometimes disagree about hyphenated compound words, so be consistent in your own writing. Generally, follow these rules:

A. Hyphenate compound modifiers preceding a noun, but do not hyphenate those modifiers if they follow a noun.
   He was a well-dressed man.
   The man was well dressed.

Always hyphenate technical terms that require hyphenation:

The pulse is positive-going.
ohms-per-volt rating
T-type low-pass filter

B. Do not hyphenate compound modifiers if one of the words ends in *-ly*.

He was a professionally dressed man.

C. Hyphenate all compound numbers between twenty-one and ninety-nine and some fractions.

one-tenth	two-thirds
one half	one quarter

2. Hyphens are sometimes used to divide a word at the end of a line. Remember that hyphens must be placed between syllables (use your dictionary to verify syllables), and one-syllable words cannot be divided. It is common practice not to begin a line with short syllables. Instead, write the entire word on the next line. Divide hyphenated words only at the hyphen. Avoid two consecutive lines ending in a hyphen.

Wrong: The difficult and time-consum-
         ing lab gave me a ter-
         rible headache.
Correct: The difficult and time-
         consuming lab gave me a
         terrible headache.

3. Hyphens, like dashes, can be used to highlight information.

To simplify circuits for analysis, we can use many methods:
-Superposition
-Thevenin's theorem
-Norton's theorem
-Millman's theorem

Exercise 1   *Correctly punctuate the following sentences by adding parentheses, dashes, brackets, hyphens, or slashes. Be prepared to defend your answer, especially for notation setting off parenthetical remarks.*

1. The announcement stated, "I want all the machines that are broke *sic* sent back to the lab."
2. The test results will indicate go no go
3. His last job I haven't the slightest idea of his other jobs was in the production department.
4. I tried every two number combination between thirty five and sixty five.
5. We received the replacement on 4 20 00.
6. Mr. Harcort the manager and Mr. Leninson the job foreman arrived for the appointment.
7. A capacitor and an inductor can be connected in an inverted L configuration.
8. A band pass filter it will follow a typical band pass response curve allows a certain band of frequencies to pass.
9. She is paid on the basis of 40 hours week.
10. My goals are 1 finish my education, 2 work for a large company, and 3 start my own business.

Apostrophes

The apostrophe (') has two common uses: in contractions and in possessives.

Contractions

Contractions are combinations of two words, similar to some abbreviations. However, rather than using only the first letters, we combine the two words to form one new word. Usually we keep the beginning or all of the first word and combine it with the last part of the second word.

should not	shouldn't
I would	I'd
she has	she's
she is	she's

Notice that the apostrophe is added exactly where the letter or letters are dropped. The only common exception is *won't,* the contraction of *will not.*

Contractions are used more often in informal writing, where the two full words would sound awkward and formal. In formal reports, use contractions as seldom as possible. If you do use them, be sure that the words are contracted correctly.

Note: Apostrophes are used to indicate any intentionally dropped letters or numbers:

the class of '87 (1987)
we're comin' (coming)
o'scope (oscilloscope)

Exercise 1 *Write the contractions for the following pairs of words.*

1. he would _____
2. is not _____
3. I have _____
4. it is _____
5. they are _____
6. are not _____
7. they have_____
8. cannot _____
9. you are _____
10. should not _____

Exercise 2   *Use each contraction from Exercise 1 in a sentence. Proofread each sentence to make sure that you can correctly substitute the two uncontracted words.*

1. _____

2. _____

3. _____

4. _____

5. _____

6. _____

7. _____

8. _____

9. _____

10. _____

Possessives

Possessive apostrophes indicate ownership. The apostrophe changes a noun to an adjective. Read the following sentence.

Sam drove his father's car.

We know that the car Sam drove belonged to his father. We have turned *father* (a noun) into an adjective to describe a certain car. When we turn a noun into a possessive adjective, we add an apostrophe plus an *s* to the end of the noun. Usually the sentences could be reworded stating the ownership as a prepositional phrase.

Sam drove the car of his father.

If you are not sure of whether or not to add an apostrophe plus *s* to the end of a word, try rewording the sentence, putting the name of the "owner" in a prepositional phrase.

Exercise 3   *Reword the following possessives stating the object owned, followed by a prepositional phrase.*

Example: neighbor's yard
Rewritten: yard of the neighbor

1. week's salary _____

2. resistor's value _____

3. student's grades _____

4. runner's time _____

5. lab's hours _____

Plural Possessives

Notice that in each case in Exercise 3, the "owner" is singular. When an owner is plural, we add the apostrophe after the plural form of the word.

the workers' schedules (the schedules of the workers)
the children's room (the room of the children)

If the name of the plural owner ends with an *s*, we don't have to add an extra *s* after the apostrophe.

> the cities' problems (the problems of the cities)
> the ladies' club (the club of the ladies)

Exercise 4   *Using the phrases from Exercise 3, rewrite each possessive phrase with a plural owner.*

> *Example:*   the neighbor's yard
> *Rewritten:*   the neighbors' yards

1. _____
2. _____
3. _____
4. _____
5. _____

Some possessives do not need apostrophes; these are the **possessive pronouns** such as *hers, his, yours, its, ours,* and *theirs*. These pronouns are usually used as adjectives, but they do not have an apostrophe.

> *Examples:*   The car is *hers*.
> *His* job is difficult.
> Hold the hammer by *its* handle. [Do not confuse *its* with the contraction *it's* (*it is*).]
> The mistake was *theirs*.

Special Cases

1. For compound words or word groups, add the apostrophe plus *s* only to the last word.

> *Example:*   someone else's idea
> Chief Executive's decision
> father-in-law's consent

2. If a singular word ends in *s*, especially a name, add another *s* after the apostrophe.

> *Example:*   Chris's turn
> the Willis's house

3. Add an apostrophe plus *s* to each individual owner.

> *Example:*   teacher's and students' experiences
> workers' and managers' responses

4. To show joint ownership, you can either add an apostrophe plus *s* to both names or only to the last name.

> *Example:*   Mr. Nguyen and Mr. Lee's business
> (the same business)

> *Or:*   Mr. Nguyen's and Mr. Lee's business
> (the same business)

Exercise 5   *Add apostrophes if necessary.*

1. The new presidents questions had to be answered.
2. The peoples choice was determined by the election.
3. The immune systems cells were destroyed.
4. The two pictures frames were identical.
5. The coaches thanked all the players parents.
6. A laws interpretation is made by the judicial system.
7. Ross wrote two different résumés for his teacher.
8. The students used Ohms law in the experiment.
9. Ruth gave her two-weeks notice.
10. The rear brakes condition was good.

Exercise 6   *Add the possessive apostrophes where needed. There are 11 apostrophes needed.*

Many handicapped people have learned how to turn their disabilities into victories. One such person is my friend, Phil, a junior in an electronics engineering technology program. Phils eyes are extremely weak, leaving him nearly blind. He is not able to see objects more than 2 inches away. He cannot read a computers keys or a tests words without bumping his nose into them. Somehow he can push his calculators keys while looking at the readout. He listens to his teachers lectures, and later he reads a classmates notes written with a dark pencil. His textbooks are recorded on cassettes, and his two lab partners assist him in all his labs safety procedures. Phils GPA is over 3.00. Last term, he was placed on the Deans List. We all admire Phils and many other handicapped students strength and determination to overcome their disabilities.

Quotations

Many technical writers find it necessary to quote other authors or speakers. A **quotation, or quote,** is information repeated or reproduced from another source. Quotes can add interest, precision, and credibility to technical material. Quoted material includes information repeated in the exact words of the original speaker or writer (**direct quotes**) or slightly reworded or summarized accounts of the original version (**indirect quotes).** Both types of quotes are cited in a footnote, endnote, or parenthetical note (refer to Appendix 3), but only direct quotes have special punctuation. Failing to cite the source of either a direct or indirect quote is considered **plagiarism,** a serious writing offense. Avoid plagiarizing by using quotations marks (in direct quotes) and citing the source (in direct and indirect quotes).

Direct Quotations

The exact words of a speaker or writer are surrounded by **quotation marks** (" ").

> "Robots are just a combination of hydraulics, mechanics, electronics, and computers."

Sometimes the speaker is included in the sentence. The quotation marks still surround the quote, but the speaker is set off by a comma.

> "Technical specialists are very expensive," notes Nancy Johns.
> "Technical specialists," notes Nancy Johns, "are very expensive."

Rules for Writing Direct Quotations

Rule 1: Begin every quoted sentence with a capital letter even if the first word of the quote is not the first word of the sentence.

> One assembly worker said, "Is that the last thing?"

Do not begin quoted words or phrases with a capital letter unless other reasons require a capital letter.

> A "dedicated robot" is one that has been permanently programmed to perform certain tasks.
> "Dedicated robots" are those that have been permanently programmed to perform certain tasks.

Rule 2:   Place the first set of quotation marks immediately next to the first word of the quote (no space).

Rule 3:   Put the closing set of quotation marks at the end of the quote and after any other punctuation. If the quote ends the sentence, put the final quotation mark outside the period or question mark.

> He asked, "How did you know?"

Exception: If the whole sentence is a question, but the quote is not a question, place the question mark after the closing quotation mark.

> Did the manager say, "Take the day off"?

If the quote is followed by other information, end the quote with a comma and the second set of quotation marks.

> "The manager just left," the worker observed.

Exception: If the quote ends in a question mark or exclamation mark, keep it in.

> "What did he say?" asked the worker.

If the speaker's name interrupts the quote, two pairs of quotation marks must be used.

> "I heard him say," responded my partner, "that we have tomorrow off."

Rule 4:   If you are including a parenthetical note (see Appendix 3) at the end of a quoted sentence, place the final quotation mark before the first parenthesis.

> "Robots aren't replacing entire shifts of workers" (Fey, p. 49).

Rule 5:   Use single quotation marks (' ') around a quote within a quote.

> Hoska reports, "I was called into plants by management and told, 'Find us a place to use a robot.' "

Rule 6:   Do not use a semicolon before a direct quote.

Rule 7:   Place the closing set of quotation marks at the end of the entire quote if it consists of more than one sentence.

Rule 8:   Use quotes sparingly or not at all. Instead, try to write ideas in your own words.

Rule 9:   Do not use more than one direct quote per page.

Rule 10: Do not quote more than one paragraph at a time.

Note: Quotes that are formal or long are sometimes introduced with a colon.

> The manual includes the following caution:
> "Do not connect the power cable to the instrument before verifying that the intended source matches the AC line configuration of the instrument."

Exercise 1   *Add quotation marks and commas where necessary. Keep each sentence a direct quote.*

1. It took a human mind to invent the wheel and build a computer said my professor.
2. He added now people can make a machine with a mind of its own.
3. He asked us what can robots do?
4. I answered I think they can cut, weld, paint, and lift things.
5. That's right he said but someday soon they will do much more.
6. Someday soon they may clean up toxic wastes, disarm bombs, or explore the ocean floor he suggested.
7. Did you say disarm bombs? I asked excitedly.
8. Yes he replied and much more.
9. I then asked what is it that makes a robot a robot?
10. Most of us agree said my professor that a robot must be two things. It must be mobile, and it must be programmable.

Exercise 2   *Write five direct quotations of your own using correct placement of quotation marks, commas, and ending punctuation.*

1. _____
   _____
2. _____
   _____
3. _____
   _____
4. _____
   _____
5. _____
   _____

Indirect Quotations

An indirect quotation states the general ideas of another speaker or author, but not the exact words. Read the following indirect quotes.

Nancy Johns noted that technical specialists are very expensive.

Troubleshooter Bob Adams reports that when a crew heard that a Puma robot was going to be installed, people feared for their jobs.

Indirect quotes may be close to a direct quotation except for an added *that,* or they may be condensed accounts of longer direct quotes. Note that no special punctuation is used for indirect quotes, which makes them easier to write. Remember, however, that indirect quotes still need to be cited in research papers either by a footnote, endnote, or parenthetical note.

Plagiarism

Many technical reports are based on the research of other people. When writing about a new or difficult subject, some writers are unable to reword their sources. Most writers understand the obligation to cite the source of a direct quote. What is less understood, however, is how to handle a passage that has been slightly reworded.

Passages that have essentially been copied with slight word or structural changes are still considered indirect quotes and should be cited.

A passage is an indirect quote if the following situations apply:

1. The order of ideas is the same. No additional information is added.

2. The words or sentence structure is changed, but the ideas are the same.

Read the following direct quote and a paraphrased, indirect quote. Both should be cited as quotations. The passage is taken from "Machine Vision," by Nello Zuech, *Robotics Today,* April 1986, p. 35.

Direct quote: "The 1986 market for machine vision should exceed $200 million, a 33% growth over 1985. If one considers products with dedicated performance envelopes, such as photomask inspection systems used in microelectronic manufacturing or off-line dimensional measuring equipment, the market will easily exceed $300 million" (Zuech, p. 35).

Indirect quote: Machine vision markets in 1986 should grow by 33% over the 1985 market, reaching an excess of $200 million. Dedicated performance envelopes such as photomask inspection systems may bring the market to $300 million (Zuech, p. 35).

Exercise 3 *Rewrite the direct quotes from Exercise 1 as indirect quotes. Remember to make the indirect quotes sound natural and smooth. The first one has been done for you.*

1. "It took a human mind to invent the wheel and build a computer," said my professor.

My professor said that it took a human mind to invent the wheel and build a computer.

2. _____

_____

3. _____

_____

4. _____

_____

5. _____

_____

6. _____

_____

7. _____

_____

8. _____

_____

9. _____

_____

10. _____

_____

Other Uses of Quotation Marks

1. Quotation marks are used to set off titles of articles, stories, songs, and poems—italicize or underline titles of books and magazines.

> The article "Shuttle Communications" is in the summer issue of *Hands-On Electronics*.

2. Quotation marks are used to show words with special or unusual meanings.

> Abraham "Honest Abe" Lincoln
> Ice (especially "dry ice") can cause tissue damage if applied directly to a burn.

Note: Do not use quotation marks for common nicknames. Once a word has been highlighted with quotation marks, continued usage of that word does not require quotation marks.

Exercise 4   *Add quotation marks where necessary.*

1. The Technically Speaking column appears in most issues of *IEEE Spectrum*.

2. One issue talked about nouns that are being used as verbs, such as *keyboard* and *messenger*.

3. Add automated reasoning and smart machine to the list of items coming into frequent use from the field of artificial intelligence.

4. One reader asked if anyone could verify the following account:

At one time the British Labor Party, in its annual convention, came within a few votes of passing a resolution that stated No one shall receive less than the average wage.

5. I then looked up Great Britain on the Internet.

Appendices

Common Symbols and Abbreviations

ELECTRONICS UNIT SYMBOLS
(from Berlin: *The Illustrated Electronics Dictionary*)

Unit	Symbol	Unit	Symbol
ampere	A	joule	J
ampere-hour	Ah	kelvin	K
ampere-turn	At	lambert	L
baud	Bd	lumen	lm
bel	B	lux	lx
coulomb	C	maxwell	Mx
decibel	dB	meter	m
degree (angle)	°	mho	mho
degree (temperature)		oersted	Oe
Celsius	°C	ohm	Ω
Fahrenheit	°F	radian	rad
kelvin	K	second	s
Electronvolt	eV	siemens	S
farad	F	tesla	T
gauss	G	var	var
gilbert	Gb	voltage	V
henry	H	voltampere	VA
hertz	Hz	watt	W
horsepower	hp	watt-hour	Wh
hour	h	weber	Wb

METRIC SYMBOLS

Prefix	Symbol	Multiplier	Prefix	Symbol	Multiplier
exa-	E	10^{18}	centi-	c	10^{-2}
peta-	P	10^{15}	milli-	m	10^{-3}
tera-	T	10^{12}	micro-	μ	10^{-6}
giga-	G	10^{9}	nano-	n	10^{-9}
mega-	M	10^{6}	pico-	p	10^{-12}
kilo-	k	10^{3}	femto-	f	10^{-15}
deci-	d	10^{-1}	atto-	a	10^{-18}

STATE ABBREVIATIONS (U.S. POSTAL SERVICE)

State	Abbreviation	State	Abbreviation
Alabama	AL	Illinois	IL
Alaska	AK	Indiana	IN
Arizona	AZ	Iowa	IA
Arkansas	AR	Kansas	KS
California	CA	Kentucky	KY
Colorado	CO	Louisiana	LA
Connecticut	CT	Maine	ME
Delaware	DE	Maryland	MD
District of Columbia	DC	Massachusetts	MA
Florida	FL	Michigan	MI
Georgia	GA	Minnesota	MN
Guam	GU	Mississippi	MS
Hawaii	HI	Missouri	MO
Idaho	ID	Montana	MT
Nebraska	NE	Rhode Island	RI
Nevada	NV	South Carolina	SC
New Hampshire	NH	South Dakota	SD
New Jersey	NJ	Tennessee	TN
New Mexico	NM	Texas	TX
New York	NY	Utah	UT
North Carolina	NC	Vermont	VT
North Dakota	ND	Virginia	VA
Ohio	OH	Virgin Islands	VI
Oklahoma	OK	Washington	WA
Oregon	OR	West Virginia	WV
Pennsylvania	PA	Wisconsin	WI
Puerto Rico	PR	Wyoming	WY

TITLE ABBREVIATIONS

Title	Abbreviation
Chief Executive Officer	CEO
Director	Dir.
Doctor	Dr.
Honorable (judge)	Hon.
Junior	Jr. (abbreviated only if following a name)
Man	Mr.
Manager	Mgr.
Married woman	Mrs.
President	Pres.
Professor	Prof.
Reverend	Rev.
Superintendent	Supt.
Vice-President	VP
Woman	Ms.

STANDARD ADDRESS ABBREVIATIONS (U.S. POSTAL SERVICE)
Use when an address contains excessive characters (over 30 per line).

Address	Abbreviation	Address	Abbreviation
Alley	ALY	Mountain	MTN
Annex	ANX	Parkway	PKWY
Avenue	AVE	Place	PL
Beach	BCH	Plaza	PLZ
Boulevard	BLVD	Point	PT
Bridge	BRG	Port	PRT
Causeway	CSWY	Ridge	RDG
Center	CTR	River	RIV
Circle	CIR	Road	RD
Corner	COR	Route	RTE
Court	CT	Spring	SPG
Creek	CRK	Springs	SPGS
Crossing	XING	Square	SQ
Divide	DV	Street	ST
Drive	DR	Summit	SMT
Estate	EST	Terrace	TER
Expressway	EXPY	Trace	TRCE
Fort	FT	Trail	TRL
Highway	HWY	Turnpike	TPKE
Lake	LK	Valley	VLY
Lane	LN	View	VW
Mount	MT	Village	VLG

STANDARD SECONDARY UNIT ABBREVIATIONS (U.S. POSTAL SERVICE) (continued)
Use when an address contains excessive characters (over 30 per line).

Unit	Abbreviation	Unit	Abbreviation
Apartment	APT	Room	RM
Building	BLDG	Space	SPCE
Department	DEPT	Suite	STE
Floor	FL	Trailor	TRLR
Office	OFC	Unit	UNIT

ABBREVIATIONS OF DAYS AND MONTHS*

Days	Abbreviations	Months	Abbreviations
Sunday	Sun.	January	Jan.
Monday	Mon.	February	Feb.
Tuesday	Tues.	March	Mar.
Wednesday	Wed.	April	Apr.
Thursday	Thurs.	May	May (no abbr.)
Friday	Fri.	June	June (no abbr.)
Saturday	Sat.	July	July (no abbr.)
		August	Aug.
		September	Sept.
		October	Oct.
		November	Nov.
		December	Dec.

*Do not abbreviate days of months in formal letters or reports, except in footnotes or bibliographies.

NUMERALS

Arabic	Greek	Roman
0	. . .	. . .
1	α	I
2	β	II
3	γ	III
4	δ	IV *or* IIII
5	ε	V
6	ς	VI
7	ζ	VII
8	η	VIII *or* IIX
9	θ	IX *or* VIII
10	ι	X
11	ια	XI
12	ιβ	XII
13	ιγ	XIII *or* XIIV
14	ιδ	XIV *or* XIIII
15	ιε	XV
16	ις	XVI
17	ιζ	XVII
18	ιη	XVIII *or* XIIX
19	ιθ	XIX *or* XVIIII
20	κ	XX
30	λ	XXX
40	μ	XL *or* XXXX
50	υ	L
60	ξ	LX
70	ο	LXX
80	π	LXXX *or* XXC
90	ο	XC *or* LXXXX
100	ρ	C
200	σ	CC
300	τ	CCC
400	ν	CD *or* CCCC
500	φ	D *or* IƆ
600	χ	DC *or* IƆC
700	ψ	DCC *or* IƆCC
800	ω	DCCC *or* IƆCCC
900	⅄	CM, DCCCC, *or* IƆCCCC
1000	. . .	M *or* CIƆ
2000	. . .	MM *or* CIƆCIƆ

GREEK ALPHABET

A α	alpha	a
B β	beta	b
Γ γ	gamma	g, n
Δ δ	delta	d
E ε	epsilon	e
Z ζ	zeta	z
H η	eta	ē
Θ θ	theta	th
I ι	iota	i
K κ	kappa	k
Λ λ	lambda	l
M μ	mu	m
N ν	nu	n
Ξ ξ	xi	x
O o	omicron	o
Π π	pi	p
P ρ	rho	r, rh
Σ σ	sigma	s
T τ	tau	t
Υ υ	upsilon	y, u
Φ φ	phi	ph
X χ	chi	ch
Ψ ψ	psi	ps
Ω ω	omega	ō

Tips for Word Processing

If you have never had a typing class or general keyboarding instruction, you might feel handicapped when you are expected to turn in typed reports, letters, résumés, and memos. Finding the letters on the keyboard will be the biggest problem until you practice. Be sure that you allow yourself extra time for slow typing.

There are other things that you need to know when you begin typing. Some terms that are used in this appendix (or in the user documentation included with your word processor) need to be defined:

- **Keyboard:** The input device with keys for letters, numbers, and punctuation marks.

- **Mouse:** The input device with one or more buttons. Click the mouse buttons to place the cursor at a certain point in a file, or to select a menu or button in the program window. Laptops may have a track ball, joystick, or finger pad that acts as a mouse.

- **Indent:** Press the **Tab** key to move in 5 to 10 spaces at the beginning of a paragraph, used especially when you will double-space a report.

- **Double-space:** Leaving an empty line between each printed line or between single-spaced paragraphs.

- **Click:** Click the left mouse button one time.

- **Double-click:** Click the left mouse button twice within 1 second.

- **Right-click** (on PCs only): Click the right mouse button one time.

- **Select:** Click with the left mouse button.

- **Title bar:** The dark bar at the top of an application window that usually contains the application and file name. It appears dark when the window has focus.

- **Menu bar:** Rows of words (also called commands) that open lists (menus) of functions when you click them. Menu items with arrows on the right side or three periods have submenus (also called secondary menus or cascading menus) that open when you click them or move the mouse over the menu item. Some menu items open windows (also called dialog boxes) for more selections.

- **Toolbar:** Rows of buttons (icons) below the menu bar that activate functions when you click them. Some word processors have several toolbars available to turn on or off. Most allow users to customize their toolbar to add buttons for functions they use frequently and remove buttons for functions they rarely use. For example, in Mocrosoft Word, click **View** > **Toolbars** to see the menu of available toolbars. Toolbars that are checked (click to check or uncheck) are currently visible.

- **Task bar:** The row at the bottom of the screen that shows all the running programs. If you minimize (hide) a program (by clicking the minus button in the title bar), the program continues to run but won't show on the screen. The name of the program continues to appear as a button in the task bar. To maximize (show) the program again, double-click the program button in the task bar.

- **Online Help:** The information panels that open when you click the **Help** button or a question mark (**?**) in a title bar. For example, in Word, if you click the **Help** menu item, a menu opens with all the types of help available with the program. Among the items are Microsoft Word Help and What's This?

- **What's This? Help:** The What's This? command (sometimes called context-sensitive help) opens a small window on the word-processor screen with information about that area of the screen. For example, in Word, if you click **What's This?,** you start the context-sensitive help program. Then, when you move the mouse, you see a question mark next to the mouse arrow. Click an area of the screen, and if help is available, a panel appears on the screen with information or instructions on how to use the area. Press any key to stop the program.

- **Minimize, Restore, and Close buttons:** Use the three buttons on the far-left side of the title bar to control your application. Click the Minimize button (-) to minimize your program window. The program name appears in the task bar, and it continues to run in the background. To maximize it again, double-click the name in the taskbar. Click the middle button (it toggles between two small windows and one large window) to change the size of your word-processor window, which is useful when you want to view two windows at one time. Click the two small windows to reduce the size, and click the one larger window to increase the size of the window. Click the Close button (**X**) to exit the program—you might be prompted to save your files before the program closes.

Punctuation and Spacing

A keyboard has a variety of keys for letters, numbers, and punctuation marks. You can also insert symbols and foreign letters using key combinations or lists of letters and symbols.

For example, in Microsoft Word, press and hold **Alt** + **Ctrl** and press **C** to produce the symbol for copyright: ©. Or you can click **Insert** > **Symbol** to open the Symbol window. From this, you can select a foreign letter, such as ö, and click **Insert.**

Use the following general rules for punctuation and spacing:

Rule 1:   Indent 5 to 10 spaces at the beginning of each new paragraph. Indent each paragraph when you double-space a file. Indent following a bullet symbol when writing a bulleted list. Be consistent in the number of spaces you indent.

Rule 2:   Use conventional spacing before and after punctuation marks:

- Place any punctuation mark next to the word that precedes it (no space).
- Put quotation marks immediately next to the first and last words of the quote, but outside other punctuation.
- Leave one space after commas and semicolons.
- Leave two spaces after a colon or a period, question mark, or exclamation mark at the end of sentences.
- Do not space after the first period of abbreviations such as A.M., N.Y., U.S., and M.D.
- Do not leave a space before or after a hyphen, as in "seven-year-old boy."
- Use two hyphens (no spaces before or after them) to indicate a dash (I enjoy summer—June through September—more than any other season).

Example: We were thrilled to have a paid, three-day vacation! Is the Martin Luther King, Jr. holiday—the third Monday in January—celebrated everywhere in the United States?

Rule 3: Use wide enough margins so that there is some white space around the text. Typical (default) margins are one inch on top, bottom, right, and left sides. You can change the default margins (and other values) using the menu options. For example, to change margins in Microsoft Word, select **File > Page Setup,** and adjust the margins in the Page Setup window.

- Six lines are approximately 1 inch.
- Ten spaces are approximately 1 inch.
- Each sheet of typing paper (or printer paper) is about 66 lines down and 80 spaces (columns) across.

Rule 4: Use consistent spacing and margins for each report, letter, memo, and résumé. For example, double-space long documents for easier reading. Single-space letters, memos, e-mails, and short reports, with a double space between paragraphs. If you want to become a "power-user," learn how to use templates or styles for formating entire reports and documents quickly. For example, Microsoft Word includes features to format font, font size, and paragraph and line spacing in the **Format** menu.

Using a Word Processor

Word processors might have unique features and procedures, but many features are standard among most word processors. If you are totally new to word processing, the following information is important to know.

Overview

Word processors are written for specific computers, such as an IBM-compatible (usually referred to as a personal computer, or PC) or Macintosh. PCs can have different operating systems, such as Windows 98 or Windows NT. Most word processors are installed from CD-ROMs onto the hard drive. You can save your files either on the hard drive, a diskette, or a zip disk. Most files created on a Mac cannot be imported to a PC—check the user manual. But files created in a word processor on one Windows operating system can usually be opened in the same application on another Windows operating system. For example, if you write a file using Word 97 on a Windows 98 operating system, you can open the file on a computer with Word 97 for Windows NT.

Like most software applications, word processors are constantly being released as new versions, with new features, options, and bugs. Some development companies offer telephone technical support only on a fee basis. Always check the online technical support (locate the Web site in the user manual or search for it using an Internet search engine) before calling a company with a problem. Or join a users' group or online forum to learn tips, problems, and work-arounds from other users.

Disks

Treat disk carefully—it is fragile and has a limited life span. Some things that will destroy disks quickly are heat (including direct sunlight), moisture, dust or hair, magnetic fields, or crushing. The best way to carry disks around is in a hard, plastic case or in protective pockets.

Menus

Most word processors are menu-driven, which means that you select letters or numbers from a menu, or list of functions, such as **Type, Print,** or **Save.**

Getting Help

All programs include online help, which you can access from a menu or by pressing the **F1** key. Usually this will open a series of panels on your screen. Some might include links to other screens or topics. You can usually open a table of contents or index to search for a topic. To search, type the closest word you can think of to start an alphabetical search for the topic. Sometimes it takes creative thinking to find the exact topic.

If a window has a **Help** button or a question mark in the title bar, click it to open online help for the window. Or use the **What's This? Help** to open information about an area of the screen.

Accelerator Keys (or Hot Keys)

Many programs provide shortcuts, key combinations, and function keys (**F1, F2,** and so on) to activate a function without using the mouse and the menus—some people prefer keyboard commands. For example, in Microsoft Word, you can press and hold the **Ctrl** button and press **P** to open the **Print** window. To learn the accelerator keys, open the menus and look on the right side of the menu items. Using the previous print example, if you open the **File** menu, you will see **Ctrl + P** next to the **Print** command.

Most programs also include a Quick Reference card in the software package that lists the accelerator keys and other shortcuts for the program. Some accelerator keys have become standard among word processors, and it is hoped that this trend will continue.

Saving a File

Save often. Power outages and computer failures can mean losing hours of work unless you save periodically. The working computer memory is temporary. The disk or hard drive memory is permanent. If you lose power, you can retrieve your file, but it might lose everything you added since the last time you saved it.

The first time you save a file, you must use the **Save As** command. In the **Save As** window, you accept or change the location, name, and file type. Some word processors have their own file types, identified by a three-letter extension. Do not change the extension unless you plan to port the file to another program. Do not use unusual characters in the file name—generally, stick to letters and numbers, but check the documentation for the program for exact limitations.

Either select a specific drive and folder in which to store the file, or note the default file location in which you will save the file. If you lose a file (for example, if you didn't notice where it was saved), you will have to search for it using the **Find** feature. Or you can open it from the list of last-opened files (for example, at the bottom of the **File** menu), and then use the **Save As** window to show the current location of the file. Once a file is saved, you can use the **Save** menu item (such as **File > Save**), toolbar button, or accelerator key (**Ctrl + S**) to save the file periodically.

Once you have named a file, the name will appear in the catalog list or directory of the disk. If you give your new file the same name as an old file and save it in the same folder, it will overwrite the old file.

Automatic Word Wrap

This feature means that you do not have to press the **Return** key as you type close to the right margin. The computer will automatically "wrap" the last word around to the next line, called a soft return. Press the **Enter** key only when you want to start a new paragraph, called a hard return. Hard returns can make editing and revising a difficult process, so it's best to use the word-wrap feature.

If you want to double-space the printed copy of the file, use the formatting options to double-space for you, and you can continue to use word-wrap. For example, in Microsoft

Word, first select all the text that you want to double-space. Then select **Format > Paragraph.** Under **Line Spacing,** select **Double.** When you click **OK,** your file will automatically reformat with double spacing.

Editing

All word processors provide tools (menus and buttons) to revise and enhance your typing. You can change the font, font size, and text attributes (such as bold or underlining) of your file. Word processors include many basic and advanced editing features. The most useful are the **Undo** and **Redo** commands, which undo your last action or restore your last **Undo,** multiple times.

Most programs also contain shortcuts for moving the cursor around on the screen, spelling checks and grammar checks, auto-correct (for words that you frequently misspell or long names or phrases that you type frequently), easy formatting options, and quick setup for tables or multicolumn layouts. You can put text in boxes and put borders around pages. Many programs also provide the ability to include hyperlinks and Web site addresses (URLs) that are active if the reader views the file on a computer. Take some time to learn these shortcuts and use the tools included with your word processor—they can make you more efficient and productive.

Printing

You can sometimes select whether to print one-sided or two-sided (also called duplex printing), portrait (vertical) or landscape (horizontal), or multiple copies. Click **Print Preview** (if available) to view a thumbnail of each page before printing. You can print on regular paper or acetate (to project as a transparency) by changing the paper stock in the printer.

Conclusion

All word processors are purchased with a user manual, online help, and other types of printed reference information to answer specific questions. Some have workbooks, tutorials, multimedia movies (sometimes called a Quick Tour), or demonstration disks (demos) to give guided practice in using the program.

Most documentation today is written so that even inexperienced computer operators can easily understand basic operating rules. These manuals have been written by professional technical writers—who once learned basic technical writing just as you are now doing. Refer to these materials to learn more about the features, options, and functions available with your word processor. Then let the computer do the mechanical part for you, and you can focus on the content.

Sample Reports

This appendix illustrates four different reports. Included are:

- Instructions
- Descriptive Report
- Formal Lab Report
- Formal Proposal

Sample Report: Instructions

> **To:** Professor K. Jones
> **From:** Billy Malone
> **Date:** 9/21/99
> **Subject:** Satellite Dish Installation
>
>
> The purpose of this memo is to give simple but complete instructions on how to install a home satellite dish system.
>
> Note: The installation is not difficult, but I recommend that you have some basic wiring and construction skills. If you feel comfortable with the following jobs, then you should have no problem with installing a satellite dish system:
>
> - Installing a ceiling fan
> - Assembling a basketball goal
> - Installing a light switch
> - Installing a garage door opener
>
> Equipment and materials that you will be using for this project are:
>
> - Satellite dish and receiver i.e. RCA, Dish Network
> - Home Installation Kit (can be obtained at the dish retailer)
> - Drill and wood drill bits
> - Screw driver
> - Adjustable wrench
> - Pliers
>
> **Warning! Avoid power lines!**
>
> Take extreme care to avoid contact with overhead power lines, lights and power circuits. Contact with these items can be fatal.
>
> **Caution:**
>
> It is extremely important to ground the dish to a central point in the house. A nearby lightning strike can easily damage an ungrounded dish, the receiver, and your TV.

Caution:

If you are using a ladder to install the dish, be sure not to underestimate the weight of the satellite. The dish may cause you to lose your balance.

The major steps in installing a satellite dish include the following:

- Locate the position to install the dish.
- Route cable through house to receiver.
- Connect cable to receiver.
- Adjust and level dish to line up with satellite.
- Program receiver to satellite transmission.

I.   Locate the position to install the dish.

Determine the mounting location of the dish. The dish must have a clear view to the satellite, which means that between the dish and the satellite there can be no obstructions.

a. Make sure there are no trees, now or in the future. (*See figures 1 and 2*)
b. Check for buildings or structures in the area.
c. Make sure there are no obstructions to block view.
d. Mount the disk on a sturdy surface i.e. (side of the house or chimney)

Note: Detailed instructions on proper mounting of the dish can be found in the home installation kit.

Figure 1. Correct line of sight

Figure 2. Incorrect line of sight

II. Route cable through house to receiver.

Note: The length of the coaxial cable must be 112 inches or shorter. If not, an amplifier must be used.

a. Find the shortest distance to the receiver.
b. Mount the grounding block close to the entrance of the cable.
c. Drill a hole in the wall slightly larger than the coaxial cable.

d. Connect the cable to the grounding block and then push the other end of the cable through the hole in the wall.

Connect cable to receiver.

a. Route the cable through the house to the back of the receiver.
b. Connect the cable to the jack on the back of the receiver labeled _Satellite_. (*See Figure 3*)
c. Use silicone to seal the hole in the wall that was drilled for the cable.

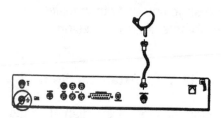

Figure 3. Satellite jack located on the back of receiver.

Adjust and level dish to line up with satellite.

a. Use a level to make sure the dish is mounted at a 90-degree angle to the mounting surface. (*See figure 4*)
b. Set the indicator on the side of the dish to 32 degrees. (*See figure 5*)
c. Position the dish to face southeast and then tighten the mounting bolts.

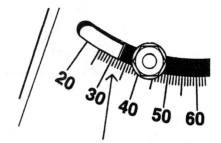

Figure 4. Correct position of mounted dish.   Figure 5. Adjustable angle slide for dish.

Program receiver to satellite transmission.

a. Turn on the satellite receiver and a final adjustment screen will be displayed.
b. Follow the step-by-step instructions on the screen to sharpen the signal from the satellite to the receiver. (*See figure 6*)
c. Select the programming package that you would like to purchase.

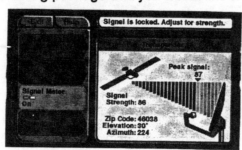

Figure 6. Step by step instruction screen.

Conclusion

The installation of the satellite system is not too complicated. If you follow the procedures and are aware of the safety precautions, you should have no problem with this project. Remember to follow the following steps in sequence.

1. Locate the position to install the dish.
2. Route cable through house to receiver.
3. Connect cable to receiver.
4.   Adjust and level dish to line up with satellite.
5. Program receiver to satellite transmission.

*Illustrations courtesy of RCA Satellite System Installer Guide.

Sample Report: Descriptive Report

THE GREENHOUSE EFFECT

by

JERRY LOPEZ

Ms. A. Rutherfoord
ENG-131
January 12, 1989

414 Boxwood Drive
Decatur, Georgia
January 12, 1989

Ms. Andrea Rutherfoord
Department of General Education
State Institute of Technology
Atlanta, Georgia 30345

Dear Ms. Rutherfoord:

I am submitting this report to meet the requirements of Communication Skills (ENG–131). In this report, I have examined the greenhouse effect, with emphasis on the predictions made by climatologists concerning the future impact of current changes in the atmosphere.

The background of the phenomenon has been obtained from encyclopedias, but the primary source of significant data is from the *Global 2000 Report to the President.* Written in 1982, this document addresses the concerns of former President Jimmy Carter. It was his commitment to our quality of life and the preservation of life that spurred scientists to examine the effects of various changes in our environment.

I believe that this report will be useful in stressing the fragile balance which allows life on earth to continue. Awareness is the first step in correcting and eliminating life-threatening practices.

An abstract follows which explains my approach in this report.

Sincerely,

Jerry Lopez

Jerry Lopez

CONTENTS

ABSTRACT

The greenhouse effect is a behavior in which the atmosphere acts as a greenhouse glass roof to control and protect the earth's temperature from the sun's lightwaves. New particles which are introduced into the atmosphere threaten to destroy the delicate balance necessary for the greenhouse effect to sustain plant and animal life on earth.

INTRODUCTION

In recent years, there has been a lot of publicity over the greenhouse effect. This publicity may lead some to believe that the greenhouse effect is a product of man, but that is simply not true. The greenhouse effect is what makes the earth's atmosphere temperate enough to support human life. In a television broadcast of "Carl Sagan's Cosmos," the narrator discussed the possibility that changes in the greenhouse effect caused the extinction of the dinosaurs. In order to gain a circumspective view of this life-sustaining phenomenon, it is necessary to examine the cause-and-effect relationship that makes up the greenhouse effect, the components of the relationship, and predictions that have been drawn by scientists regarding the greenhouse effect.

CAUSE AND EFFECT

The greenhouse effect is due to the ability of molecular elements, suspended in the earth's atmosphere, to block ultraviolet wavelengths and trap infrared wavelengths. This causes the average global temperature to stay between the freezing point and boiling point of water.

The sun emits energy in many different wavelengths. Some of this energy (ultraviolet wavelengths) is not good for carbon-based life on earth. The atmosphere blocks these harmful emissions. It is our atmosphere that lets sunlight (infrared wavelengths) through to the surface of the earth which creates heat. The heat, however, cannot pass out of the atmosphere easily. As a result, the earth is warmed and life is sustained through a balance of sunlight and atmospheric particles (*World Book*, p. 783).

Thus the atmosphere acts like the glass roof of a greenhouse, which creates a climate necessary to sustain the life within the greenhouse. The introduction of particles not normally found in the atmosphere can potentially cause an imbalance within the greenhouse effect, thus endangering plant and animal life on earth.

COMPONENTS OF THE GREENHOUSE EFFECT

The components of the greenhouse effect are of major importance to man. The particles known as ozone are found in the upper portion of the atmosphere. These particles tend to block the ultraviolet light emitted by the sun. If ultraviolet light were to reach the earth's surface in abundance, it would cause most carbon-based life to die. A small increase in ultraviolet light is suspected to be a cause in the increased number of cases of skin cancer in the human population.

Water and carbon dioxide are the major elements in the warming effect of the earth. Changes in these levels could possibly cause a loop reaction. For example, an increase in carbon dioxide could cause a warming trend which would then cause an increase in the melting of the polar ice caps. Water vapor has a warming effect, thus causing a further warming.

THE FUTURE

Dinosaurs became extinct long before human beings first appeared, yet modern man's habits threaten the greenhouse balance which supports life today. People cannot control nature, but they can hopefully gain the wisdom needed to at least avoid causing the complete annihilation of life on earth.

In 1977, then-President Jimmy Carter commissioned a group of scientists and officials to study the probable changes in the world's population, natural resources, and environment through the end of the century. They produced *The Global 2000 Report to the President* which summarized the responses of 24 major climatologists from seven countries. One scenario resulting from the survey predicted a large global warming by the end of the century, which would bring with it a greater likelihood of continental drought in the United States.

The human race uses huge amounts of fossil fuels today. One by-product of burning fossil fuels is carbon dioxide, one of the elements which interacts in the greenhouse effect. Carbon dioxide levels in the atmosphere have increased by 5% in the past 20 years (*Global 2000*). "Since the total amount of carbon in the earth-atmosphere system is constant, the carbon being added to the atmosphere pool must come from a nonatmospheric carbon pool somewhere within the system" (*Global 2000*, p. 261).

Some scientists suspect that the additional carbon is the by-product of burning fossil fuels. They feel that this increase threatens to tip the delicate balance in the greenhouse effect necessary to sustain life on earth.

It is impossible at this time for scientists to predict what will happen to our climate in the future, but, as the *Global 2000* report goes on to say, "Many experts nevertheless feel that changes on a scale likely to affect the environment and the economy of large regions of the world are not only possible but probable in the next 25 to 50 years" (p. 257).

CONCLUSION

With this new information, we can conclude that the human race can create its own fate. Can we gain the wisdom and take the necessary steps to ensure our survival, or will we ignore our adverse effects on nature's balance and cause our own extinction? Education has, in the past, led the human race to achieve and triumph over conditions beset upon it by nature. Once again, education will be the answer.

BIBLIOGRAPHY

Council on Environmental Quality and the Department of State. *The Global 2000 Report to the President.* New York: Penguin Books, 1982.

"Greenhouse Effect," *Academic American Encyclopedia*, Vol. 9. Danbury, CT: Grolier Inc., 1986.

"Sun," *The World Book Encyclopedia*, Vol. 18. Chicago: World Book-Childcraft International, 1981.

Sample Report: Formal Lab Report

COMMUNICATIONS LAB #7
LOW-FREQUENCY HETERODYNING

by

SHARON LINDELL
REG OLSON
CARL TALBERT

Professor H. Erickson
CS-301
March 4, 1989

EQUIPMENT USED

Oscilloscope
Function Generator
Digital Multimeter
DeVry Console 80

MATERIALS USED

LM3900 Integrated Circuit (1)
1N914 Diode (1)
10-kΩ Resistors (3)
1-MΩ Resistor (1)
2.2-MΩ Resistor (1)
0.002-μF Capacitor (3)
0.2-μF Capacitor (1)

PURPOSE

The purpose of this experiment was to investigate the theory and operation of low-frequency heterodyning circuits.

THEORY

When two signals, each of a different frequency are mixed together, as shown in Figure 1, the result will be the difference in frequencies of the two signals. This output signal can then be fed into a low-pass filter with a cutoff frequency that allows the desired detected signal to pass through but filters out any undesired portion of the spectrum.

PROCEDURE

The circuit of Figure 1 was constructed on a breadboard. Next, the static voltage levels at each pin of the LM3900 IC were measured. Then, two function generators were each set to a frequency of 20 kHz and an amplitude of 50 mV(p-p) and fed into the inputs of the circuit as shown in Figure 1. The frequency of input 1 was changed to 19.5 kHz and the output was observed on an oscilloscope. This output was then sketched, indicating the amplitude and frequency. Also, the amplitude was varied between 50 and 100 mV(p-p). The output for this was again observed. Next, input 2 was reset to 50 mV(p-p) and the frequency of input 1 was varied slightly above and below 20 kHz. Finally, input was disconnected from the circuit and the bandwidth of the circuit was determined by finding the frequencies (upper and lower) at which the output amplitude is 0.707 of the amplitude of the signal at 20 kHz.

RESULTS

Output Readings of
Steps 2 and 3.
f_{out} = 666.67 Hz
V_{out} = 4.5 V(p-p)

Output Reading
  of Step 4.
f_{out} = 333.33 Hz
V_{out} = 7 V(p-p)

Output Readings
  of Step 5.
Above 20 kHz f_{out} = 80 Hz
Below 20 kHz f_{out} = 689 Hz

Output Readings
  of Step 6.
Upper Cutoff:
  f_{out} = 66.7 kHz
  V_{out} = 3.2 V(p-p)
Lower Cutoff:
  f_{out} = 4.5 kHz
  V_{out} = 3.2 V(p-p)

DISCUSSION

With input 2 set at 19.5 kHz, the output signal was the frequency difference of the two input signals. The output signal was recorded to be 4.5 V(p-p) at 666.67 Hz. Referring to the results section, it is evident that the statement made in the theory section was supported. This is substantiated by the fact that 20 kHz − 19.5 kHz is approximately equal to the 666.67-Hz output. The error can be attributed to calibration of the function generators, the fact that they were not in phase with each other, and the fact that the two input capacitors were not exactly equal in value. The output waveform can be examined in Figure 2.

When the amplitude of input 2 was changed to 100 mV(p-p), the output amplitude increased to 7 V(p-p) and the output frequency decreased to 333.33 Hz. This change resulted in a halving of the frequency and a doubling of the output voltage. For the waveform sketch, see Figure 3.

Adjusting the frequency of input 1 caused a shift from the original frequency of 666.67 Hz. From this it can be stated that the greater the deviation from ±20 kHz, the larger the output frequency. This supports the fact that the heterodyne process does indeed take the difference of the two input signals and translates the spectrum of the one that is chosen to represent the message. The frequency response, with the high and low cutoff frequencies, can be observed in Figure 4.

CONCLUSION

In this experiment, we evaluated a low-frequency heterodyning circuit that produced the difference between two input signals. From the results section, it is readily observed that varying one of the inputs affects the output of the entire circuit. This circuit is very useful in radio broadcasting, where signals are easier to modulate at one frequency but must be broadcast at another frequency. This process can be accomplished very easily by using a heterodyne circuit.

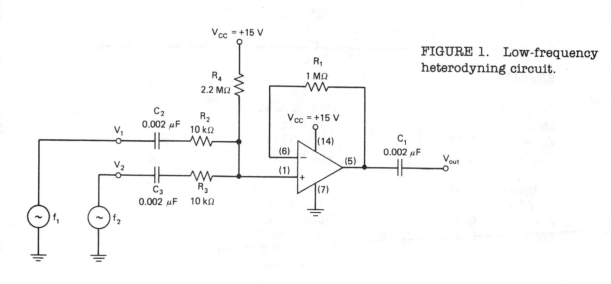

FIGURE 1.   Low-frequency
heterodyning circuit.

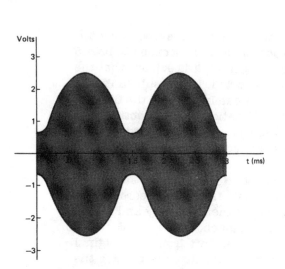

FIGURE 2.   Output waveform
of steps 2 and 3.

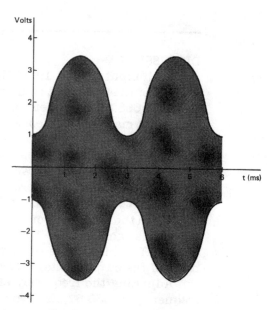

FIGURE 3.   Output waveform
of step 4.

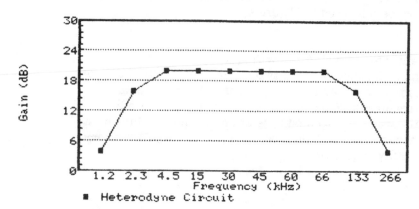

FIGURE 4.   Frequency
response of the output.

Sample Report: Formal Proposal

Computer Information Systems and Engineering Technology

Recommendation Report for the Purchase of Cellular Service and Phone

Prepare for: K. Jones

Prepared by: Matt Mikulak
Matt Langford
Anthony Davis
DeAnthony Norwood
Dawnalee McKenna; Project Manager

Date: October 2, 1999

This report details the results of our investigation of five cellular companies, the services they provide, and the phones they offer. It also gives a recommendation for one of these companies and shows how their services can benefit our company.

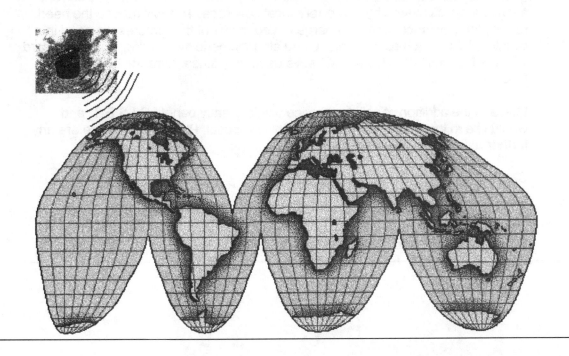

C.I.S.E.T.

Computer Information Systems
and Engineering Technology
460 N Decatur Rd
(770) 284-0911

Kyle Jones
Engineering Supervisor
C.I.S.E.T.

Dear Mrs. Jones:

This report includes our findings and recommendations for a local cellular service provider, a cellular option package, and a cellular phone.

We discussed the results of our research as a group on September 15, 1999 and determined that Powertel offered the best cellular package and phone to meet the present needs of C.I.S.E.T. We are recommending the Powertel Personal Power 50 plan combined with the Nokia 6190 phone. All of C.I.S.E.T.'s current service area is covered by Powertel's local coverage. This will reduce the need for any long distance roaming charges. Also, many of the features we need in a cellular phone, such as call waiting and alphanumeric messaging, come standard with the Nokia 6190. This too will save us money since there will be no extra feature charges.

I believe the addition of cellular service would greatly benefit C.I.S.E.T. and would like to thank you for considering this proposal. Please call if there are any further details you would like investigated.

Sincerely,

Dawnalee P. McKenna
Project Manager/ Information Systems Engineer

TABLE OF CONTENTS

LIST OF ILLUSTRATIONS

Executive Summary

The purpose of this report is to provide information on the best cellular phone and service offered. C.I.S.E.T cannot respond to customers fast enough; this decreases profits. In order to increase customer response C.I.S.E.T needs to provide its employees with cellular phones. Our research team carefully and extensively evaluated five different cellular phone companies and phones. We decided that the NOKIA 6190 is the best phone that meets the needs of the company and Powertel is the best service.

Evaluative Criteria

The research team of Davis, McKenna, Mikulak, Langford, and Norwood researched five different cellular service provider and five different types of cellular phones. Most of the phone providers offered a rebate if you signed up for their service and used one of their phones. However, we researched each phone and service separately. The criteria for choosing a service provider and cellular phone is as follows:

- Cost
- Warranty
- Types of Phones
- Coverage Area
- Option Packages

Advantages of Cellular Phones

If C.I.S.E.T issues each of their field workers a cellular phone, the workers morale and job satisfaction will increase which in turn increases productivity. By incorporating the use of cellular phones the workers will have:

- Faster responses to customer call-ins
- Fewer accidents because of unnecessary travel
- Increased job completion

Recommendation and Costs

The NOKIA 6190 with Powertel's service best meets the needs of C.I.S.E.T, and offers lower prices than its competitors. The cost of the phone is $159.95 and the cost of the service is $50.00 per month with a $50.00 activation fee. The total cost of using cellular service per year, including the phone, would be $759.95.

DISCUSSION SECTION

C.I.S.E.T has been having difficulties with communicating between our field representatives and our home office. This has affected our productivity, delayed the completion of several important jobs, and caused some customer dissatisfaction.

In the discussion section we will talk about five cellular services. We will also discuss one phone offered by each of these services. These phones were chosen to best meet the needs of C.I.S.E.T. The services and phones that were investigated and chosen as best are as follows:

Service	Phone
Powertel	Nokia 6190
AirTouch	Qualcomm QCP860
Sprint PCS	Samsung SCH 2000
Bell South	Bosch World 718
Motorola	IDEN i1000 plus

Powertel

Type of Phone

Powertel and the Nokia 6190 phone allow you to communicate anywhere in the United States. It has many features and accessories. Powertel and the Nokia 6190 phone best meet the needs of our company. By using the combination of these two products we will insure that we receive a reliable service at a price that is best suited to C.I.S.E.T.'s budget.

Nokia 6190

Figure 1
Source: Powertel
http://www.powertel.com

Features:

- Under 5 ounces
- Voice Mail
- Numeric paging
- Caller ID
- Caller ID block
- Call waiting
- 1st incoming minute free Supports English, Spanish, and French

3

- Answer calls by pressing any key
- Signal strength and battery power indicator
- Fixed Antenna 5-line LCD Display
- 250 name and number memory locations
- Easy last 8 number redial
- Personal security code and keypad lock
- Caller ID
- Receive and send alphanumeric messages
- Message and voice mail waiting indicator
- Sleep mode for extended battery life
- Improved security-privacy and clone protection
- Data capable(Optional accessories required)
- Calendar
- 35 different ring tones
- 4 games
- Caller Groupings

Accessories:

- 4 battery options; Li-Ion 900mAh, NiMh 900 mAh; NiMH Vibra 900 mAh, or the Ultra extended Li-Ion Battery 1500 mAh.
- Complete car kit
- Compact car kit
- Compact desktop charging stand
- Headset car kit
- Rapid cigarette lighter charger
- Rapid travel charger
- Belt clip
- Leather carry sleeve
- Mobile holder
- 4 battery options; Li-ion 900mAh, NiMh 900 mAh, NiMH Vibra 900mAh, or the Ultra extended Li-Ion Battery 1500 mAh. The 1500 mAh is the best: 8 hours of talk time, 5.7 ounces,and 2 weeks standby time.
- Complete car kit
- Compact desktop charging stand
- Headset car kit
- Rapid cigarette lighter charger
- Rapid travel charger

- Belt clip
- Leather carry sleeve
- Mobile holder
- Memory - view the last 5 calls missed, the last 8 numbers dialed, and the last 5 calls received.
- 35 ring tones
- 4 games
- Navi - Key it allows for simple one hand use

Option Package

Powertel offers six local rate plans with the ability to ugrade and add options to each. In addition, Powertel offers 50-State Rates^SM. These option packages allow you to call anywhere in the United States for one monthly fee. The option package that best meets our immediate needs is the Personal Power 50. The details of this package are:

- 500 anytime minutes
- One time activation fee of $20
- Free long distance calls to anywhere in the U.S. - within service area
- 15 cents a minute long distance calls to anywhere in the U.S. - outside service area
- $50 per month
- $20 activation fee

Coverage Area

Local coverage includes Georgia, Alabama, Tennessee, Kentucky, South Carolina, and some areas in Flordia, Arkansa, Misouri, Indiana, Illionois, Louisana and Mississippi. Long distance plans cover the entire U.S.
See Appendix A (p. 22) for local coverage area map.

Warranty

Powertel offers a 24-month warranty on the Nokia 6190 phone.

Cost

The cost of the Powertel option package is $50 a month. This package includes 500 minutes of airtime monthly. Any usage more than 500 minutes monthly is billed at the rate of $0.35 a minute. The one time cost of the Nokia 6190 is $184.94, and there is a onetime activation fee of $20.

Benefits

Some of the benefits of Powertel and the Nokia 6190 are:

- A large local coverage area
- Ability to Upgrade, Downsize, and personalize.
- A toll free number for customer service 24 hours a day (1-888-NOKIA2U).
- Nokia has services in over 130 countries.
- Powertel will replace your phone if lost or stolen at no extra charge.
- All plans include:voice mail, built-in numeric paging, caller ID, caller ID block, call waiting, free 911 calls, first incoming minute free, free 611 customer service calls

Disadvantages

Powertel charges an additional 35 cents per minute for any airtime in excess of your contracted 500 minutes.

AirTouch Cellular

Type of Phone:

Qualcomm QCP860 Digital Phone

The Qualcomm QCP860 Digital Phone offers a 12-month warranty and costs $129.95. The following are the specifications for the phone in Figure 2:

Qualcomm QCP860

Figure 2
Source: AirTouch Cellular
http://www.airtouch.com

Features:

- 2.5 hours of talk time
- 4 days of standby time
- Data/Fax Capable
- Ability to add or change external batteries during a call without interruption
- Smart Key Menu System
- Dual Mode
- Weight 4.23 ounces (120 grams) *includes battery
- Internal Li-Ion battery

Accessories:

- 99 phonebook memory locations
- Customizable phonebook using SmartKeys
- Messaging system for voice mail and paging
- Caller ID
- Large, easy to read LCD display
- Four language options: English, French, Spanish, and Portuguese
- For an additional cost, the "Thin Phone" has 2 hot swappable piggy back batteries which give up to 10 hours of talk time and up to 17 days of standby time in CDMA digital mode using both the internal and external battery
- Available in colors

Option Package ($50/month):

- 500 anytime minutes
- One time activation fee ($20)
- Free long distance anywhere in the US

Coverage Area:

Local service includes the Metro-Atlanta area and parts of Alabama and South Carolina. Long distance service includes the entire United States.
See Appendix A (p. 23) for local coverage area map.

Warranty:

Air Touch offers a 12-month warranty for all digital phones

Cost:

The cost of the AirTouch option package is $50 a month. This package includes 500 minutes of airtime monthly. Any usage more than 500 minutes monthly is billed at the rate of $0.35 a minute. The one time cost of the Qualcomm QCP860 is 129.95, and there is a onetime activation fee of $20.

Benefits

AirTouch offers a variety of option packages. This would be beneficial if our future needs changed.

Disadvantages:

Additional airtime minutes are charged at a higher rate than most of the other service providers. AirTouch charges a $20 activation fee. Local coverage area is smaller than that of Powertel.

9

Sprint PCS

Type of Phone:

Sprint PCS offers large variety of phones and accessories. Their phone that best meets the needs of C.I.S.E.T. is the Samsung 2000.

Samsung SCH 2000

Figure 3
Source: Spint PCS
http://www.sprintpcs.com

Features:

- Convenient size and weight: measures (5.7"x2.1" and weighs 7oz.)
- Battery provides 3 hours of talk time and 50 hours of standby.
- Voice activated dialing
- Vibrating ring
- Ringer mute
- Large buttons
- Comes with desktop charger
- voicemail
- Numeric paging
- First incoming minute free
- Caller ID
- Call waiting
- Three way calling
- Phone is dual-band. Digital and analog

10

Accessories:

- Extended battery offers 4 hours of talk time and 95 hours of standby
- Cigarette lighter adapter
- Dual slot desktop charger that allows phone one extra battery to be charged at the same time
- Hands free car kit
- Leather carrying case

Option Package:

Sprint PCS offers a wide range of option packages. The package that best meets our current needs includes:

- 500 anytime minutes
- No activation fee
- Free long distance on all calls made in the Sprint home areas
- Voicemail, numeric paging, first incoming minute free, caller ID, call waiting, and three-way calling at no additional charge
- Monthly charge 0f $49.99
- Sprint add a phone - allows a second user with a separate Sprint PCS phone number to share the minutes on one service plan.(14.99 per month)
- Equipment replacement - protects the phone and selected accessories from loss, theft, and damage.($3.25 per month)
- Off peak option - add 500 minutes of off peak airtime per month - Off peak times are Monday - Thursday 9pm. - 7am. Friday - Monday 7pm. - 7am.($7.99 per month)
- Text messaging - allows text messages, up to 100 characters, to be received on your phone.($1.99 for 30 messages 25 cents each additional message or $9.99 for 500 messages 10 cents each additional message)

Coverage Area:

Sprint has the smallest local coverage area of the five services we investigated. The local coverage area is the Metro-Atlanta area. Long distance coverage includes various cities throughout the United States.
See Appendix A (p. 24) for local coverage area map.

Warranty:

Samsung offers a one-year warranty on their phone. Sprint offers a replacement plan for $3.25 a month.

Cost:

The cost of the option package is $50 a month. This package includes 500 anytime minutes of airtime monthly. Any usage more than 500 minutes is billed at the rate of $0.25 a minute. The onetime cost of the Samsung SCH-2000 is $100, and there is no activation fee.

Benefits:

- No activation fee
- No contract

Disadvantages:

- Sprint PCS coverage is only available in select cities and the features such as Voice mail, and numeric paging can only be accessed on this network
- Once outside the Sprint home area, the phone is subject to long distance and roaming charges.

BellSouth Mobility

Type of Phone

BellSouth offers a selection of eight different phones ranging in price from $49.95 to $699.95. The phone that is best suited to our needs is the Nokia 6190.

Bosch World 718

Figure 4
Source: BellSouth Mobility
http://www.Bellsouthdcs.com

Features:

- Illuminated display
- Battery charge indicator
- Signal strength Indicator
- Voice message indicator
- Text message indicator
- 27 Selectable ringer tones/melodies
- Any key answer
- Easy to use menu system
- Built-in phone directory
- Call Waiting
- Call hold
- Call forward

13

- Caller Line Identification
- Speed Dialing
- Built-in calculator
- Keypad Lock
- Multiple Languages
- Enhanced full rate speech coding
- Text message capable
- Short message receive/send
- Display list of outgoing, missed or answered calls
- Data/Fax capable

Accessories:

- Handsfree car kit
- Additional accessories available from Bosch distributor

Option Package

BellSouth offers a wide range of option packages and add-ons. These packages rang in price from $25 a month to $243.30 a month. The package that best meets our needs is the Digital Advantage 450 with the addition of the DCS Executive Package with VIP Mobile Memo. This Digital Advantage 450 package includes:

- 450 anytime minutes
- Call Waiting
- MegaZone coverage area

The added digital feature package DCS Executive with Mobile Memo includes:

- VIP Mobile Memo with pager notification
- Message wait indicator
- Caller ID
- Call forwarding
- 3-Party calling
- No answer transfer
- Detailed billing
- Limited lifetime warranty
- Emergency roadside service

Coverage Area

Local coverage area is more than 22,000 square miles. This includes continuous local coverage from Macon and Metro Atlanta to the Alabama, Tennessee, and South Carolina borders. Long distance roaming areas include the United States, Australia, Bahrain, Belgium, Denmark, Finland, France, Germany, Hong Kong, Ireland, Italy, Macau, Netherlands, New Zealand, Norway, Poland, Portugal, South Africa, Spain, Sweden, Switzerland, Taiwan, Thailand, Turkey, and the UK (England, Scotland, Wales and Northern Ireland). See Appendix A (p. 25) for local coverage area map.

Warranty

BellSouth offers a two-year limited lifetime warranty with many of their digital phones including the Nokia 6190. The digital feature package, DCS Executive with Mobile Memo, includes a limited lifetime warranty.

Cost

The cost of the Digital Advantage 450 package is $45 a month. This package includes 450 minutes of airtime monthly. Any usage more than 450 minutes monthly is billed at the rate of $0.37 a minute. The one-time cost of the Nokia 6190 telephone is $159.95, and there is no activation fee.

Benefits

- BellSouth uses GSM (Global System for Mobile communications) technology. At present, this is the world standard in digital technology and is used by more than 200 service providers in over 100 countries. This would allow for international roaming should the need arise (Requires a Bosch World 718 Dual-Band phone). GSM technology allows for better clarity, less static, and the highest level of call security.

- BellSouth also offers a wide range of option plans and services. This would allow us to select the services that we need and reject the ones we do not.

- BellSouth has its strength coverage area in the southeastern United States. This is currently an advantage for us since our service area is in the metro Atlanta area.

Disadvantages

- All plans require a minimum 12-month service agreement.
- MultiSaver primary, 2nd and 3rd lines require a minimum of a 24-month service agreement on all lines.
- BellSouth has services throughout the world, but their primary focus is on the southeastern United States. This may prove to be a disadvantage for C.I.S.E.T. should we expand beyond our current service region.

16

Motorola

<u>*Type of Phone*</u>

Motorola offers both standard digital phones and their iDEN series phones. The iDEN series phones function as a microbrowser, a two-way radio, and a digital phone. The Motorola phone that best meets our current needs is the iDEN i1000 PLUS.

IDEN i1000 PLUS

Figure 5
Source: Motorola
http://www.Motorola.com

Features:

- Lightweight - 5.4 ounces
- Built-In microbrowser to access White and Yellow pages, directions and stock quotes
- Built-In speakerphone

17

- Mobile office
- See-through cover
- Time and date display
- ViraCall alert
- Voice mail indicator
- Message indicator
- Turbo Dial
- T9 Text input
- Multi-Language Support
- 60 hours of standby time
- 180 hours of continuous talk time
- Large 4 line backlit display
- Continuous signal and battery strength indicator
- Missed call indicator
- Built-In fraud and cloning protection
- 100 member private call directory
- 30 programmable talk groups with name tags
- Call alert queuing
- One-Touch private or group call access
- Quickstore private Ids
- Messaging service with one touch call back
- Time and date stamp on messages

Accessories:

- Adjustable headset earpiece microphone with audio adapter
- Headset earpiece
- Microphone with audio adapter
- Lightweight headset
- Lithium Ion slim battery
- Lithium Ion standard battery
- Plastic holster soft leather case
- Desktop dual pocket charger
- Handsfree car adapter
- Standard travel vehicular battery

Option Package

Motorola offers cellular service through various providers throughout the United States. We investigated Nextel, which is one of the two providers in our local area. Nextel's Performance 600 plan is their option package that best meets our current needs. This plan includes:

- 600 local minutes
- Unlimited numeric pages

Coverage Area

Nextel's local coverage area includes the Metro-Atlanta area with continuous coverage to Florida. Long distance coverage includes various cities throughout the United States. See Appendix A (p. 26) for local coverage area map.

Warranty

Nextel offers no extended warranty options.

Cost

The cost of Nextel's Performance 600 plan is $59.95 a month. The Motorola iDEN i1000 PLUS digital phone is $309.99

Benefits

Offer plans that include cellular service only, as well as other plans to incorporate all your business needs. This might be beneficial if our needs increase in the future.

Disadvantages

Smaller local coverage area than most of the other services. Package plans, rates and additional airtime is higher than some of the other service providers.

Cost Analysis Table

Cellular Provider	Phone	Type of Phone	Cost of Phone	Cost of Service (per month)
Powertel		Nokia 6190	$159.95	$50.00
AirTouch		Qualcomm QCP860	$129.95	$50.00
Sprint		Samsung 2000	$100.00	$50.00
BellSouth		Bosch World 718	$249.95	$45.00
Motorola		IDEN i1000Plus	$309.99	$69.95

Conclusion and Recommendation

Conclusion

The purpose of our research was to better connect our field representatives with both the home office and our clients. The best solution to this problem would be a cellular phone and cellular service. With features such as voicemail, caller id, numeric paging, and call waiting, our field representatives could be reached anytime.

We also had to take into consideration the coverage area so that our reps could be reached anywhere. This would eliminate the problem that we have with our current pager system. The coverage area must also be large enough to eliminate long distance and roaming charges for our traveling field representatives.

Recommendation

Based on our research we recommend Powertel for our cellular service and the Nokia 6190 for our cellular phone. The service provided by Powertel gives us the largest local coverage area, which should eliminate the need for long distance and roaming charges. Also, if C.I.S.E.T. should expand and need a larger coverage area, Powertel is the only cellular service that offers a fixed rate contract for long distance service. The airtime of the 500-minute package that we are recommending can be used anytime, anywhere. This will save us money by not having to pay peak rate prices for daytime usage. We feel that these features, combined with Powertel's large selection of option packages, will ensure that we will be satisfied both now and in the future.

Also based on research, we are recommending the Nokia 6190 phone. This phone supports the features that we feel best suit our needs. The Nokia 6190 also has the longest talk and standby battery life of any phone we investigated. This will reduce the need to purchase added charging stations.

Despite the onetime $20 connection fee and the one-year contract, we feel that the Nokia 6190 phone and Powertel's Personal Power 50 service package best covers our needs for the lowest price.

Spelling and Misused Words

FREQUENTLY MISSPELLED WORDS

aberrant, aberration
abscissa
absorb, absorption
acceptor
accessible
accommodate
accuracy
achromatic
acknowledgement
across
admissible
aerosol
afterward
age, aged, aging
air-cooled
air-dried
align, alignment
a lot (*not* alot)
alpha
ammeter
ampere
ancillary
anonymous
aperture
arc, arced, arcing
asterisk
asymmetric
audible
audio-frequency
auxiliary
axes (*pl. of* axis)
bandwidth
bargain
battery
beneficial
benefited
beta
beveled

biased
binary
Bohr-Sommerfeld atom
breadboard
bypass
by-product
cannot
cassette
catalog
channeled
characteristic
chassis (*s. and pl.*)
circuit
coherent
cohesive
collapsible
collectible
combustible
compatible
comprehensible
comprise
condensable or condensible
conductor
cooperate
coordinate
corollary
coulomb
coworker
crises (*pl. of* crisis)
criteria (*pl. of* criterion)
crystal
curvilinear
cylinder
data (*pl. of* datum)
defensible
diagnosis
diagrammed
dielectric

diesel
diffusible
dimensional
disassemble
disk, diskette
eccentric
efficient
electrode
electrolyte
electronic, electronics
eligible
embarrass
emitter
engineer
exhaustible
expandable
fascinate
feasible
flux
Fourier
frequency
fulfill *or* fulfil
fundamental
fusion
gauge
geometric
germanium
guaranteed
gyrator
harmonic
height
henry, henries
hertz (*s. and pl.*)
horizontal
hybrid
hysteresis
illogical
imaginary

FREQUENTLY MISSPELLED WORDS (*Continued*)

immediately
impedance
incident
indispensable
inductance
insertion
instantaneous
instrument
insulation
integrated
intensity
intermittent
interruption
inversely
isolation
isotropic
joule
judgment *or* judgement
junction
kelvin
Kirchhoff
knowledge
leakage
leisurely
length
license
linear
liquid
logarithm
logarithmic
luxury
magnetic
majority
maneuver
match
meant
mercury
microprocessor
minority
modulation
multiplex
multiplier
mutual
necessary
neutralize
neutron
nickel
nuclear
occasional
occur
occurrence
ohmmeter
omission

omitted
opportunity
optimistic
oscillate
oscilloscope
parallel
paraphrase
particularly
pastime
peculiar
permanent
permeability
permeance
permissible
pleasant
polarized
possibly
potentiometer
privilege
procedure
professor
prognosis
programmable
programmed *or*
    programed
questionnaire
quiescent
radiation
received
reciprocal
recommend
reference
rehearsal
repetition
resistance
resistivity
resistor
resonance
resonant
rotary
rough
satellite
saturation
scarcity
schedule
schematic
secretary
sensitivity
separate
siemen
silicon
sincerely
socket

solder
spectrum
squelch
statistics
straight
strategy
summarized
susceptible
synchronous
technical
technician
technique
tendency
therefore
thermionic
thermostat
transducer
transformer
transient
transistor
transmitter
transponder
tubular
tweeter
ultrasonic
ultraviolet
universal
until
useful
usually
vacuum
variable
vector
velocity
vertical
vibration
video
visibility
voltage
volume
wattmeter
wavelength
wire-wound
woofer
x-axis
X-band
X ray (noun)
X-ray (verb)
y-axis
Y-junction
z-axis
Zener

MISUSED OR CONFUSED WORDS

a, an, and
accept, except
advise, advice
affect, effect
alternate, alternative
analog, analogous
analysis, analyze
anonymous, unanimous
access, excess
all ready, already
all together, altogether
angel, angle
born, borne
brake, break
caliber, caliper
causal, casual
command, commend
complement, compliment
conscience, conscious,
    conscientious
continual, continuous

decent, descent, dissent
desert, dessert
device, devise
dyeing, dying
elapse, lapse, relapse
elicit, illicit
eminent, imminent
ensure, insure
envelop, envelope
expect, suspect
fair, fare
farther, further
from, than
ladder, later, latter
least, lest
lose, loose, loss
moral, morale
morality, mortality
knew, new
know, no
of, off

official, officious
past, passed
perfect, prefect
perpetrate, perpetuate
personal, personnel
peruse, pursue
picture, pitcher
precede, proceed
quiet, quit, quite
recent, resent
sign, sine
stationary, stationery
suppose, supposed to
then, than
they're, there, their
thorough, though, thought,
    tough
to, too, two
threw, through
use, used to
wear, were, we're

Irregular Verbs

Present	Past	Past Participle (Used with *have, has,* or *had*)
be, is, am, are	was, were	been
beat	beat	beaten
become	became	become
begin	began	begun
bend	bent	bent
bid (offer)	bid	bid
bid (command)	bade	bidden
bind	bound	bound
bite	bit	bitten
bleed	bled	bled
blow	blew	blown
break	broke	broken
bring	brought	brought
build	built	built
burst (*not* bust)	burst	burst
buy	bought	bought
cast	cast	cast
catch	caught	caught
choose	chose	chosen
cling	clung	clung
come	came	come
cost	cost	cost
creep	crept	crept
cut	cut	cut
deal	dealt	dealt
dig	dug	dug
dive	dove, dived	dived

Present	Past	Past Participle (Used with *have, has,* or *had*)
do	did	done
drag	dragged (*not* drug)	dragged
draw	drew	drawn
drink	drank	drunk
drive	drove	driven
eat	ate	eaten
fall	fell	fallen
feed	fed	fed
feel	felt	felt
fight	fought	fought
find	found	found
fly	flew	flown
forget	forgot	forgotten, forgot
freeze	froze	frozen
get	got	got, gotten
give	gave	given
go	went	gone
grow	grew	grown
hang (execute)	hanged	hanged
hang (suspend)	hung	hung
have	had	had
hear	heard	heard
hide	hid	hidden
hit	hit	hit
hold	held	held
hurt	hurt	hurt
keep	kept	kept
know	knew	known
lay (place)	laid	laid
lead	led	led
leave	left	left
lend	lent	lent
let	let	let
lie (speak falsely)	lied	lied
lie (recline)	lay	lain
lose	lost	lost
mean	meant	meant
prove	proved	proved, proven
read	read	read
rise	rose	risen
run	ran	run
say	said	said
see	saw	seen

Present	Past	Past Participle (Used with *have, has,* or *had*)
sell	sold	sold
set	set	set
shake	shook	shaken
shine (glow)	shone	shone
shine (polish)	shined	shined
shoot	shot	shot
show	showed	shown, showed
shrink	shrank	shrunk
sit	sat	sat
speak	spoke	spoken
spin	spun	spun
spring	sprang, sprung	sprung
swear	swore	sworn
take	took	taken
teach	taught	taught
tell	told	told
think	thought	thought
throw	threw	thrown
troubleshoot	troubleshot	troubleshot
wear	wore	worn
win	won	won
write	wrote	written

Job Application

This appendix illustrates a typical job application.

APPLICATION FOR EMPLOYMENT PRE-EMPLOYMENT QUESTIONNAIRE AN EQUAL OPPORTUNITY EMPLOYER

PERSONAL INFORMATION

NAME (LAST NAME FIRST)				SOCIAL SECURITY NO.	
PRESENT ADDRESS	APT. NO.	CITY		STATE	ZIP
PERMANENT ADDRESS	APT. NO.	CITY		STATE	ZIP
ARE YOU 18 YEARS OR OLDER? ☐ YES ☐ NO	PHONE				

DESIRED EMPLOYMENT

POSITION		DATE YOU CAN START	SALARY DESIRED

ARE YOU EMPLOYED NOW? ☐ YES ☐ NO   IF SO MAY WE INQUIRE OF YOUR PRESENT EMPLOYER? ☐ YES ☐ NO

EVER APPLIED TO THIS COMPANY BEFORE? ☐ YES ☐ NO	WHERE?	WHEN?
EVER WORKED FOR THIS COMPANY BEFORE? ☐ YES ☐ NO	WHERE?	WHEN?

REASON FOR LEAVING

NAME OF LAST SUPERVISOR AT THIS COMPANY

WHO REFERRED YOU TO THIS COMPANY?
☐ EMPLOYMENT AGENCY   ☐ NEWSPAPER ADVERTISING   ☐ FRIEND
☐ STATE EMPLOYMENT OFFICE   ☐ COLLEGE PLACEMENT SERVICE   ☐ WALK IN   ☐ OTHER

EDUCATION

SCHOOL LEVEL	NAME AND LOCATION OF SCHOOL	NO. OF YEARS ATTENDED	DID YOU GRADUATE?	SUBJECTS STUDIED
GRAMMAR SCHOOL				
HIGH SCHOOL				
COLLEGE				
TRADE, BUSINESS OR CORRESPONDENCE SCHOOL				

GENERAL

SUBJECTS OF SPECIAL STUDY OR RESEARCH WORK

SPECIAL TRAINING

SPECIAL SKILLS

Adams
9288

(Jan. 1992)

FORMER EMPLOYERS
LIST BELOW LAST THREE EMPLOYERS, STARTING WITH THE MOST RECENT ONE FIRST.

NAME OF PRESENT OR LAST EMPLOYER				
ADDRESS	CITY		STATE	ZIP
STARTING DATE	LEAVING DATE		JOB TITLE	
WEEKLY STARTING SALARY	WEEKLY FINAL SALARY	MAY WE CONTACT YOUR SUPERVISOR? ☐ YES ☐ NO		
NAME OF SUPERVISOR		TITLE		PHONE
DECRIPTION OF WORK				
REASON FOR LEAVING				

NAME OF PREVIOUS EMPLOYER				
ADDRESS	CITY		STATE	ZIP
STARTING DATE	LEAVING DATE		JOB TITLE	
WEEKLY STARTING SALARY	WEEKLY FINAL SALARY	MAY WE CONTACT YOUR SUPERVISOR? ☐ YES ☐ NO		
NAME OF SUPERVISOR		TITLE		PHONE
DECRIPTION OF WORK				
REASON FOR LEAVING				

NAME OF PREVIOUS EMPLOYER				
ADDRESS	CITY		STATE	ZIP
STARTING DATE	LEAVING DATE		JOB TITLE	
WEEKLY STARTING SALARY	WEEKLY FINAL SALARY	MAY WE CONTACT YOUR SUPERVISOR? ☐ YES ☐ NO		
NAME OF SUPERVISOR		TITLE		PHONE
DECRIPTION OF WORK				
REASON FOR LEAVING				

REFERENCES

BELOW, GIVE THE NAMES OF THREE PERSONS YOU ARE NOT RELATED TO, WHOM YOU HAVE KNOWN AT LEAST ONE YEAR.

	NAME	ADDRESS	BUSINESS	YEARS ACQUAINTED
1				
2				
3				

SERVICE RECORD

BRANCH OF SERVICE	DISCHARGE DATE RANK

HAVE YOU BEEN CONVICTED OF A FELONY WITHIN THE LAST 5 YEARS?	☐ YES ☐ NO

IF YES, EXPLAIN. (WILL NOT NECESSARILY EXCLUDE YOU FROM CONSIDERATION)

AUTHORIZATION

"I CERTIFY THAT THE FACTS CONTAINED IN THIS APPLICATION ARE TRUE AND COMPLETE TO THE BEST OF MY KNOWLEDGE AND UNDERSTAND THAT, IF EMPLOYED, FALSIFIED STATEMENTS ON THIS APPLICATION SHALL BE GROUNDS FOR DISMISSAL.

I AUTHORIZE INVESTIGATION OF ALL STATEMENTS CONTAINED HEREIN AND THE REFERENCES AND EMPLOYERS LISTED ABOVE TO GIVE YOU ANY AND ALL INFORMATION CONCERNING MY PREVIOUS EMPLOYMENT AND ANY PERTINENT INFORMATION THEY MAY HAVE, PERSONAL OR OTHERWISE AND RELEASE THE COMPANY FROM ALL LIABILITY FOR ANY DAMAGE THAT MAY RESULT FROM UTILIZATION OF SUCH INFORMATION.

I ALSO UNDERSTAND AND AGREE THAT NO REPRESENTATIVE OF THE COMPANY HAS ANY AUTHORITY TO ENTER INTO ANY AGREEMENT FOR EMPLOYMENT FOR ANY SPECIFIED PERIOD OF TIME, OR TO MAKE ANY AGREEMENT CONTRARY TO THE FOREGOING, UNLESS IT IS IN WRITING AND SIGNED BY AN AUTHORIZED COMPANY REPRESENTATIVE."

DATE   SIGNATURE